Earthquake Hazard Impact and Urban Planning

AF780029

Series Editor:
Thomas A. Birkland

About the Series

Recent years have shown that all manner of disasters have become increasingly damaging, dangerous and complex. Recent examples include the 2004 South Asian tsunami, Hurricane Katrina in 2005, and the Sichuan earthquake of 2008.

Disasters like these have, for years, been understood as neither strictly man-made or strictly natural; most disasters are a complex blend physical forces and social organization. And some disasters, such as the Madrid bombings of 2004 and the London bombings of 2005, are man-made, but their effects are vast, with structural, social, and environmental ramifi cations that reverberate locally, nationally, and globally. We cannot explain the full complexity of these events using single-discipline approaches.

This series will therefore reflect and promote the increasingly interdisciplinary nature of hazards and disaster research. Its reach will be wide because hazards research is very broad, encompassing such varied fields of study as research on industrial accidents, research on public health, biosecurity issues and "homeland security", just to name a few. And its reach will include an wide array of facets because scholars from the social, natural, and behavioral sciences have concluded that we cannot fully understand hazards and disasters without an appreciation for the social, natural and man-made environments involved.

In addressing the technical, social and environmental issues of hazards and disasters, this series will include up-to-date, comprehensive texts on key hazards and disasters at the local, regional, national, and global levels. It will cover numerous issues critical to successful research and practice and will consider these issues through an international lens. And it will involve all fi elds connected to the study of hazards and natural disasters, including disciplines found within the natural sciences, physical sciences, social sciences, and engineering. The series' coverage will therefore be broad, but it will be centered on the interactions between society, technology, and the natural environment in the hazards and disasters field.

Proposals for this book series may be sent to the Series Editor, Thomas A. Birkland, Kretzer Professor of Public Policy, North Carolina State University, at tom_birkland@ncsu.edu or the Publishing Editor, Fritz Schmuhl, at Fritz.Schmuhl@springer.com

Editorial Board:

JoAnn Carmin, Massachusetts Institute of Technology, Cambridge, MA, USA
Nicole Dash, University of North Texas, Denton, TX, USA
Daniel Nohrstedt, Crisis Management and Research (CRISMART), Swedish National Defense College, Stockholm

For further volumes:
http://www.springer.com/series/8583

Maria Bostenaru Dan • Iuliana Armas
Agostino Goretti
Editors

Earthquake Hazard Impact and Urban Planning

Springer

Editors
Maria Bostenaru Dan
Department of Urban and Landscape Design
Faculty of Urban Planning
Bucharest
Romania

Agostino Goretti
Civil Protection Department
Seismic and Volcanic Risk Office
Rome
Italy

Iuliana Armas
Department of Geomorphology
Faculty of Geography
University of Bucharest
Bucharest
Romania

ISBN 978-94-024-0778-5 ISBN 978-94-007-7981-5 (eBook)
DOI 10.1007/978-94-007-7981-5
Springer Dordrecht Heidelberg New York London

© Springer Science+Business Media Dordrecht 2014
Softcover reprint of the hardcover 1st edition 2014
This work is subject to copyright. All rights are reserved by the Publisher, whether the whole or part of the material is concerned, specifically the rights of translation, reprinting, reuse of illustrations, recitation, broadcasting, reproduction on microfilms or in any other physical way, and transmission or information storage and retrieval, electronic adaptation, computer software, or by similar or dissimilar methodology now known or hereafter developed. Exempted from this legal reservation are brief excerpts in connection with reviews or scholarly analysis or material supplied specifically for the purpose of being entered and executed on a computer system, for exclusive use by the purchaser of the work. Duplication of this publication or parts thereof is permitted only under the provisions of the Copyright Law of the Publisher's location, in its current version, and permission for use must always be obtained from Springer. Permissions for use may be obtained through RightsLink at the Copyright Clearance Center. Violations are liable to prosecution under the respective Copyright Law.
The use of general descriptive names, registered names, trademarks, service marks, etc. in this publication does not imply, even in the absence of a specific statement, that such names are exempt from the relevant protective laws and regulations and therefore free for general use.
While the advice and information in this book are believed to be true and accurate at the date of publication, neither the authors nor the editors nor the publisher can accept any legal responsibility for any errors or omissions that may be made. The publisher makes no warranty, express or implied, with respect to the material contained herein.

Printed on acid-free paper

Springer is part of Springer Science+Business Media (www.springer.com)

Preface

Earthquakes are common phenomena that since time immemorial have been permanent threats and challenge for the human being. The violent movement at the Earth's surface induced by the seismic waves deeply marked the humanity consciousness, as demonstrated by abundant folklore of legends, proverbs and taboos. Some of large earthquakes deeply marked the course of humankind, in such a degree that led to the disappearance of civilizations.

People are becoming progressively more vulnerable to disasters of all types, including earthquakes, and this is true in case of both poor and rich countries (Hurricane Katrina and Fukushima nuclear disaster are typical examples in two of the topmost developed countries).

Natural hazards are unavoidable, but they do not necessarily become natural disasters. The critical issue put up to society nowadays is the tremendous increase of the vulnerability of human communities and infrastructures. If any previous major event happened today, it would cause much more consequences than it did at the time. This is a direct consequence of a rapid urbanization process, a tremendous increase of population and infrastructure, economic developments, expansion of real-time communication, industrial interdependence that make the system more sensitive to earthquake impact. The most affected communities and regions are those where no proper planning and environmental practices were implemented.

Intensive risk is mainly a characteristic of large cities or densely populated areas where the level of vulnerability to natural hazards has increased rapidly. For this reason, disaster management has assumed great importance in recent times. Repeatedly warning signs appear worldwide: the devastating earthquakes occurred within the past years (Indonesia, Kashmir, Sichuan, Haiti, Japan) are some of the most costly (both in terms of human lives and economic loss) natural disasters in human history.

Let us take a single example: the tragic consequences of poor local housing infrastructure (particularly schools and hospitals). The tremor with magnitude 6.9 that took place on December 7, 1988 at 11:41 local time (07:41 UTC) in the Spitak region of Armenia killed at least 25,000 people. It has been estimated that if the earthquake had occurred 5 min later, the number of deaths would have been considerably diminished, because the children would have left the schools' unstable buildings. During the earthquake that hit the Pakistan on the 8th of October 2005 ($M_w = 7.6$) about 7669 schools collapsed during morning classes and about 18,095 students and 853 teachers and educational staff died then (Source: Asian Development Bank and World Bank, 2005). Another tragic case was reported on May 12, 2008 in Sichuan (China) when during an earthquake of magnitude 7.9 over 7,000 inadequately engineered schoolrooms collapsed.

The main topic of the present book refers to the role of urban planning in earthquake disaster management, as one of the crucial elements determining the vulnerability level to earthquakes. However, this approach is often neglected in earthquake risk mitigation strategies and preventing measures. The new point of view is to transfer the focus from building scale to city scale. This broader view implies mechanisms that are interdisciplinary, with public, private and civil society participation involving all concerned entities within an urban area.

Lack of control and planning in urbanization can have tragic consequences. Demographic pressure (due to the inflow of rural population to the larger cities) results in poorly developed shanty-towns in marginal and risky areas and development of illegal, uncontrolled and unplanned urban infrastructure.

Social and environmental factors can contribute also to vulnerability. Frequently, the socially disadvantaged population has to use short-term survival strategies without any long-term prospect or investment. Since individuals in the dense populated city become more and more dependent on the system, real psychological traumas can be induced in large portions of population by system disruptions.

Disaster risk comprises different types of potential losses which are often difficult to quantify. Thus, each destroying earthquake brings into our attention the vulnerabilities, defects and critical points in our system facing the seismic wave impact. Experience from passed events shows that although earthquakes can lead to enormous toll in human lives and cause billions of dollars of damage to property, infrastructure and services, many of the catastrophic consequences of these earthquakes were preventable. Taking into account the fact that such events are rare (at local scale at least), it is essential to consider the lessons learned from previous disasters, and take advantage of this information in order to incorporate adequate preventive measures to reduce the vulnerability to similar phenomena in the future.

At the same time, in dealing with disasters, administration cannot rely on normal procedures to implement appropriate responses since disasters are non-routine events that require non-routine response. Thus, the recommendations are to appeal to multihazard approaches, interdisciplinarity and multicriteria decision making in order to reach considerable disaster risk reduction.

The editors have taken care to bring together new approaches in all the principal aspects of the earthquake risk management: vulnerability, planning, risk assessment and recovery. Case studies, mostly from Italy and Romania, are investigated and innovative methodologies and current challenges and recommendations are found particularly useful for further research from the papers' presentation.

Finally, I appreciate this book as an important step forward in providing knowledge and tools to enhance society capacity to cope with disasters, to handle the situations of crisis, to promote a culture of earthquake safety and prevention.

Senior Researcher in Seismology, Mircea Rădulian
National Institute for Earth Physics,
Măgurele, Romania

Editors' Note

The idea of this book grew out of a long term collaboration between the editors, at the World Housing Encyclopedia, at some project proposals, and respectively at two nationally funded projects in Romania "Multihazard and vulnerability in the seismic context of the city of Bucharest" and "Spatial and temporal patterns of urban vulnerability", with which it adhered to the TU0801 and IS1104 COST actions, as well as common organisation of sessions from the series "Natural Hazards' Impact on urban areas and infrastructure" (the series continues with different conveners). Funding came also from the Marie Curie European Reintegration Grant PIANO "The innovation in the plan of the current floor: Zoning in blocks of flats for the middle class in the first half of the 20th century".

This first volume focuses on earthquake hazards, with some papers on the general problem, while a second volume is envisaged for hazards related to water.

The volume is focused mainly on Romania and Italy, where the editors are based, and reflects also the Marie Curie Actions funded mobility of the main editor.

We would like to acknowledge the following reviewers:

Marco Mucciarelli, Antonino Marvuglia, Nicolas Lachance-Bernard, Ioan Ianos, M. A. Grado-Caffaro, Ioan Sorin Borcia, Rohit Jigyasu, Sergiu Petrea, Andreea Duţu, Bogdan Srdanovic, Sonia Giovanazzi, Alex Barbat, Stefan Shanov, Dumitru Stanica, Jose Antonio Puppim de Oliveira, Meine Pieter van Dijk, Michael Wurm, Derya Maktav, Ioana Siminea, Sever Georgescu, Radu Andrei.

We also wish to express thanks for the publishing assistance of the series editor, Prof. Thomas Birkland, of the publishing editors Tamara Welschot and Fritz Schmul and of the editorial assistant Judith Terpos, for kind guidance and patience during the long production time of this volume.

Bucharest, July 2012

Contents

About the Editors

Dr. Dipl.-Ing. Maria Bostenaru Dan graduated in architecture with a specialization in urbanism, at the Universität Karlsruhe/Germany in 1999. She holds certificates in additional studies in "Working with multimedia" and "Project Management" from the same university. Her scientific title is from the "Ion Mincu" University of Architecture and Urbanism, where she is employed as a research scientist. She was awarded three Marie Curie Fellowships by the European Commission, two of them in Pavia, Italy (two and half years), a German Research Foundation scholarship in Karlsruhe, Germany (three years) and a Canadian Centre of Architecture support grant (one month). Recently she was awarded a postdoctoral residential grant from Dumberton Oaks Harvard Institute (USA) and a DOMUS scholarship by the Hungarian Academy of Sciences, where she is member of the external public body. She was an editorial board member from 2003–2006 at the "World Housing Encyclopedia" (USA). She is a committee member on three COST actions and the ESF NeDiMAH network and at ICOMOS ISCARSAH. In the framework of the first COST action, she spent two month at the University of Algarve, Portugal, and in the second she spent them at the Lisbon University Institute. With respect to the NeDiMAH network, she has recently organised a cross-working group workshop in Bucharest. She is an expert evaluator for Marie Curie Career Integration Grants, a funding scheme of the European Commission, COST actions, as well as for cultural projects of the Administration of the National Cultural Funds, Romanian Ministry of Culture. Maria BOSTENARU DAN has nearly 200 publications and presentations, including ISI journals, seven authored books in German and English, one former edited book at Springer, a currently co-edited one, an edited historic manuscript and been a member of collectively written books such as on urban regeneration, 21st century housing in Romania and historic photography. She is editorial board member for one Romanian and two international journals, guest editor for three ISI international journals and reviewer for even more journals and conferences. She acted as session organiser for the series from which this book emerged, and organised sessions also at other conferences.

Prof. Dr. Iuliana Armaş is a full professor in Natural Hazards and Physical Geography at the Faculty of Geography, University of Bucharest, Romania.

She graduated summa cum laude with a Bachelor's degree in Geography and a Bachelor's degree in Psychology from the University of Bucharest,Romania, in 1990 and 2008, respectively. She obtained her PhD in Geography from the University of Bucharest in 1995, with a geomorphologic research on a drainage basin.

In the last ten years, she has attended several Short Term Fellowships and training courses, including Applied Geomorphology (Köln, Germany); Remot Sensing and GIS in Applied Earth Sciences (ITC, The Netherlands, ESRI-GIS Romania); Multihazard and Risk Assessment (ITC, The Netherlands); MultiRISK (Bonn, Germany); Radar Remote Sensing (ROSA/ESA/DLR).

Prof. Armaş has more than twenty years of research and teaching experience in the field of geomorphology, landslides and seismic hazard, vulnerability and risk assessment, and risk perception. She designed and developed curriculums in Natural Hazards, Risk Perception and Physical Geography at the University of Bucharest and delivered teaching and training classes in Physical Geography, Natural Hazards and Geomorphology.

She conducted, as Principal Investigator, eight national research programs including a wide range of earth science topics and environmental problems of regional and national importance, being involved as a team member in many other research projects. She is MC member of the COST actions "Semantic enrichment of 3D city models for sustainable urban development" (2009–2012) and "The EU in the new complex geography of economic systems: models, tools and policy evaluation" (since 2012).

Her research has been published in 14 books and over 100 papers, reports, and monographs. Since 1997 she participated in the most important national and international conferences in her field of expertise. In 2005, 2006 and 2007 she received the University of Bucharest awards for her research work and results. Prof. Armaş is a member of a number of professional associations: the European Geosciences Union, the Society for Risk Analysis, the Romanian Geographical Society and the Association of Geomorphologists, also occupying leading positions (1996–2000-Vice-president in the Romanian Cartographic Association; 2000–2004-Romanian Geographical Society). She was session co-organizer at the European Geosciences Union in 2010 (NH9.12), and was invited as an expert in the EU-project CapHAZ-Net, at the workshops on Vulnerability and Risk Perception of Natural Hazards and on Risk Communication. She is a member in the editorial board of two Romanian research journals, five school-books (where she is also co-author), and peer reviewer for several Romanian and international ISI-indexed journals.

In 2006, Prof. Armaş founded and is the Director of the *Center for Risk Studies, Spatial Modeling, Terrestrial and Coastal System Dynamics* (www.geodinamic.ro), University of Bucharest, having more than 10 permanent employees, senior researchers, master and Ph.D students, who are currently involved in five research projects financed by the Romania National Authority of Scientific Research.

Dr. Agostino Goretti Civil Protection Department, Seismic and Volcanic Risk Office, Via Vitorchiano 4, 00189 Rome, ItalyBorn in Rome in 1961. Took his degree in Structural Engineering in 1985 and Ph.D in 1991 at Rome University, "La Sapienza", Involved from 1987 to 1994 in the research activities of the Department of Structural and Geotechnical Engineering of Rome University, "La Sapienza". Involved since 1987 in the structural design, strengthening and retrofitting of bridges and buildings. Involved since 1987 in academic and professional teaching. Contract professor from 2000 to 2008 at Faculty of Architecture (Rome University "La Sapienza") and Engineer Faculty (Rome III University). Invited as speaker in several international workshop and courses. Coordinator and lecturer at the Course "Post-earthquake Building Safety and Damage Assessment" at Rose School, Pavia. Participation since 1988 to the most important national and international conferences in the field of Structural Engineering with emphasis on Building and Bridge Repair and Strengthening; Vulnerability and Seismic Risk Evaluation. Author of more than 80 papers among national and international congress, journals and technical reports.

Working since February 1996 at National Seismic Survey, Department of Technical National Surveys, Presidency of Ministry Council. Involved in technical-scientific activities, mainly in the field of damage assessment and building and lifelines vulnerability assessment. In 2002 the National Seismic Survey has been moved to the National Civil Protection and changed its name in Seismic Risk Office. Participation to the reconnaissance team and coordination of the damage assessment after the following earthquakes 26.4.1996 Sila Greca (Italy); 27.9.1997 Umbria-Marche (Italy); 9.9.1998 Pollino (Italy); 11.3.2000 Lazio (Italy); 29.10.2002 Etna (italy); 31.10.2002 Molise (Italy); 6.4.2009 Abruzzo (Italy). Participation to national research programs, such as TREMA, SAVE, Pic Urban II Crotone Project. Member of the Working Group for S. Giuliano microzonation. Member of the Technical Committee supporting the national plan for school building seismic safety. Participation to international research programs, such as RADIUS, WHE, ENSeRVES, LESS-LOSS, TRIPOD, STEP, DrHouse. Visit to Greece and Turkey within the Expert Exchange Program funded by High Institute for Emergency Planning in the field of post-earthquake safety inspections. Short Term JSPS Fellowship (45 days) in Japan at Building Research Institute and National Institute for Land and Infrastructure Management, to conduct a research entitled "Post-earthquake usability and damage evaluation of reinforced concrete buildings designed not according to modern seismic codes". Member of the UN Project RADIUS (Risk Assessment Tools for Diagnosis of Urban Areas against Seismic Disaster). Member of the EERI/IAEE Project "Encyclopedia of Housing Construction Types in Seismically Prones Areas of the World". Participation in the EERI project concerning damage data collected after Molise 2002 earthquake Participation in the EERI-USGS PAGER project concerning real-time world-wide seismic damage assessment. Cooperation with the World Bank in the need assessment after the Sumatra 2009 earthquake. Participation as EU expert to the Haiti Post Disaster Needs Assessment in 2010. EU Expert Exchange on seismic risk (15 days) with France and French Territories in 2011.

Chapter 1
Earthquake Hazard Impact and Urban Planning—An Introduction

Maria Bostenaru Dan, Iuliana Armaş and Agostino Goretti

Abstract Classically, the discipline dealing with earthquakes is called 'earthquake engineering' and considered to be a branch of structural engineering. Thus, in case of projects dealing with strategies for earthquake risk mitigation, approaches from architecture, urbanism and geosciences other than seismology are often ignored. The contribution of architecture and urbanism to natural hazards mitigation is however important as:

- Retrofitting interventions are performed many times on heritage buildings, or, more generally, on buildings with cultural value,
- Today interventions are needed not only at building scale, but at city scale.

For implementing risk reduction strategies, urban planning plays a key role, interventions can only take place with the support of the affected people, their acceptance being influenced by the quality of landmarks (after Lynch 1960, 'The Image of the city') built by architectural objects which are bearers of memory.

This work approaches the impact of earthquakes, with a broader view on multihazard on urban areas. It is different from other similar works because it gives a particular importance to urban planning and architecture issues, such as integrated conservation. The majority of papers included in the book assert the impact of hazards on historical sites in case of earthquakes which shook the world and focusing on the need of rebuilding. Other papers emphasize pioneering approaches such as the Abbruzzo 2009 earthquake approach on which more research is still needed, the usage of social housing for reconstruction need and the outreach of urban planning issues towards public education and civil protection fields.

M. Bostenaru Dan (✉)
Department of Urban and Landscape Design, "Ion Mincu" University
of Architecture and Urbanism, Academiei 18–20, 010014 Bucharest, Romania
e-mail: maria.bostenaru@iaim.ro

I. Armaş
Faculty of Geography, University of Bucharest, 1 Balcescu Bd.,
010041 Bucharest, Romania

A. Goretti
Seismic and Volcanic Risk Office, Civil Protection Department,
Via Vitorchiano 4, 00189 Rome, Italy

M. Bostenaru Dan et al. (eds.), *Earthquake Hazard Impact and Urban Planning*,
Environmental Hazards, DOI 10.1007/978-94-007-7981-5_1,
© Springer Science+Business Media Dordrecht 2014

Keywords Risk · Hazard · Vulnerability · Social vulnerability · Multidisciplinarity · Transdisciplinarity · Interdisciplinarity · Urban planning

1.1 Definition of Terms

1.1.1 Risk

The concept of risk can be looked at both subjectively and objectively. An objective risk is calculated as part of an evaluation process that identifies probable losses, while a subjective risk is a personal perception linked to different viewpoints and potential threats (the relativist-constructivist perspective in social sciences). The last approach assumes that the concept of risk is a social product, a result of the modern anomie society which is heavily dependent on technology. As the German sociologist Niklas Luhman suggested in 1993, the concept of risk incorporates the concept of decision-making. The 'risk' is a potential future loss due to a specific decision (Luhman 1993).

The word risk was first used in the seventeenth century in the field of naval insurance to describe the possibility of incurring adverse outcomes. During the eighteenth century the concept was developed in Pascal's theory of probability and Bernoulli's theory of arbitrary occurrence, and in the nineteenth century the concept of risk was integrated into economic studies related to business strategy development (or business continuity planning). During the twentieth century, the science of natural disasters defined the notion of risk as the probability of being affected by a hazard as a community.

With reference to extreme events affecting communities, local authorities distinguish between a tolerable risk and a disaster or catastrophe involving severe dysfunctions in a specific society.

In the context of current human society evolution, which increases the complexity of territorial systems, any given risk has a dual nature, being also a result of the accompanying positive risk assumed by society. The risk of a community exposed to a hazard can be evaluated as a formula of loss predicting the probability of possible outcomes, being also a way of measuring the probable extent of a 'threat'.

To quantify the risk, we have to multiply the temporal and spatial probability of occurrence of a specific hazard intensity with the amount of elements-at-risk and their value. The risk elements are people, buildings, lands and their uses, infrastructure and services. The risk's magnitude depends on the vulnerability of risk elements and their degree of exposure. In recent decades, more attention was given to the damage potential of natural hazards. Many approaches and conceptual thinking calculate the risk of property and human life, which jointly represent the collective risk of an area. However, an area of wide implementation as well as an adequate level of detail in risk assessment is still largely missing.

Risk analysis constitutes the support of a decisional process based on specific measures aiming to limit and reduce the outcome of hazards.

1.1.2 Hazard

The word 'hazard' (the origin of the word is the Arabic 'az-zahar' which means a game of chance) has a negative connotation in geosciences implying the possibility of determining processes and situations, actions or non-actions that have the potential to bring damage and loss for a human society in a specific area and at a specific time.

The hazard is always understood as a threat and not the event itself. Although it can be predicted to influence a specific area at a given time with different levels of intensity, it remains an uncertainty as to when and where exactly the hazard will occur.

Many times the hazard is identified with extreme events regardless of their cause, natural or man induced. The hazard also includes hidden unknown elements representing possible future threats. The hazard can manifest itself as a singular event or a combination of events or sequentially included in the cause and effect process.

Hazards can be characterized by a specific geographical location, intensity or magnitude, frequency and probability of occurrence. They have a dynamic trend (connected to a specific magnitude and periodical cyclicity), so they can be quantified through their magnitude-frequency connection, through probabilistic modelling and recorded by historical archives. Any territory can be defined by a specific hazard.

Hazards can develop into risks only if they become potentially damaging to the interest of a human community, vulnerable and exposed to a specific naturally occurring event. If this extreme event happens in an uninhabited area it does not represent a risk. Additionally if the natural phenomenon maintains itself within tolerable limits it is perceived as a resource and not a hazard. The tolerable limits vary according to the socio-economic and natural factors specific to each human community. Research in earth sciences has put emphasis on determining the hazard potential by examining, modeling and assessing a variety of individual processes. This book focuses only on seismic hazard and urban vulnerability to the seismic event, aiming to develop better risk management tools for densely inhabited areas.

1.1.3 Vulnerability

1.1.3.1 Vulnerability of the Built Environment

The concept of vulnerability is applied in a broad field and at different spatial levels. Seismic vulnerability is the pre-disposition of elements at risk, such as buildings, infrastructures, people, organisations, etc. to be affected, damaged or destroyed by the seismic event. Physical vulnerability, sometimes called primary vulnerability, usually refers to physical damage, while functional vulnerability refers to the function abruption. Secondary vulnerability refers to the consequences of the physical damage, such as the repair or the replacement cost of a damaged building. The methodologies for vulnerability assessment can be variously divided. The subdivision depends obviously on the nature of exposed elements (i.e. buildings, bridges, lifelines, tunnels, etc.) and on the spatial scale or resolution for analysis (i.e. individual building, urban, regional or national scale for territorial analysis buildings, etc). When considering special buildings or critical infrastructures, the physical vulnerability assessment is generally performed by using seismic time histories for the input motion, together with detailed structural models. For large scale vulnerability evaluation, it is common to establish typologies of exposed elements and evaluate the vulnerability of representative elements. Seismic intensity is generally expressed in terms of macroseismic intensity (e.g. European Macroseismic Scale EMS98) or Peak Ground Acceleration, spectral acceleration or spectral displacement at the natural period of the structure which may be more objective measures of the earthquake's severity. It is common to assess vulnerability through damage probability matrices or building fragility functions. The latter ones provide the probability to reach or exceed a certain degree of damage given a specific building type and a seismic intensity.

1.1.3.2 Social Vulnerability

In the context of disasters, the use of the concept in social sciences was first introduced by O' Keefe et al. (1976), in exploring the key role played by socioeconomic factors in creating a weakness in responding to, and recovering from, the effects of extreme natural events.

The first model of vulnerability aimed to determine biophysical vulnerability, namely conditions making people vulnerable to natural disasters (Burton et al. 1993; Anderson 2000). Several elements trigger dependence in this case, such as: probability of a hazard occurrence, a specific magnitude and area extent, proximity to the hazard, rapidity of onset and its duration. This approach gained usage, as the detection of two types of vulnerability, biophysical and social, was considered independent (Cutter 1996). The biophysical aspect is also considered in some research perspectives (UNISDR 2004) to be built up by two components: physical and environmental.

The second model aimed to determine the social vulnerability (e.g., Bohle et al. 1994; Blaikie et al. 1994; Cutter et al. 2003; Rygel et al. 2006), and have raised most controversial aspects (Hewitt 1997; Bankoff 2003).

The third model was built on the concept developed by Cutter in the first model (1996): 'hazards of place' where exposures and resilience were integrated, depending on particular places or regions (Kasperson et al. 1995). This theoretical approach integrated by several studies was the starting point for the ESPON Hazard Model (Kumpulainen 2006). This last model successfully includes the aims of the previous two: both biophysical vulnerability and social aspects (McEntire 2000; Weichselgartner 2001).

Newer literature emphasises the critical importance of social dynamics of disasters and the barriers in the science-policy interface (Dowity and Allen 2011; Kasperson and Berberian 2011).

In current research, vulnerability is broadly defined as the 'potential for loss' (e.g., Mitchell 1989, 2001; Petak and Atkisson 1982; Susman et al. 1983; Thywissen 2006), or as a predictive variable which designates the potential of being harmed relative to the elements at risk and event intensity (e.g., Kates 1985; Dow 1992; Ionescu et al. 2009). Unfortunately, vulnerability still means different things to different people, and despite all the theoretical knowledge we still lack practical applications.

1.1.3.3 Risk Management

Risk management represents the full range of procedures and tasks that ultimately lead to the implementation of policies and risk mitigation measures. Depending on social and technical feasibility and cost/benefit considerations, investing in preventive measures is justified in areas frequently affected by disasters. In the context of public awareness and education related to disaster risk reduction, changing attitudes and behaviour contribute to promoting a 'culture of prevention'.

All the activities are formulated within the open-ended process of the Emergency Management Cycle that includes Response, Recovery, Preparedness and Mitigation.

Response is the provision of assistance or intervention during or immediately after a disaster to meet the life preservation and basic subsistence needs of those people affected.

Recovery includes the decisions and actions taken after a disaster with a view to restoring or improving the pre-disaster living conditions of the stricken community, while encouraging and facilitating necessary adjustments to reduce disaster risk. Recovery (rehabilitation and reconstruction) affords an opportunity to develop and apply disaster risk reduction measures.

Preparedness includes the activities and measures taken in advance to ensure effective response to the impact of hazards, including the issuance of timely and effective early warnings and the temporary evacuation of people and property from threatened locations.

Mitigation includes structural and non-structural measures, or tangible and non-tangible actions, undertaken to limit the adverse impact of natural hazards, environmental degradation and technological hazards.

Structural measures refer to any physical construction to reduce or avoid possible impacts of hazards, which include engineering measures and construction of hazard-resistant and protective structures and infrastructure. Seeing the risk as the convolution of Hazard, Vulnerability and Exposure, structural measures mitigate the risk throughout the reduction of the vulnerability. Structural measures reduce most the seismic risk, while, for example, insurances just transfer the risk. Non-structural measures refer to policies, awareness, knowledge development, public commitment, and methods and operating practices, including participatory mechanisms and the provision of information, which can reduce risk and related impacts.

Structural measures include both improving standards for new buildings and infrastructures (Seismic code, Seismic hazard, Microzonation, Land use & Urban planning, Controls) and strengthening existing buildings and infrastructures. In case of new buildings the total construction cost increases in the range of 5–10 % if seismic provisions are taken into account. The strengthening of existing buildings can range between 15 % and 40 % of the replacement cost. This is the main reason why the strengthening of a large number of existing buildings is seldom implemented.

The political payback is another reason: immediate for new construction, between 10 and 30 years for existing building. This explains why structural measures are implemented just after destructive events, when the risk is perceived at most by society.

Since large scale programs require massive financial investments and produce benefits in the long term, public authorities are in charge of mitigating the risk in public buildings and sometimes in the private sector as well. One example of structural measures is the Italian national plan for school seismic risk mitigation that was implemented after the 2002 San Giuliano di Puglia earthquake when 27 children died in the collapse of the primary school. The cost for the seismic upgrading of all the schools in Italy was estimated to be about 13,000 mio. €, that is 0.5 % of GDP. An initial prioritisation reduced the needs to 4,000 mio. €. As for today (Law 289/2002, art. 80) two excerpted plans have been implemented, for a total of 450 mio. € funded by the Italian State and 1,600 school buildings strengthened. Another plan, also funded by the Italian State, devoted to seismic mitigation of school buildings (OPCM3728/2008) has a budget of 20 mio. € per year, allowing the full retrofit of about 40 buildings per year. However the most important ongoing mitigation plan was funded after the 2009 L'Aquila earthquake, when 308 people died. About 1,000 mio. € have been allocated in the years 2010–2016 not only for public and private buildings strengthening, but also for bridge strengthening.

Obviously Italy is not the only country implementing such plans. In France the national plan 'Plan séisme', aimed to seismic risk prevention, started on 2005. Two years later, on January 2007, the 'Plan Séisme Antilles' started as well (http://www.plan-seisme-antilles.org/), being the most seismic area of France. It is an interdepartmental program aimed to reduce vulnerability to seismic risk through intangible and tangible actions. The State planned to allocate about 340 mio. € in the period 2007–2013 for the strengthening or reconstruction of public buildings, such as Emergency management buildings, Educational buildings, Hospitals and health buildings, Social housing, Infrastructure and lifelines. A special mention on the upgrading of the primary schools for which an annual program of approximately 20 mio. € is expected.

Similarly the Californian plan for seismic risk mitigation of Hospitals (1994–Senate Bill 1953) is intended to bring all the hospitals in compliance with the code (operational after major earthquakes) by the year 2030. If the deadline is not reached, the project will have to be abandoned. According to OSHPD, 56 % of the hospitals in California have not yet achieved the safety requirements. The estimated cost of the program was 14 billion $ in 1990 and no funding was provided by the State. The cost of the program increased to 42 billion $ in 2002 (RAND estimate) and continues to increase due to increases in construction costs. In 2007 (Senate Bill 306) a seven year extension with respect to the 2013 seismic safety deadline was allowed but hospitals were required to be replaced, if needed, by January 1, 2020, instead of 2030.

Finally an example of seismic risk mitigation program for residential buildings is the one implemented and funded by the Umbria Region, Italy (Regional Law n. 18/2002). The financial contributions are up to 50 % of the cost of the structural works correlated to that of finishing correlated, with the maximum of 20,000 Euro for each dwelling. The program addresses residential masonry buildings only. Similarly the seismic risk mitigation program for residential buildings in Eastern Sicily, Italy, is a program funded by the Italian State (Law 433/1991) with 120 mio. of Euro. Buildings can have access to funding according to a priority list based on the indicator given by the average number of expected victims in the building per year. The financial contribution ranges from 225 €/sm to 280 €/sm including VAT and technical fees, with a 50 % reduction for dwellings used as temporary housing.

1.2 Challenges

The United Nations General Assembly declared 2008 as the International Year of Planet Earth to increase awareness of the importance of Earth sciences for the achievement of sustainable development. According to Gociman (2000) the idea of natural and man-made hazard management aiming at sustainable development starts from Greek mythology, and is associated with the meaning of primordial elements, including the earth. Most primordial elements (water, fire, earth, air) have dual significance; they give life through their characteristics that lead to the development of mankind and take away life through the hazards connected to them. Geological hazards such as earthquakes, addressed in this book, volcanic eruptions and landslides are connected to earth's elements.

Social sciences and the arts have been much neglected in the study of earthquakes, which is considered an engineering field. The book's challenge is to develop a multidisciplinary view on increasing resilience techniques to earthquake hazards available to decision makers. All citizens, researchers, teachers, and students must be involved and educated.

The book aims to reach both junior and senior researchers and intends to prove the need to develop university curricula in the field, which is a major challenge.

Each one of four sections of the book addresses the issue of decision making, sometimes at the level of participatory planning, as well as the role of specialists from urban planning, geography, engineering and seismology/earth science, while engineering sciences include civil engineering and remote sensing/geo-information. In short, the book is a long interdisciplinary dialogue between these fields, rarely considered and researched together.

Another challenge is to equally address both heritage and new planning, in the twenty first century focused on new planning, which in our opinion should rely on lessons from the past.

1.3 Multidisciplinarity, Interdisciplinarity, Transdisciplinarity

A holistic approach to natural hazards, in this particular case of earthquakes coming from natural sciences, engineering, social and economic sciences and the arts, has never been aimed for so far. In this book we have contributing architects, geographers, civil engineers, remote sensing/geoinformation engineers, seismologists (physicians and geologists), and ecologists.

Although earthquake engineering was traditionally considered a branch of civil/structural engineering, research has to be open to natural sciences, related to hazards, and to socio-economic sciences, related to the consequences, and to the arts, related to communication and perception. The Earthquake Engineering Research Institute had a pioneering approach to this issue by encouraging the contribution of disciplines from all these fields, unlike traditional Earthquake Engineering Associations.

A successful approach to the role of urban planning in earthquake mitigation was developed at the beginning of the twenty first century in a multidisciplinary frame of the European project RISK-UE. The Urban System Exposure method was developed: "system components and their functionality, then the 'elements-at-risk', their indicator-quantified exposure, and the vulnerability factors were analysed, leading to the identification of problems and opportunities. The spatial distribution of strategic or vulnerable elements was mapped. The elements were ranked according to their role in the system during normal, crisis and recovery periods. After appropriation of the problems posed by seismic risk, 'management-plans' and 'plans-of-action' are proposed for strategic implementation by decision-makers through land-use decisions." (Bostenaru 2005). But this is only one stage in the cycle of reactions to earthquake impact. In Bostenaru (2005) we identified models for reconstruction planning, preparedness planning, mitigation planning and resilience planning, which may be correlated to the disaster management cycle phases (response, recovery, ex-ante disaster planning, preparedness), but, first of all, we aim to define the role of planning, which this book is addressing. (Urban) Planning must take place after risk assessment in parallel to decision making, as suggested by the outline of the book. According to Gociman (2000) the disaster management follows the following cycles:

- Crisis (from the impact of the disaster over intervention in the first 24 hours, and about 4 weeks of rehabilitation),
- Post-disaster reconstruction (divided into 6 months spans) which ends at the moment when development starts,
- Development, which ends with prevention and preparedness activities, before the disaster strikes again.

This way a spiral results, the disaster leading to economic regression, with the development in the aftermass leading to a higher developed state than before the disaster. In the contribution by Cerqua and Rapicetta in this book the disaster management phases were correlated to the participative approach. According to Cerqua and Rapicetta all phases must contain participative components: the response (emergency after the impact), the recovery (rehabilitation), the ex-ante disaster planning (prevention, reconstruction and mitigation), and preparedness (preparedness itself and warning). We note that the phases proposed by Cerqua and Rapicetta do not contain resilience planning, but all the other types proposed by Bostenaru (2005), in a paper which builds on multicriteria decision making and also follow a participative approach. Resilience planning might be the key for participation. What was developed in RISK-UE corresponds to this level of resilience planning, and correlates urban planning with seismology. This leads to the role of urban planning we outline in the following section. The participative approach addresses the role of social sciences in our book.

The series to which this book belongs highlights the 'management' side, and the contribution of socio-economic sciences and humanities to the field. For this reason the focus centers on such aspects, less than on the technical ones. The first book in the series also deals with policies to reduce environmental hazard impact.

Regarding the arts we may refer to the representation of catastrophes in photography but also in other artistic projects addressing the (keeping of) memory. In 2009 a conference was held in Ghent ("Tick your catastrophe!" http://www.catastrophe.ugent.be/, proceedings available 2011, Le Roy et al. 2011) and a book on the topic of nineteenth century photography of natural and man-made disasters, from the collections of the Canadian Centre for Architecture (Blau and Kaufman 1989), is to follow another successful Springer book from the collections of Kozak and Cernak (2010) on drawings on natural disasters in the modernity. To the artistic approach we respond by analysing arts' integration into public spaces following disasters, to create a memory of the event, at the planning stage.

The role of economic sciences should be the subject of further studies, as we will outline in the conclusion.

1.4 The Role of Urban Planning

One of the tasks of urban planning is protecting citizens from the effects of natural hazards. Another task is protecting the built heritage.

The image of planning has changed several times during the twentieth century. A major change was adoption of the functionalist model, developed in the first half, and implemented mainly around mid-twentieth century. How this model still affects us in earthquake reconstruction is a topic dealt with in this book.

Another planning change regards the procedures, more specifically the introduction of participative planning, in the framework of which, strategic planning developed. This resulted from critique of the functionalist model. We will return to this topic when talking about decision structures, since participative planning is a decision model: the citizens voice gains importance. In strategic planning the final output of the planning is not foreseen from the beginning, but it is shaped gradually while implemented, it is a process. How this philosophy applies to earthquake reconstruction is discussed in this book. Recently EEA Grants provided an opportunity to hold citizen education projects in Romania with the help of NGOs. In the framework of such an event we worked in a group on how to influence decision makers to support earthquake retrofit of seismic risk class I buildings in Bucharest, Romania. The advocacy planning strategy developed identified the tactics, the goal, the allies (sustaining parts), the target and the resources for different steps, which were: petition, audience, open letter, public coffee shop, signatures list, online mapping of buildings, design competitions, photo competition and exhibition (Bostenaru 2012). Further research on the role of participation in earthquake mitigation, not just reaction to unfortunate events, is needed. Also, the public participation methods have changed in the twenty first century due to information and communication technologies within this field. Examples include public participation GIS- PPGIS, which uses spatial and visualisation tools such as satellite imagery, and digital and sketch maps to change geographic involvement and awareness at a local level, which is based on crowd sourcing for data collection. This field should be considered because data for different analysis is often most difficult and costly to obtain and research can be hampered when it lacks data. Web 2.0 software provides an opportunity for exchange, following the way confined masonry guidelines developed for Haiti over different Internet forums, also supporting Building Information Modelling.

This marks the transition to the third change in planning in the last 100 years, which is that induced by information technology tools. Instead of artistic drawing on paper now computer tools are available for spatial planning and urban modeling. This allows on one side approaching not only the building scale, but the city scale, when assessing vulnerability or risk scenarios, allowed by the rapid spread of the power of computers which could not be foreseen to this extent when the Report on the Ecological City was adopted in 1996 and highlighted the role of GIS and other scenarios. However, more development is still to be wished, between the enrichment with numerical data which allows statistical analysis and the visualization power of GIS, which has not yet reached that of the artistic expression of drawing. Several papers in the book successfully display how urban planning can be supported by results achieved through the analysis of data by means of GIS.

1.5 Organisation of the Book

The book is organised in four sections. They follow the succession of analyzing earthquake hazard, earthquake vulnerability and earthquake risk. While hazard, exposure and vulnerability, the input for risk and impact studies, are approached in a condensed way, more space is dedicated to the study of 'risk', being strongly linked to 'impact' the focus of this book. Thus we have differentiated between risk assessment and risk management. Finally risk and impact studies serve as decision making tools. Urban planning can be seen also as another decision making tool. And although economic aspects are present at all stages, the economic efficiency analysis is a decision making tool which will be approached during further future research.

After introduction, the first section, including three papers, addresses urban hazard, exposure and vulnerability. Each paper addresses one of the following issues: urban hazards through the work of geologists in Bucharest's City case study, urban exposure through the work of geodesists specialized in remote sensing in a chapter aimed at the exposure of population, in Istanbul's City Case study, and the case of an architect addressing urban geography. Thus this section presents contributions from different geosciences fields.

The second and the third section address 'risk', as an expression of earthquake impact.

Section II consists of four chapters and is dedicated to 'Earthquake risk assessment'. The first chapter, written by seismologists, is about local site effects in a Romanian case study from the Danube Delta. The second chapter is from the field of geophysics, and addresses an Italian case study: the damage pattern in case of rich cultural heritage. The third chapter is again a Romanian case study, from Bucharest's City and presents a GIS multicriteria analysis, from the point of view of geography and seismology, including also social vulnerability, building a basis for further decision making tools. The fourth chapter is the case of a city located in the strongly seismic Southern Italy, presenting the results of a European project, which links civil engineering and other disciplines. The aim of the paper is to define urban strategic structures, which are different from the mentioned "Urban System Exposure". This section focuses on case studies from Romania and Italy, and includes multidisciplinary issues ranging from engineering seismology to remote sensing, this time applied from cultural heritage to civil engineering. The issue of population is addressed through the social vulnerability perspective.

The third section addresses risk management. It includes three papers. The first paper, from Spain, addresses a general approach to measuring risk governance. It is a methodology paper, not a case study. The other two papers are from the point of view of the architect intervening in case of a disaster. The first deals with an on-going issue from Bucharest—planning the urban strategy before disaster strikes, while the other addresses the role of architects in immediate reconstruction: the emergency housing, with a specific case study from Romania. Although, of course prevention is most desirable, the contribution of urban planning to disaster management started historically, well before the twentieth century, with reconstruction. It was shown with the historic earthquake resilient typology presented in the first section. With the emerging discipline of urban planning, the first paper includes urban planners next to architects, to whom the task of urban planning belonged before.

The papers in section IV, written by architects, are dedicated to participative planning in Italy. Participative planning means the democratisation of planning through inclusion of citizens. The first paper gives the framework for participative planning in case of natural disasters, while the second presents a case study of the reconstruction after the l'Aquila earthquake put in a global context, bearing in mind Kolontar's and Devecser's case from Hungary and Corbeni from Romania.

The book closes with a chapter devoted to conclusions and recommendations for future actions.

The case studies overweigh the methodological papers, but nevertheless new methodologies are introduced and exemplified in these case studies. Contributors come from both academia or governmental structures, or are independent researchers.

References

Anderson MB (2000) Vulnerability to Disaster and Sustainable Development: A General Framework for Assessing Vulnerability. In: Pelke R Jr, Pielke R Sr (eds) Storms, Vol 1. Routledge, London, pp 11–25

Bankoff G (2003) Cultures of disaster: society and natural hazards in the Philippines. Routledge Curzon, London

Blaikie P, Cannon T, Davis I, Wisner B (1994) At risk: natural hazards, people's vulnerability, and disasters. Routledge, London

Blau E, Kaufman E (eds) (1989) Architecture and its image: four centuries of architectural representation: works from the collection of the Canadian Centre for Architecture, Montréal: Centre canadien d'architecture/Canadian Centre for Architecture; Distributed by the MIT Press, Cambridge

Bohle HG, Downing TE, Watts MJ (1994) Climate change and social vulnerability: the sociology and geography of food insecurity. Global Environ Chang 4:37–48
Bostenaru Dan M (2005) Multidisciplinary co-operation in building design according to urbanistic zoning and seismic microzonation. Nat Hazards Earth Syst Sci 5:397–411
Bostenaru Dan M (2012) Participatism and disaster management, Argument, 4, in print
Burton I, Kates RW, White GF (1993) The environment as hazard, 2nd edn. Guilford Press, New York
Cutter SL (1996) Vulnerability to environmental hazards. Prog Hum Geog 20(4):529–539
Cutter SL, Boruff BJ, Shirley WL (2003) Social vulnerability to environmental hazards. Soc Sci Q 82:242–260. doi:10.1111/1540-6237.8402002
Dow K (1992) Exploring differences in our common future(s): the meaning of vulnerability to global environmental change. Geoforum 23(3):417–436
Dowity RA, Allen BL (eds) (2011) Dynamics of disaster: lessons on risk, response, and recovery. London, Earthscan
Gociman CO (2000) Tipologia hazardului si dezvoltarea durabila. Generalitati, concept, problematica, Editura Universitară „Ion Mincu", Bucharest
Hewitt K (1997) Regions of risk: a geographical introduction to disasters. Longman, London
Ionescu C, Klein RJT, Hinkel J, Kumar KKS, Klein R (2009) Towards a formal framework of vulnerability to climate change. Environ Model Assess 14:1–16
Kasperson JX, Kasperson RE, Turner BL (eds) (1995) Regions at risk: comparisons of threatened environments. United Nations University Press, Tokyo
Kasperson RE, Berberian M (eds) (2011) Integrating science and policy: vulnerability and resilience in global environmental change. London, Earthscan
Kates RW (1985) The interaction of climate and society. In: Kates RW, Ausubel JH, Berberian M. Climate impact assessment, SCOPE 27. Wiley, New York, pp 3–36
Kozák J, Cermák V (2010) The illustrated history of natural disasters. Springer, Dordrecht
Kumpulainen S (2006) Vulnerability concepts in hazard and risk assessments. In: Schmidt-Thomé P (ed) Natural and technological hazards and risks affecting the spatial development of European Regions. Geological Survey of Finland Special Paper 42:65–74
Le Roy F, Wynants N, Hoens D, Vanderbeeken R (eds) (2011) Tickle your catastrophe!: imagining catastrophe in art, architecture and philosophy (studies in performing arts and media). Academia Scientific, Ghent
Luhman N (1993) Risk: a sociological theory. Aldine de Gruyter, New York
McEntire DA (2000) Sustainability or invulnerable development? Proposals for the current shift in Paradigms. Aust J Emerg Manage 15(1):58–61
Mitchell JK (1989) Hazards research. In: Gaile GL, Willmott CJ (eds) Geography in America. Merrill, Columbus, pp 410–424
Mitchell JK (2001) What's in a name?: issues of terminology and language in hazards research (Editorial). Environ Hazards 2:87–88
O'Keefe P, Westgate K, Wisner B (1976) "Taking the naturalness out of natural disasters." *Nature*, Vol 260. pp 566–567
Petak WJ, Atkisson AA (1982) Natural hazard risk assessment and public policy: anticipating the unexpected. Springer, New York
Rygel L, O'Sullivan D, Yarnal B (2006) A method for constructing a social vulnerability index: an application to hurricane storm surge in a developed country. Mitig Adapt Strat Glob Change 11:741–764
Susman P, O'Keefe P, Wisner B (1983) Global disasters: a radical interpretation. In: Hewitt K (ed) Interpretations of calamity. Allen & Unwin, Boston, pp. 264–283
Thywissen K (2006) Core terminology of disaster reduction. Measuring vulnerability to Natural Hazards Towards disaster resilient societies. United Nations University Press, J. Birkmann
UN/ISDR (2004) Living with Risk. UN. http://www.unisdr.org/eng/about_isdr/bd-lwr-2004-eng.htm
Weichselgartner J (2001) Disaster mitigation: the concept of vulnerability revisited. Disaster Prevention Manage 10(2):85–94

Part I
Urban Hazard, Exposure and Vulnerability

Chapter 2
Active Fault Systems and Their Significance for Urban Planning in Bucharest, Romania

Dumitru Ioane, Mihail Diaconescu, Florina Chitea and George Garbacea

Abstract In urban planning activities, besides a detailed seismic zonation, knowledge of the areal distribution of active faults, and particularly of faults crossing the city in heavily built districts, may be valuable in numerous cases. This chapter aims to identify active faults or fractures in Bucharest, a large city developed over thick layers of Quaternary unconsolidated sediments. To achieve this goal, the research is based on an integrated analysis of all existing relevant data coming from geology, geomorphology, geophysics and seismology. Another purpose is to understand the role of faults in the spatial variability of seismic damaging effects on buildings in view of a better future urban planning. The N-S fault system seems to be involved in neotectonics, affecting the geomorphology between Bucharest and Ploieşti cities. It could represent an active transcrustal fault, the deepest seismic event recorded within Bucharest at subcrustal depth being located on its southern prolongation. Within Bucharest the presence of faults affecting Pleistocene sediments was previously illustrated using borehole data, shallower vertical displacements of Quaternary formations along faults being interpreted in this study. The existing seismic reflection data recorded within Bucharest was used to interpret shallow faults considered as active or capable faults crossing Quaternary sedimentary deposits.

Keywords Tectonics · Neotectonics · Geomorphology · River network · Romania

2.1 Introduction

Active fault systems may have significant influences in slow ground displacements during neotectonic vertical or/and horizontal movements, or sudden ones, as a consequence of high magnitude earthquakes.

D. Ioane (✉) · F. Chitea · G. Garbacea
Faculty of Geology and Geophysics, Department of Geophysics,
University of Bucharest, 6 Traian Vuia str., 020956 Bucharest, Romania
e-mail: d_ioane@yahoo.co.uk, dumitru.ioane@g.unibuc.ro

M. Diaconescu
National Institute for Earth Physics, Măgurele, 12 Călugăreni str.,
077125 Ilfov, Romania

M. Bostenaru Dan et al. (eds.), *Earthquake Hazard Impact and Urban Planning*,
Environmental Hazards, DOI 10.1007/978-94-007-7981-5_2,
© Springer Science+Business Media Dordrecht 2014

In large cities such as Bucharest, Romania, developed over quite thick sequences of Quaternary unconsolidated sediments, there were no real possibilities of geologically mapping faults or fault systems, the geological structures being deeply buried. Due to their specific structure, the unconsolidated sediments do not preserve traces of active faulting due to neotectonic processes or seismic events. In geotechnical and hydrogeological wells or in the city subway tunnels such tectonic features are not easily observed by specialists, the resulting geological cross sections representing many times only changes in geological facies and no faults or fractures.

An integrated analysis of all existing relevant data coming from geology, geomorphology, geophysics and seismology may represent a first stage of a study aiming at locating active/capable faults or fractures in the Bucharest city area and understanding their role in the spatial variability of seismic damaging effects on buildings in view of a better future urban planning.

2.2 Geographic Location, Geomorphology and Rivers Network

2.2.1 Geographic Location

Bucharest, the capital of Romania, is located in the northern part of the Romanian Plain (Fig. 2.1), between the Danube River (60 km southward), the Carpathian Mountains (100 km northward) and the Black Sea shore (250 km eastward).

The city covers an area exceeding 20 km in diameter; its geographical coordinates may be roughly considered as 44°43, in latitude and 26°10, in longitude.

2.2.2 Geomorphological Data

The topography in the Bucharest area is quite flat, the relief decreasing gently southeastward, toward the Danube. The difference in relief elevations within Bucharest ranges from 50 m (SE) to 100 m (NW), following the main topographic regional trend from river Danube to the Carpathians.

Geomorphological data in the area where Bucharest is situated show that starting from the city northern limit there is a sudden change of about 10 m in the topographic elevations on a lineament trending N–S, between Bucharest and Ploieşti cities. The area located west of this lineament is characterized by higher altitudes as compared to that situated eastward.

At the regional scale, the geomorphological boundary between the southern plain and the northern hilly region follows two main trends, as shown by the digital terrain model (http://srtm.csi.cgiar.org 2011), displayed in Fig. 2.2: (1) NE–SW from the East Carpathians bend zone to Ploieşti city and then again, from Bucharest toward Danube river; (2) N–S between Ploieşti and Bucharest cities.

Fig. 2.1 Geographic location of Bucharest city, Romania

The high topography area located especially in the western part of Bucharest is considered to represent a remnant of uplifted Middle Pleistocene rocks, affected by erosion processes due to both Dâmboviţa and Argeş rivers (Lacatusu et al. 2008).

As an interesting geomorphological feature, the Dâmboviţa riverbanks are asymmetrical, the south-western riverbank being much higher (15 m) as compared to the north-eastern one (2 m). Within Bucharest, this asymmetry was also observed at the Colentina riverbanks.

2.2.3 Rivers Network

Palaeohydrological studies showed that during Middle Pleistocene-Lower Holocene the rivers and their tributaries modeled the relief of this area, by both erosion and sediments deposition processes.

The main rivers crossing Bucharest are Dâmboviţa and Colentina. Their river courses trend NW–SE and display the same asymmetry regarding tributaries—both

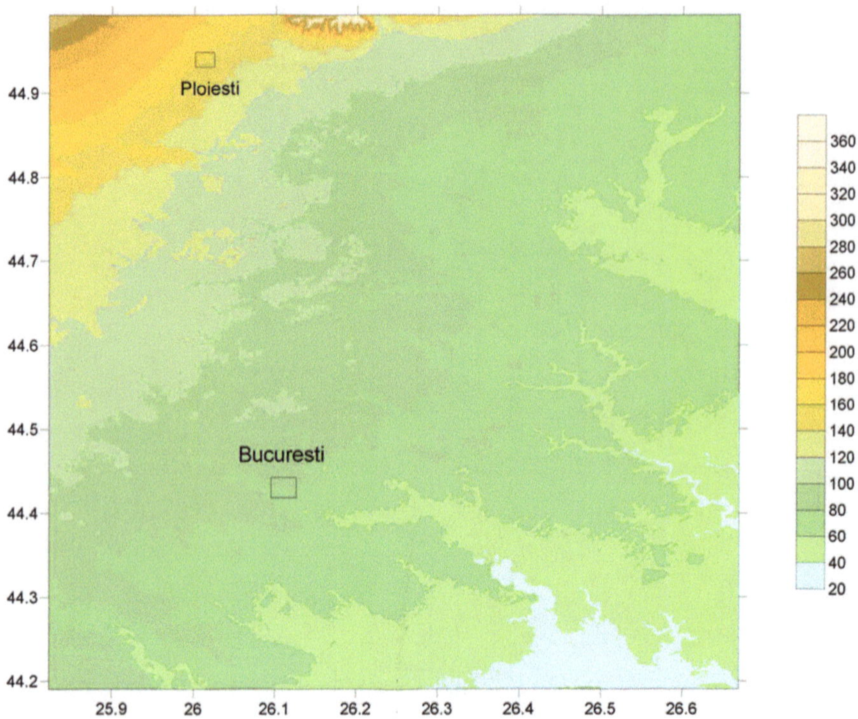

Fig. 2.2 Regional geomorphology between Bucharest and Ploiesti cities (topographic heights in meters)

rivers receive only left side tributaries, a feature obviously related to the riverbanks asymmetry already mentioned (higher right bank as compared to the left bank.

Colentina river, which crosses the northern part of the city, displays a meandering course especially at the northern Bucharest outskirts. Since numerous peat bogs developed between the river meanders, during the first half of the twentieth century these areas were transformed into lakes: Străuleşti, Băneasa, Herăstrău, Floreasca, Tei, Fundeni and Pantelimon.

Dâmboviţa river used to have a meandering course especially when crossing the Bucharest central part (Oltean 2005). Its course was modified into a straight one during the nineteenth century within the city limits and its tributaries may no longer be observed. Some of its former meanders were transformed into lakes: Carol and Tineretului lakes on the right bank and Cişmigiu Lake on the present day left bank (Mândrescu et al. 2008).

The main directions of the rivers network, analyzed at a regional scale, are as follows (Fig. 2.3):

- NW–SE: Mostiştea, Colentina, Dâmboviţa, Sabar, Argeş;
- W–E: Câlniştea, Argeş, Ialomiţa;
- NE–SW: tributaries of Ialomiţa, Mostiştea and Colentina

Fig. 2.3 Rivers network in the region where Bucharest is located

The two rivers which once favored the defense of medieval Bucharest, Dâmboviţa and Colentina, cross from NW to SE in the present time city area. Both river valleys displayed once large meanders within the old city centre.

Older city maps show small streams on the left side of the Dâmboviţa; the one that trends NNE–SSW started south of Victoria Square and the one that trends NNW–SSE started in Ioanid Park and crossed Rosetti Square. The Colentina river still keeps some of its tributaries (Pipera and Saulea streams) which trend NNE–SSW and NW–SE, respectively (Lăcătuşu et al. 2008).

2.3 Geological, Tectonic and Neotectonic Data

2.3.1 Geological Setting

Bucharest is situated on the Wallachian sector of the Moesian Platform, west of the Intramoesian Fault (Sandulescu 1984).

The crystalline basement is here situated at depths ranging between 6 and 8 km (Polonic 1998) and was not reached by boreholes.

The sedimentary cover was built during four main sedimentary sequences: Palaeozoic, Permian-Triassic, Lias-Upper Cretaceous and Middle Miocene-Holocene (Paraschiv 1979).

The Upper Cretaceous-Badenian unconformity is gently deepening between the Danube river and Bucharest, but starting from Bucharest it goes rapidly deeper toward the Carpathians.

Within Bucharest city limits the Upper Cretaceous sedimentary deposits were found in boreholes in its southern part at about 1,000 m, while in the northern part of the town they were intercepted at more than 2,000 m.

The thickness of the Middle Miocene-Holocene sequence increases rapidly in Bucharest from south to north, from 452 m at Berceni area to 2053 m at Otopeni area (Lăcătuşu et al. 2008), a distance of about 20 km.

The Quaternary unconsolidated geological formations have thickness variations ranging between 150 m at the city southern limit (Măgurele area) and more than 250 m (Otopeni area).

The Lower Pleistocene deposits (Beds of Frăteşti) consist of gravels (pebbles of quartzite, gneiss, micaschist, granite, conglomerate) and sands, separated by clay intercalations. This formation is continuously deepening northward: 80 m depth in the city southern part and 180 depth at its northern limit). The map at their upper surface in the Bucharest area shows NE–SW lineaments of sudden deepening (Mândrescu et al. 2008).

The Middle Pleistocene consists of sand and clay beds alternances (Beds of Uzunu).

The Upper Pleistocene consists of the Mostiştea sands (8–20 m), the intermediate clays (5–20 m), the Colentina gravels (thickness of 4–18 m, with pebbles of micashist, gneiss, sandstone) and loess. In the southern part of Bucharest the intermediate clay beds are lacking, the Mostiştea sands being in continuity with the Colentina gravels, both reaching 20–30 m in thickness.

The loess deposits have thickness of 3–5 m between Colentina and Dâmboviţa rivers, and 15–20 m south of Dâmboviţa river. They include within Bucharest city limits 2–3 Pleistocene and Holocene paleosol levels (Mândrescu et al. 2008).

2.3.2 *Tectonics*

2.3.2.1 Regional Tectonics

Bucharest is situated in the central part of the Moesian Platform, in the vicinity of the transcrustal Intramoesian Fault, considered as a senestral strike-slip regional fault (Cornea and Lazarescu 1982). The classical tectonic and geodynamic picture of the eastern Moesian Platform considers a tectonic block having as south-western limit the Intramoesian Fault (senestral) and as north-eastern limit the Peceneaga-Camena Fault (dextral). This crustal block is displaced toward north-west and is involved in the Vrancea zone seismicity (Cornea and Lazarescu 1982).

The dominant crustal fault systems in the Moesian eastern compartment trend NW–SE, while the western compartment displays a more complex picture, with faults trending NW–SE, N–S, W–E and NE–SW.

Recent interpretations of reflection seismic data attested that the NW–SE fault system affects also the Tertiary sediments, while the ENE–WSW system is active until present as transtensional strike-slip. During the evolution of this strike-slip system the NW–SE transpressional system was reactivated (Rabagia et al. 2000).

Bucharest is closely located in the contact zone between an older eastern compartment (Dobrogean) and a younger western compartment (Wallachian) along the NW–SE trending Intramoesian Fault system. Considering the regional relationship between the Moesian Platform and the Carpathian Mountains, the city northern limit overlaps the tectonic limit between the platform and the outer zone of the Carpathian foredeep (Sandulescu 1984; Visarion et al. 1998).

The reflection seismic surveys carried out for oil exploration in the Bucharest area depicted in the sedimentary cover a regional northward deepening monocline, affected by regional longitudinal normal faults trending E–W: Petreşti-Corbii Mari, Cartojani-Ileana, Fundeni-Harleşti, Videle-Vidra (Paraschiv 1979).

It was recently considered that in the Bucharest area the intersection of two major strike-slip systems are responsible for the active tectonics and high seismicity. Important wrench zones occur south of Bucharest, involving the Dobreni and Danube faults (Rabagia et al. 2000). Regional wrench tectonics was also suggested when interpreting recent GPS data (Munteanu 2009).

2.3.2.2 Local Tectonics

Geological cross-sections based on borehole data illustrate faults affecting the Upper Miocene-Holocene sedimentary formations beneath Bucharest, or only suggest such tectonic features at the upper part of the Quaternary sedimentary deposits:

1. A normal fault, interpreted on geological borehole information, is situated south of the Dâmboviţa river, affecting at least the Pliocene geological formations on more than 400 m depth (Fig. 2.4). Considering the coal intercalations in clay beds, better developed within Upper Romanian-Lower Pleistocene sedimentary deposits, the vertical displacement of its compartments may be approximated at 200 m. Since there are no possibilities of geologically correlating the two fault compartments based on data from seven boreholes on a N–S distance of 30 km, horizontal displacements along this fault must be also taken into consideration;
2. The necessity of interpreting vertical displacements within the geological structures in the first 50 m beneath the topography is well illustrated in a geological cross-section trending NE–SW, between Giuleşti and Bragadiru areas, a distance of ca 12 km that was investigated by more than 20 boreholes (Fig. 2.5).

There are a number of reasons to interpret local faults in this very detailed geological cross-section:

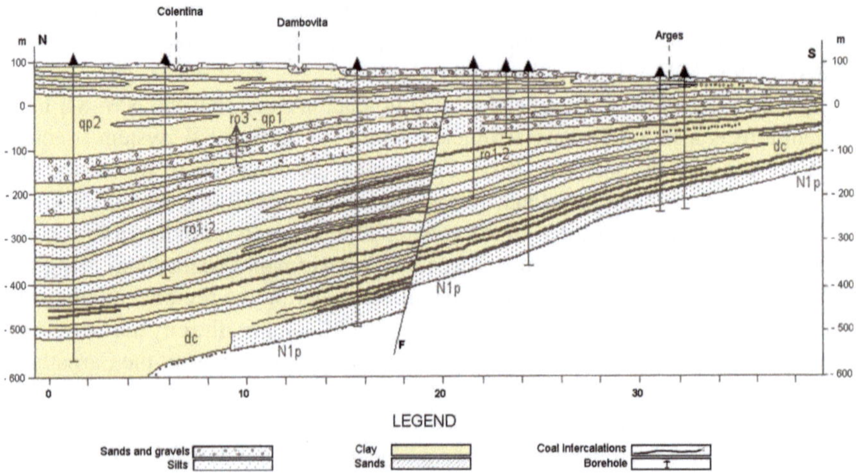

Fig. 2.4 Geological section across Bucharest city area (Lacatusu et al. 2008)

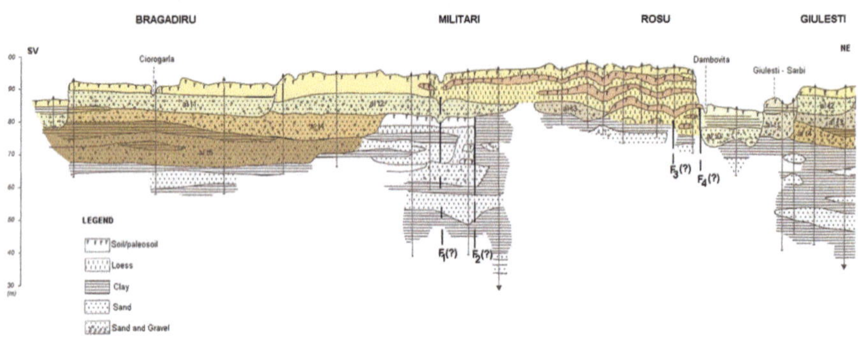

Fig. 2.5 Geological section crossing the western part of Bucharest city area (Modified after Lacatusu et al. 2008)

- The Dâmboviţa river banks display a sudden variation in surface topography (the south-western bank is higher than the north-eastern one) and in geology (the south-western bank is formed of loess and paleosols accumulated during the last 400,000 years, while the north-eastern bank is formed by alluvium deposits accumulated during the last 240,000 years). Furthermore, the boundary between the mentioned geological formations and the underlying clay bed is situated at lower levels beneath the north-eastern bank and at successively higher levels beneath the south-western bank;
- In the Militari district the alternances of loess and paleosols ends at lower levels beneath this area, thick deposits of buried loess, sands and gravels being rapidly replaced by thick clay deposits between two closely situated boreholes (Fig. 2.5).

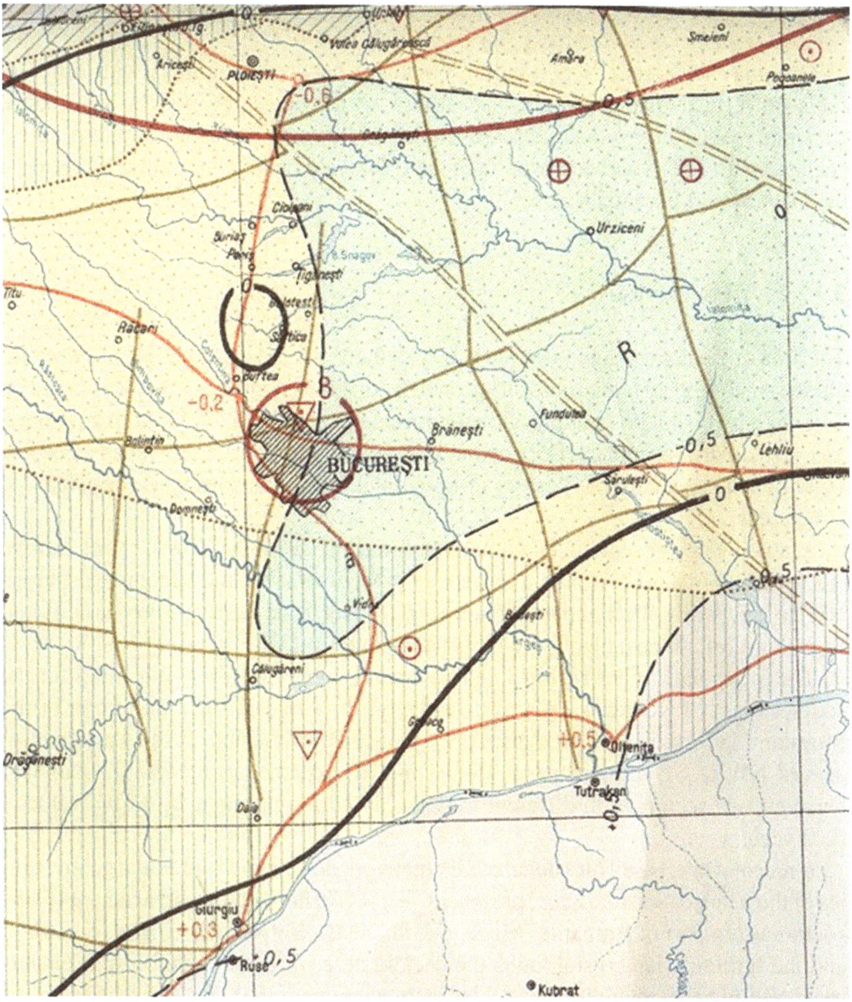

Fig. 2.6 Neotectonic data in the Bucharest city area (*Black lines*: isolines of similar rate of vertical movement; Brown lines: faults; Triangles and circles: earthquake epicenters) (Visarion et al. 1977)

2.3.3 Neotectonics

2.3.3.1 Recent Vertical Movements of Crustal Blocks

Modifications during the time of relief heights, resulted from repeated topographic leveling in a region affected by active geodynamic processes, have been utilized since the mid-twentieth century for assessing actual vertical positive (uplift) or negative (subsidence) movements of crustal blocks.

For the territory of Romania there are several maps published by different authors describing sectors with similar rates of vertical displacements (uplifting or subsiding areas on the Romanian territory). In Fig. 2.6 is presented a map of the region where

Bucharest is located, extracted from the 1: 1,000,000 scale *Map of recent crustal vertical movements in Romania* (Visarion et al. 1977).

One may observe a N–S trending boundary between two areas with different rates of vertical displacements, the eastern sector being affected by subsidence (0.5 mm/year). This boundary crosses the central part of Bucharest and may be associated with a N–S active crustal fault, also depicted on this map, situated between the Danube and Ialomiţa rivers.

2.3.3.2 GPS Monitoring

The area of Bucharest, due to its proximity to the Vrancea seismic zone, has been included since 1995 in numerous European or Romanian geodynamic projects based on GPS high precision positioning and monitoring of crustal blocks, for both horizontal and vertical displacements. Data on the velocity field have been published since late 1990, when especially the vectors of horizontal movements became consistent for large areas.

A study that presents data accumulated during 1994–2003 in Central Europe (Hefty 2004) offers horizontal displacements for Bucharest and Vrâncioaia stations NW to SE, while the Fundata station (the contact zone between the Eastern and Southern Carpathians) displays a higher horizontal velocity, trending N to S.

The interpolated GPS regional velocity field (Grenerczy 2004) illustrates for the southern part of Romania a general trend from NW to SE. The central part of the Moesian Platform in Romania displays a transit from WNW to ESE (western part) toward NW to SE (eastern part), suggesting the presence of a NW–SE boundary between two sectors with different trending and different intensity of the velocity field vectors.

A recent work, based on a detailed GPS network dedicated to the actual geodynamics of the Vrancea seismic zone, presented vectors for horizontal displacements in the south-eastern part of Romania (Mocanu et al. 2008). Between the Capidava-Ovidiu and the Intramoesian crustal faults the vectors describing horizontal displacements trend NW to SE, with higher intensities in the vicinity of the Capidava-Ovidiu Fault. West of the Intramoesian Fault the velocity field is less homogeneous and less intense, the vectors trending W to E north of Bucharest, N to S within Bucharest and NNW to SSE at Măgurele, a few kilometers south-west of Bucharest. GPS network stations situated at ca 50 km west and north-west of Bucharest illustrate a more homogeneous velocity field, the vectors trending NW to SE or NNW to SSE. In the latter case, this trend is similar to that obtained at Măgurele, at the outskirts of Bucharest.

Quantitative data on the tectonic blocks horizontal movements were recently published, based on 5 years of GPS observations. For the territory situated north-east of the Intramoesian Fault south-eastward horizontal displacements of 2 mm/year were determined, while in the Bucharest city area, located south-west of the Intramoesian Fault, southward horizontal displacements of 1 mm/year were observed (Munteanu 2009).

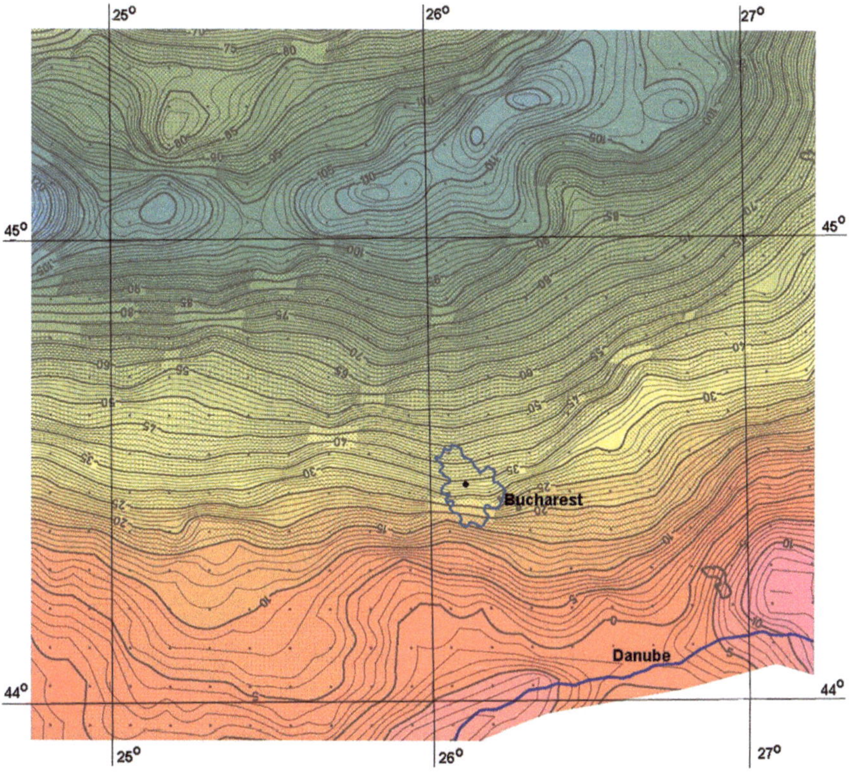

Fig. 2.7 Bouguer gravity anomaly map built on 5' × 7.5' mean data (Ioane and Ion 1992)

2.4 Geophysical Data

2.4.1 Gravity Data

The Bouguer gravity anomalies are dominated in the Moesian Platform by a rapid northward decreasing of gravity toward the Carpathians (Nicolescu and Roşca 1992; Visarion 1998), due to both sedimentary cover and crust thickening in this direction.

Smaller gravity effects, determined by fault systems, are generally covered by the high intensity anomalies due to the above mentioned geological structures. However, the Bouguer gravity anomaly map built on the 5' × 7.5' mean gravity values data set (Ioane and Ion 1992) display anomalous lineaments trending NE–SW between the latitude of Bucharest and the Carpathians, a result of the overlapping of two horizontal gradient anomalies (Fig. 2.7):

1. The W–E trending horizontal gradient anomaly, determined by the density contrasts between the sedimentary cover and the crystalline basement (W–E trending normal faults) and between the crust and the upper mantle (crustal flexure);

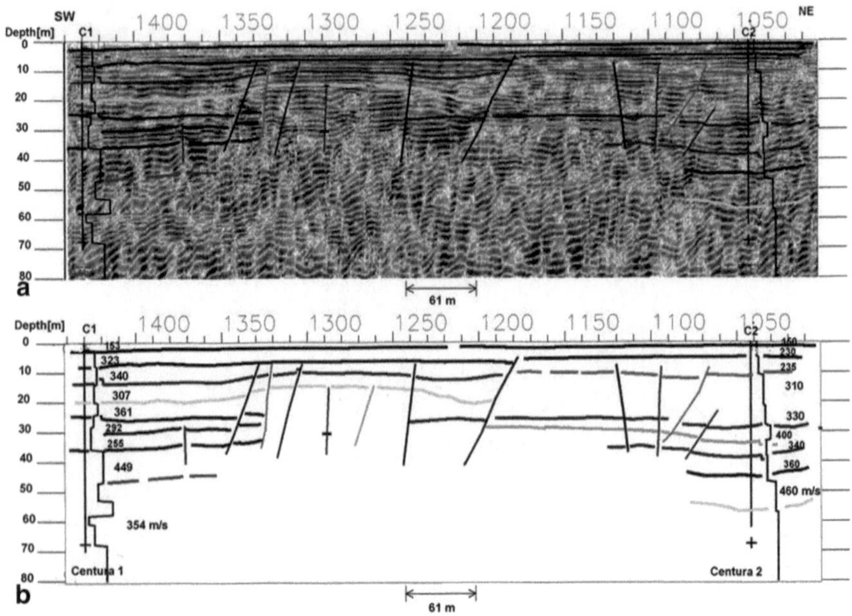

Fig. 2.8 Reflection seismic profile within Bucharest city (Bala et al. 2006)

2. The NE–SW trending horizontal gradient anomaly, determined by 'Vrancean' normal faults affecting especially the sedimentary cover and possibly the crystalline basement.

Of particular significance for local tectonics are considered the NE–SW horizontal gradient lineaments located east and west of Bucharest, the eastern one crossing the south-eastern limit of the city area.

2.4.2 Reflection Seismic Data

Shallow reflection seismic investigations were carried out within the city area for a maximum depth of 25 m. The obtained seismic velocities range between 300 and 1800 m/s for the longitudinal waves and between 90 and 230 m/s for the transversal waves. The data presented in the seismic section shown in Fig. 2.8 were recorded in the eastern part of Bucharest (Căţelu district) on a 600 m long profile, situated between two boreholes, C_1 and C_2 (Bala et al. 2006). The interpreted faults and fractures displace the near-surface reflectors up to 10 m depth beneath the surface (Fig. 2.8). A part of this micro-tectonic features were attributed to effects of unconsolidated sediments during glacial and peri-glacial time intervals (Mândrescu et al. 2008).

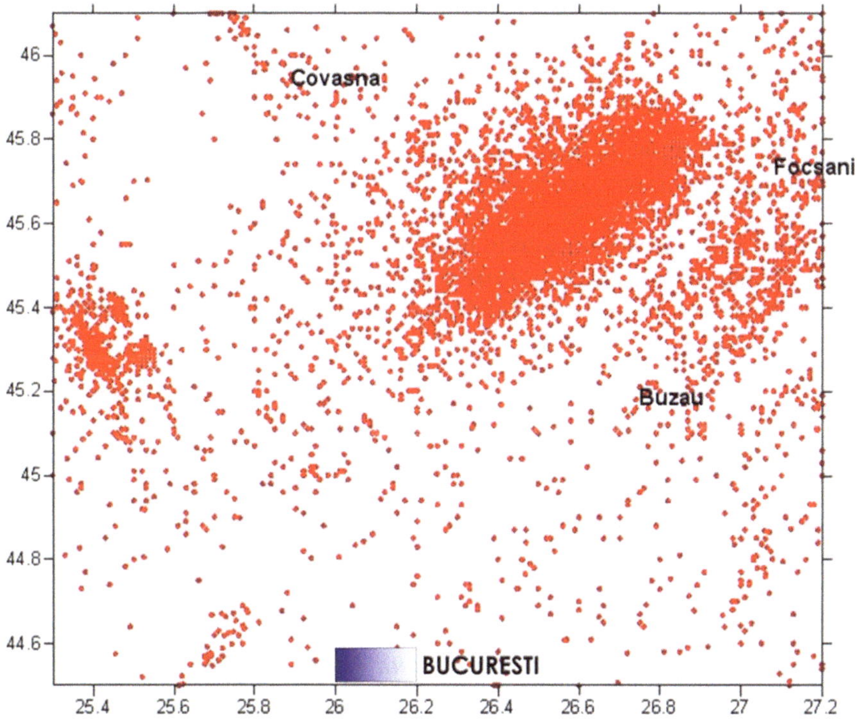

Fig. 2.9 Regional seismicity in the south-eastern part of Romania

2.5 Seismological Data

2.5.1 Regional Seismicity

Regional seismicity is mainly related in Romania to the Vrancea seismogenic zone, situated at the bending sector of the Eastern Carpathians (Fig. 2.9). The occurring earthquakes have hypocenters at both crustal and intermediate depths. It was previously suggested that the earthquake epicenters correspond within the Moesian Platform with intersections between two main fault systems: NW–SE and NE–SW (Cornea and Lazarescu 1982).

Large magnitude earthquakes occurred at intermediate depths in 1940, 1977, 1986 and 1990 during the twentieth century with destructive effects over large distances including Bucharest.

The Vrancea March 4, 1977 high magnitude earthquake was a multiple shock event, whose main shocks epicenters propagated in time from NE to SW with increasing depths, along a N40^{0}E fault plane along approximately 100 km (Enescu et al. 1982). The seismic intensity map based on information collected on the whole Romanian territory is similar to that obtained after the November 10, 1940 high magnitude Vrancea earthquake, displaying highest intensities on areas located in the Vrancea zone and in Bucharest city (Radu and Polonic 1982).

A large number of fractures and fissures were mapped by direct observation after the March 4, 1977 earthquake (Radu and Polonic 1982):

- In the Vrancea zone the main fractures were located close to already mapped N–S trending faults in the Jitia-Valea Sării area;
- In the Berca sector a NNE–SSW fracture activated muddy volcanoes;
- In the area situated north-west of Bucharest the newly created fractures between the Dâmboviţa and Argeş rivers were directed NW–SE;
- For the rest of the Moesian Platform the fractures and fissures showed N–S and W–E directions.

Besides the active NE–SW fault system, involved in the Vrancea zone seismicity with seismological significance at continental scale, the N–S trending system seems to be important for Bucharest at regional, and even local scale. The distribution of earthquake epicenters recorded between March 4 and April 30, 1977 displays events occurred at intermediate depths (NE–SW) and crustal depths (N–S and NW–SE) (Fig. 2.10—Radu and Polonic 1982).

2.5.2 Earthquakes in Bucharest and Local Seismicity

The November 1940 earthquake ($M = 7.4$) has damaged a large number of buildings in Bucharest but only one collapsed—the 13 floors Carlton block situated in the city centre. A map describing the distribution of three categories of areas with buildings damaged by this seismic event was prepared by Beles including proposals for urgent restoration (Mândrescu 1992).

The March 1977 earthquake ($M = 7.2$) caused the collapse of 32 buildings having 8–12 floors, more than 250 buildings being heavily affected. Most of the collapsed or damaged buildings were located in the central part of Bucharest and were already mentioned in the list with heavily damaged buildings after the 1940 earthquake. An evaluation of the damages determined by the 1977 seismic event, based on the analysis of 2,500 buildings was prepared and published by Mândrescu (1992).

The earthquakes that occurred in the Bucharest area, whose main characteristics are presented in Table 2.1, were selected using the ROMPLUS catalogue elaborated by the National Institute for Earth Physics, Romania. This catalog includes the existing information published in previous catalogues, such as Constantinescu and Marza (1980), Oncescu et al. (1999) and Trifu and Radulian (1991), covering the time interval 984–1997. After 1997, the catalogue was completed and continuously updated with seismic data on events produced in Romania and in its vicinity. For locating earthquakes, the ANTELOPE software was employed, the Hypoinverse 2000 module performing this task.

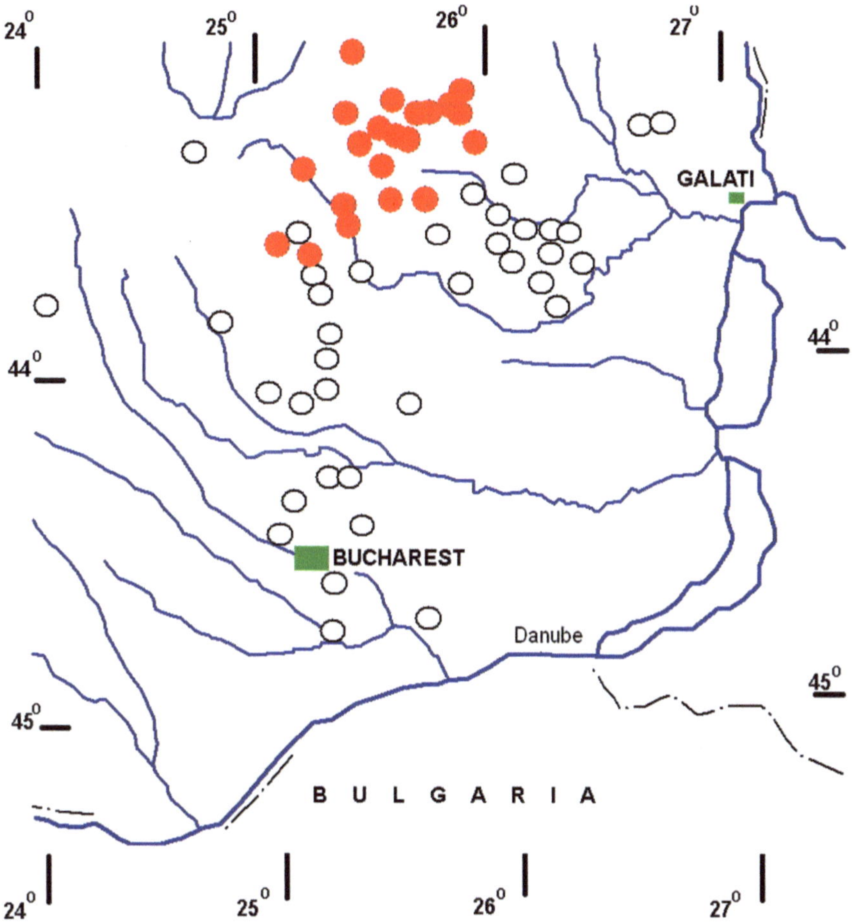

Fig. 2.10 Regional distribution of seismic events epicenters during March–April 1977 (*red dots*: intermediate seismicity; *white dots*: crustal seismicity) (Radu and Polonic 1982)

Conversion relationships in Mw of different type of magnitudes:

No	Equations	References
1	$Mw = Ms$ for lat > 43N	Oncescu et al 1999; Grunthal and Wahlstrom 2003
2	$Mw = 0.56Ms + 2.66$ if $Ms < 5.3$ for lat < 43N	Papazachos and Papazachou 1997
3	$Mw = 0.804Ms + 1.28$ if $Ms \geq 5.3$ for lat < 43N	Burton et al. 2004
4	$Mw = 1.28Mb - 1.12$	Papazachos and Papazachou 1997
5	$M_w = 0{,}52M_D + 1{,}1$ if $2{,}9 \leq M_D \leq 4{,}7$ for $h < 40$ km	Oncescu and Bonjer 1997
6	$M_S = -2{,}14 + 1{,}43\,M_L - 0{,}018\,M_L^2$ if $M_L \geq 5{,}0$	Bath 1973
7	$M_S = 1{,}59\,m_b - 3{,}97$ if $h < 40$ km then $M_w = M_S$	Gutenberg and Richter 1956
8	$M_w = 0.66I_{max} + 1.23$ if $I_{max} \geq VI$	Radu 1979
9	$M_w = 0{,}74M_D + 0{,}80$ if $2{,}6 \leq M_D \leq 5{,}0$ for $h > 40$ km	Oncescu and Bonjer 1997
10	$M_w = 0{,}92M_{GR} + 0{,}81$ if $5{,}3 \leq M_{GR} \leq 7{,}4$ for $h > 40$ km	Oncescu and Bonjer 1997

Table 2.1 Earthquakes recorded in the Bucharest area. (ROMPLUS catalogue)

No.	Year	Month	Day	Hour	Minute	Second	Latitude (°N)	Longitude (°E)	Depth (km)	Magnitude (Mw)
1	2004	9	18	3	14	24	44.16	25.56	6.4	2.8
2	2009	6	20	14	10	42	44.17	25.72	15.2	3.1
3	2003	3	10	0	13	51	44.18	26.52	5.0	2.9
4	2003	3	7	11	34	50	44.18	26.54	1.2	2.8
5	2009	6	22	23	35	17	44.20	25.73	11.9	2.6
6	1905	11	18	13	40	0	44.20	26.20	9.9	3.2
7	2008	3	2	2	24	6	44.24	26.34	12.9	2.6
8	2003	2	3	8	22	23	44.24	26.52	16.6	2.6
9	1977	4	20	21	16	15	44.26	26.27	8.2	3.0
10	1977	4	20	22	17	23	44.27	26.12	10.0	2.0
11	2007	7	12	0	15	45	44.27	26.49	18.7	2.5
12	2009	10	5	15	23	53	44.29	25.92	2.3	2.6
13	1900	6	22	14	11	40	44.30	26.30	9.9	2.6
14	1999	7	21	9	32	6	44.30	26.48	6.7	2.6
15	1988	11	19	22	23	50	44.31	26.53	10.7	2.7
16	2009	11	21	4	48	50	44.33	25.69	11.8	2.7
17	2009	9	21	21	34	10	44.36	25.93	5.2	2.3
18	2009	5	7	23	58	31	44.36	26.38	5.0	2.6
19	2008	3	7	22	29	52	44.37	26.08	20.0	2.6
20	1986	1	31	0	8	28	44.38	25.78	4.3	3.0
21	1982	6	26	21	57	42	44.38	26.20	24.5	1.6
22	2009	10	9	1	11	51	44.40	26.51	20.6	2.3
23	1907	8	13	2	12	0	44.42	26.15	15.0	2.6
24	2008	2	8	20	21	20	44.42	26.44	3.9	2.6
25	1977	3	29	17	16	28	44.43	26.33	57.6	2.5
26	1997	5	16	10	57	16	44.44	26.06	30.0	2.6
27	1998	11	3	5	14	31	44.46	26.37	12.4	2.5
28	2008	1	28	6	32	37	44.47	25.60	3.0	2.6
29	2007	12	27	23	13	35	44.49	25.60	5.0	2.3
30	1977	4	30	1	37	9	44.49	26.22	10.0	0.0
31	1982	2	20	13	52	49	44.49	26.34	30.0	1.6
32	2007	12	28	20	43	44	44.50	25.61	5.0	2.6
33	2008	1	4	20	43	35	44.50	25.62	5.0	2.6
34	1997	11	1	10	12	32	44.50	26.39	18.0	3.0
35	1985	5	11	18	24	33	44.50	26.40	5.0	1.6
36	2007	11	8	23	8	36	44.52	25.52	30.0	2.5
37	2008	1	18	9	32	26	44.52	25.62	6.1	2.6
38	1982	5	6	16	27	25	44.53	26.05	20.0	2.1
39	2007	7	20	9	15	2	44.53	26.14	15.0	2.6
40	1996	3	17	23	35	29	44.53	26.42	0.0	2.6
41	1991	8	21	4	54	44	44.54	26.17	4.7	2.6
42	2007	11	9	1	55	22	44.55	25.64	10.0	2.5
43	2008	1	16	7	26	20	44.55	25.68	1.6	2.6
44	2008	1	14	0	53	51	44.55	25.69	3.0	2.6
45	1983	8	25	7	59	4	44.55	25.89	25.0	3.1
46	1977	4	1	17	4	58	44.55	26.07	8.8	2.1
47	1988	11	20	10	43	38	44.55	26.15	6.4	2.7
48	1984	2	12	2	26	13	44.55	26.33	7.2	1.5
49	2007	12	27	4	20	58	44.56	25.68	1.7	2.9
50	2002	10	20	9	14	25	44.56	26.30	20.6	2.6
51	1977	3	28	22	20	4	44.56	26.41	24.5	2.9
52	2008	1	18	18	3	39	44.57	25.67	4.7	2.5

Table 2.1 (continued)

No.	Year	Month	Day	Hour	Minute	Second	Latitude (°N)	Longitude (°E)	Depth (km)	Magnitude (Mw)
53	2008	1	22	18	12	13	44.57	25.68	3.0	2.4
54	2007	12	30	14	56	54	44.57	25.69	3.0	2.6
55	2003	11	2	1	33	8	44.57	25.72	51.4	2.8
56	2008	1	21	0	40	8	44.58	25.69	3.0	2.5
57	1995	9	8	11	34	13	44.58	26.02	5.0	2.5
58	2009	1	31	4	30	26	44.58	26.28	16.0	2.5
59	2004	4	30	18	51	43	44.58	26.40	0.0	2.5
60	2008	1	19	16	26	36	44.59	25.68	5.8	2.5
61	2008	1	26	12	15	30	44.59	25.72	3.0	2.7
62	1985	7	27	0	35	6	44.59	25.93	20.7	1.2
63	1990	6	6	19	26	16	44.59	26.15	37.5	2.6
64	2008	1	12	1	4	3	44.60	25.74	3.0	2.5
65	2007	12	27	2	54	5	44.60	25.77	3.0	2.7
66	2003	2	19	7	22	34	44.60	26.28	50.0	2.4
67	1984	3	23	2	26	58	44.61	25.89	2.0	2.3
68	2007	12	27	8	16	13	44.62	25.70	4.6	2.8
69	2008	1	18	14	56	52	44.62	25.72	3.0	2.6
70	2005	9	27	14	50	41	44.62	25.76	3.0	2.5
71	2005	9	27	14	50	41	44.62	25.76	15.5	2.9
72	2007	4	27	0	23	46	44.62	26.19	4.4	2.5
73	2007	12	27	12	16	44	44.63	25.71	9.7	2.7
74	2008	1	5	10	58	17	44.63	25.74	3.0	2.7
75	2008	1	5	8	23	30	44.63	25.75	3.0	2.6
76	1977	4	6	11	21	23	44.63	26.27	20.0	2.8
77	2008	1	20	14	12	42	44.64	25.77	3.0	2.5
78	2008	1	20	14	12	42	44.64	25.77	0.7	2.7
79	1977	3	21	20	34	34	44.64	26.15	4.8	1.9
80	2008	1	24	1	2	1	44.65	25.75	3.0	2.5
81	2008	2	19	4	24	31	44.65	25.81	3.9	2.7
82	2008	1	26	4	4	58	44.66	25.73	9.8	3.1
83	2002	7	4	9	26	30	44.66	25.92	20.2	2.2
84	1998	1	24	7	41	43	44.66	26.41	8.3	2.4
85	2008	1	22	23	18	43	44.67	25.69	3.0	2.6
86	2008	1	11	14	56	37	44.67	25.72	13.0	2.9
87	2008	1	16	17	22	23	44.67	25.75	8.2	2.8
88	2008	1	21	0	54	58	44.67	25.78	3.0	2.6
89	2007	12	29	7	50	17	44.68	25.78	3.0	2.7
90	2008	1	26	5	39	31	44.68	25.87	3.0	2.6
91	1997	2	6	12	38	37	44.68	26.35	15.0	2.7
92	2007	12	27	11	37	44	44.69	25.77	1.2	2.8
93	1996	10	28	13	54	53	44.69	26.32	7.4	2.7
94	1977	4	1	7	26	51	44.69	26.35	2.4	2.1
95	1993	4	6	18	6	28	44.69	26.50	4.9	2.5
96	1977	4	13	1	6	8	44.70	26.33	70.6	1.9
97	2007	4	24	3	58	24	44.71	26.25	16.9	3.1
98	1993	7	14	11	55	32	44.71	26.31	20.2	2.5
99	1996	12	21	18	21	18	44.71	26.53	0.0	2.5
100	1995	4	27	3	54	36	44.72	25.88	19.4	2.5
101	2007	7	29	9	30	35	44.73	26.33	9.1	2.8
102	1992	12	3	8	27	59	44.73	26.45	3.9	2.4
103	1992	12	3	8	27	59	44.73	26.45	0.0	2.6
104	2001	7	23	11	59	36	44.74	26.19	30.0	2.6
105	2007	7	5	0	32	59	44.74	26.32	10.4	2.9

Considering the local seismicity in Bucharest, a detailed representation of local seismicity with crustal low magnitude seismic events shows seismic epicenters within the city limits, or close to its outskirts (Fig. 2.11). The hypocenters' depths generally range between 5 and 35 km (crustal earthquakes), but most of them occurred between 5 and 27 km (Fig. 2.12). The deepest seismic event, which occurred at 70 km depth, is located within the central part of the city. The earthquakes' magnitude have so far ranged between 2 and 3.2.

New observations using an improved regional seismological network evidenced, on a lineament trending NE–SW, a consistent seismogenic zone situated between Răcari and Cartojani localities, within the north-western vicinity of Bucharest. The other epicenters, which look quite scattered in this map, cannot be easily associated with known faults. However, they may be linked at a regional scale to the 'Vrancean' NE–SW trending fault systems.

The Răcari-Cartojani seismic lineaments do not represent a significant danger for the densely populated Bucharest area, the greatest magnitude being so far $M_w = 3.1$, for an event recorded on January 26, 2008.

The highest magnitude of the whole analysed area was determined for an earthquake with $M_w = 3.2$ that occured at 10 km depth, recorded on November 18, 1905.

2.5.3 *Local Seismicity and Urban Damages*

After the 1940 earthquake and the still unexplained collapse of the Carlton building in Bucharest centre, interest in seismic protection in civil engineering planning was continuously increasing. Since 1953 the seismic protection in Bucharest has been based on seismic microzonation maps.

The effects of the 1977 earthquake in Bucharest showed significant discrepancies between the areal distribution of the main damages and the existing seismic zonation. The highest expected seismic intensities were previously evaluated along the Dâmboviţa and Colentina rivers (Mândrescu 1972), while major damages to buildings were observed mainly in the city centre.

To explain the specificity of the March 4, 1977 earthquake, the following observations were considered: (1) multiple shock seismic event; (2) south-westward displacement of the rupture; (3) gradual increase of the generated energy.

A statistical evaluation of the highly damaged buildings after March 4, 1977 showed that the Magheru boulevard sector was less affected as compared to the eastern part of the city centre, this fact being mainly associated by civil engineers with errors in designing or constructing those buildings.

However, the concentration of the highest damages in the centre of Bucharest in an area of ca 5 km in diameter suggests other causes, besides construction errors. Among possible causes such as unknown deep or shallow geological structures, such as the Frăteşti gravels and associated aquifer, faults that are considered to be involved in locally increasing the seismic intensity (Sandi and Perlea 1982).

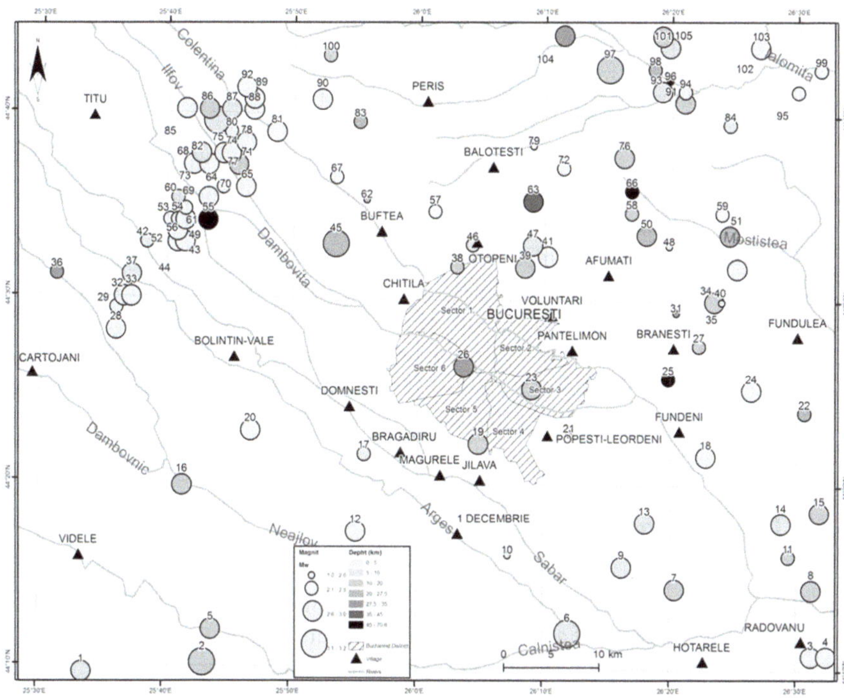

Fig. 2.11 Earthquake epicenters in the Bucharest city area

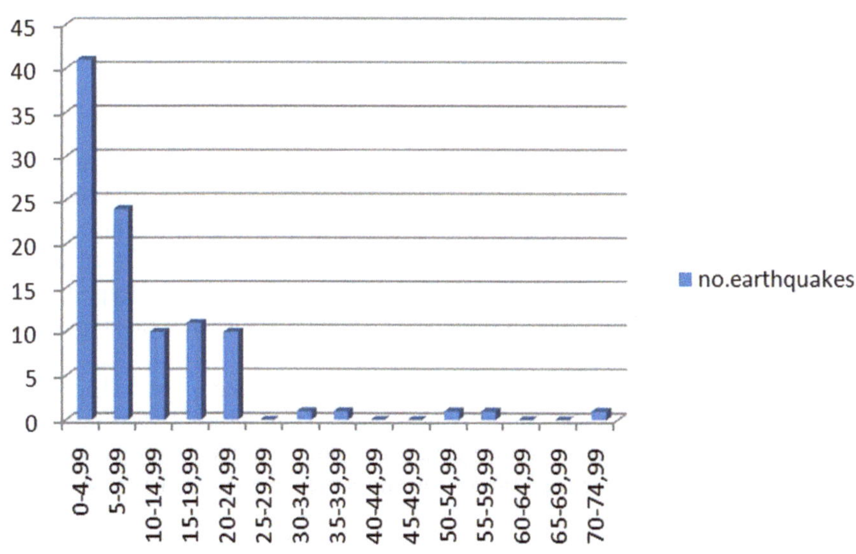

Fig. 2.12 The depth versus number of earthquakes plot for seismic events that occurred in the Bucharest area

2.5.4 *Earthquake Focal Mechanisms in the Bucharest Area*

To calculate earthquake focal mechanisms the HASH 1.2 software (www.usgs.gov) was used, a Fortran 77 code that computes earthquakes' focal mechanisms from P wave motion. For a graphic representation and interpretation of focal mechanism parameters the Earthquakes Focal Mechanism software (Wolfram Demonstration Project—Scherbaum et al. 2010) was employed.

Selected data on the focal mechanisms computed for earthquakes recorded in the Bucharest area are presented in Table 2.2. As it may be observed the selection of seismic events is quite consistent, with recordings at 8–21 seismological stations, the computed focal mechanisms showing mostly reverse faults and in a few cases, strike-slip solutions. A detailed analysis and interpretation of the kinematics of fault systems will be carried out in the next stage of this study.

2.6 Data Interpretation

Considering the reluctance of specialists in civil engineering or geotechnical works in accepting the presence of active and/or capable fault systems within the Bucharest area in Quaternary geological formations, various data coming from geomorphology (rivers networks, tectonics, neotectonics, geophysics and seismology) through an integrated interpretation attest to the presence of such tectonic features as were selected in this paper.

2.6.1 *Geomorphological Data*

The sudden change in the topography along a N–S trending lineament between Bucharest and Ploieşti cities (Fig. 2.2), which is in good agreement with the neotectonic information for actual crustal vertical movements (Fig. 2.6), illustrate the presence of a N–S regional fault representing the western boundary of a subsiding area (0.5 mm/y). It could represent an active transcrustal fault, the deepest seismic event recorded within Bucharest (depth = 70 km) being located on its southern prolongation.

The geomorphology in the region where Bucharest is located, which shaped the topographic limit between plain and hilly areas, suggest the presence of two important fault systems: (1) NE–SW, similar to the trend of the main fault lines detected by means of high seismicity in the Vrancea seismic zone; (2) N–S, as discussed above.

Table 2.2 Focal mechanism catalogue—Bucharest city area. Latitude 44.20–44.70/Longitude 25.50–26.60

No.	Year	Month		hh:mm:ss	Lat.N	Long.E	Depth (km)	MD	PLANE 1			Fault type	PLANE 2			Fault type	NS
									Strike	Dip	Slip		Strike	Dip	Slip		
49	2007	12	27	4:20:58	44,56	25,68	1,7	3	281	76	149	Reverse right-lateral oblique	19	60	16	Left-lateral strike-slip	12
75	2008	1	5	8:23:30	44,628	25,75	3	2.9	63	62	51	Reverse left-lateral oblique	304	47	140	Reverse right-lateral oblique	8
74	2008	1	5	10:58:16	44,626	25,74	3	3	139	49	119	Reverse right-lateral oblique	278	48	61	Reverse left-lateral oblique	9
86	2008	1	11	14:56:37	44,67	25,72	13	3.4	31	67	150	Reverse right-lateral oblique	133	63	26	Reverse left-lateral oblique	13
43	2008	1	16	7:26:20	44,551	25,68	1,6	2	9	84	159	Reverse right-lateral oblique	101	69	6	Reft-lateral strike-slip	8
87	2008	1	16	17:22:23	44,672	25,75	8	3	277	61	138	Reverse right-lateral oblique	31	54	36	Reverse left-lateral oblique	8
82	2008	1	26	4:04:57	44,664	25,73	9	3	264	24	36	Reverse left-lateral oblique	140	76	110	Reverse right-lateral oblique	14
39	2008	2	19	3:17:52	44,643	25,77	0,7	3.1	232	63	119	Reverse right-lateral oblique	1	39	46	Reverse left-lateral oblique	9
18	2009	5	7	23:58:31	44,358	26,383	5	2.9	189	17	92	Reverse	7	73	89	Reverse	8
5	2009	6	22	23:35:17	44,204	25,734	11,9	2.8	244	66	28	Reverse left-lateral oblique	142	65	153	Reverse right-lateral oblique	21
17	2009	9	21	21:34:10	44,364	25,928	5,2	2.3	290	47	148	Reverse right-lateral oblique	43	67	47	Reverse left-lateral oblique	18
12	2009	10	5	15:23:53	44,289	25,922	2,3	2.8	68	36	125	Reverse right-lateral oblique	207	61	67	Reverse left-lateral oblique	8

2.6.2 Rivers Network Data

As a consequence of modification in the topographic slope due to active subsidence processes east of the Bucharest-Ploieşti lineament, the straight courses of the Ialomiţa and Prahova rivers changed their direction from NW–SE to W–E. They also changed their course style from straight to meandering.

The riverbanks asymmetry of Dâmboviţa and Colentina when crossing Bucharest may reflect normal faulting at deeper levels, the downlifted compartments being the north-eastern ones.

The NW–SE trending of the rivers crossing Bucharest in the vicinity of the Intramoesian Fault may be interpreted as faults belonging to the Intramoesian fault system, largely developed between the Mostiştea and Argeş rivers.

The meandering course of the Dâmboviţa (modified during the nineteenth century) and Colentina rivers suggests a neotectonic subsiding area, with local influences determined by faults trending mainly N–S and NE–SW.

2.6.3 Geological, Tectonic and Neotectonic Data

Dâmboviţa represents a normal fault line, with episodes of strike-slip displacements, at least since the Upper Pliocene-Lower Pleistocene, since the coal beds observed in the borehole located south of the river were not found in the borehole situated north of it (Fig. 2.4).

West of Dâmboviţa the intermediate clay beds separating the Mostiştea sands and the Colentina gravels are lacking, indicating different sedimentation regimes in the Upper Pleistocene in two adjacent sectors separated by the Dâmboviţa line.

In the Militari-Roşu area (western part of Bucharest) four faults were interpreted as affecting the shallow geological formations starting at the depth of 10 m (Fig. 2.5) for reasons described in Sect. 2.3.2.2, and benefiting from a dense network of boreholes.

In the area of Bucharest most important for the goals of this study we consider the N–S trending fault system, an active system that develops from the Black Sea shoreline to river Olt. It had significant influences on the Danube river course between Feteşti and Galaţi towns, the old 'Danube Fault' trending N–S.

2.6.4 Geophysical Data

The NE–SW direction of gravity anomalous lineaments described and interpreted for the first time in this paper (Fig. 2.7) are determined by a regional fault system related to the tectonic processes associated with the Vrancea seismic zone. Their close spatial correlation with high/moderate seismicity lineaments (Fig. 2.9) suggests that they are involved in actual active tectonics. Most important for Bucharest are those affecting the city area, crossing either its south-eastern outskirts or its north-western vicinity.

The seismic reflection data obtained within Bucharest city limits may be considered of high significance for this study. In the paper published by Bala et al. (2006) shallow faults are interpreted on a reflection seismic profile performed in the eastern part of the city. Such faults, illustrated at a few meters depth, may be considered as active faults crossing Quaternary sedimentary deposits (Fig. 2.8).

2.6.5 *Seismological Data*

Regional active faults may be interpreted on the NE–SW high seismicity map evidenced in the Vrancea zone and their south-western prolongations toward Bucharest (Fig. 2.9).

In a regional sense, a N–S fault system was interpreted by seismic epicenters activated by the March 4, 1977 high magnitude Vrancea earthquake (Fig. 2.10).

The active fault depicted by low magnitude and crustal depths seismic events between Răcari and Cartojani, trending NE–SW, represents a south-western continuation of a crustal 'Vrancean' lineament clearly developed between Buzău and Focşani. This active segment is limited by two NW–SE faults located along the Colentina and Argeş river courses (Fig. 2.11).

The earthquake epicenters that look scattered within the Bucharest area are mainly associated with NE–SW prolongations of Vrancea type active faults, with active segments of NW–SE faults related to the river courses (Colentina, Dâmboviţa, Argeş) or active/capable segments of the N–S fault system.

The focal mechanisms computed for local Bucharest seismicity show mostly reverse and strike-slip faulting (Table 2.2). Considering the observations reported at regional scale by Rabagia et al. (2000) the role of strike-slip faulting and wrench tectonics seems to be most important in Moesian Platform neotectonics, this statement being also valid for the Bucharest city area.

2.6.6 *Integrated Interpretation*

To locate fault and fracture systems that may be related in Bucharest to local seismicity and understand their significance for urban planning is obviously a difficult task due to two major causes: (1) the geological structures do not outcrop, preventing geological and tectonic mapping; (2) the infrastructure and anthropogenic noise of a highly populated city do not allow the geophysical investigation and monitoring of the active/capable faults overlapped by quite thick Quaternary sediments.

Under such circumstances this first part of the study was mainly dedicated to the regional frame, by selecting and reinterpreting information from all relevant and available parts of geosciences: geomorphology, hydrology, geology, tectonics, neotectonics, geophysics and seismology.

Integrating the knowledge derived from this work the following statements may be derived:

- The Bucharest city area is situated at the junction of several fault systems (NW–SE, NE–SW, E–W, N–S) involved at various times in the geological evolution of this territory until the present. The N–S fault system seems to be involved in neotectonics, affecting the geomorphology between Bucharest and Ploieşti and representing a boundary for the eastern subsiding compartment (see Sect. 2.3.3.1);
- The most important fault systems for the seismicity of the Bucharest area are NE–SW and N–S, with enhanced seismicity at active faults junction. The active segments of the NW–SE system are mostly involved in strike-slip displacements and regional wrench tectonics;
- Geological information from boreholes and interpretation of reflection seismic measurements are in favor of shallow faulting in the city area, the recent to actual tectonics involving the Quaternary sediments;
- Local seismicity of the Bucharest area is represented by scattered low magnitude crustal earthquakes, as well as by deeper ones, such as the one that occurred at 70 km depth.

2.7 Urban Planning in Bucharest

Besides the usual geotechnical and seismic zonation techniques considered when developing or restoring a city district, the local active tectonics must be always taken into account.

For Bucharest the results of this integrated geophysical and geological study may attract the interest of geotechnical, civil engineers and architects for such information, useful when planning new buildings and restoring damaged ones. The data may be normally obtained by means of near-surface geophysics and high resolution seismological observations, but as seen in this paper, other kinds of data may be also relevant at regional or local scales: geomorphology, rivers network, shallow geology, local tectonics and neotectonics.

Additional information utilized in the final interpretation for urban planning purposes, aiming at understanding the local tectonics of the city centre, includes: (1) the distribution of collapsed or heavily damaged buildings, the latter being marked in Bucharest with large red dots on their exposure to the street; (2) the detailed topographic features, preserving the trace of original river banks of Dâmboviţa river, with influences of local tectonics (Fig. 2.13).

The Bucharest centre map displaying the collapsed (blue dots) and heavily damaged buildings (red dots) illustrates a scattered image, which is mostly developed N–S and in a less extent, W–E.

When mapping in the central area the areal distribution of less damaged buildings or sectors without obvious damages, elongated N–S and NNW–SSE areas of collapsed and highly damaged areas were contoured, separated by less damaged or lack of damages sectors. The axes of these 'anomalous' areas, presented in Fig. 2.13, suggest a detailed system of active/capable faults or fractures which may be involved

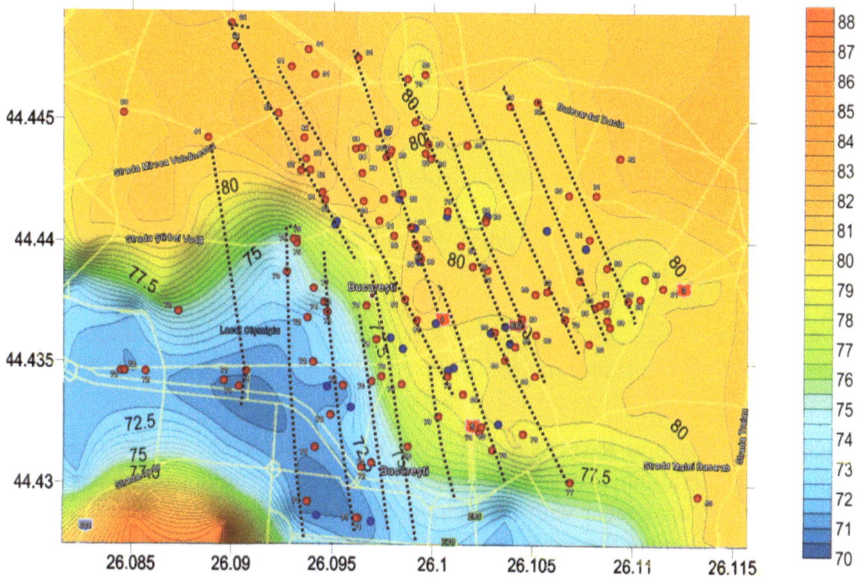

Fig. 2.13 Map of collapsed/ heavily damaged buildings in Bucharest after March 4, 1977 earthquake (*blue dots*—collapsed; *red dots*—heavily damaged), overlapping the city centre topography, with interpreted local faults/fracture systems

in the low magnitude Bucharest area seismicity, or locally enhance the effects of high magnitude Vrancea earthquakes. Two distinct groups of such tectonic features were evidenced, a southern one trending N–S and a northern one trending NNW–SSE. At this stage of our understanding these faults or/and fractures may belong to the regional N–S fault system, discussed within this study, the slight change of direction of the northern ones being due to wrench tectonics processes associated with NW–SE strike-slip faults situated along the river lines.

The detailed topographic map of the city centre (Fig. 2.13) illustrates two main features: (1) a segment of the former course of Dâmboviţa, with its most important meander transformed into Cişmigiu Park (including Cişmigiu Lake); (2) a plateau located west of Cişmigiu Park, where the old city centre was developed and the highest seismic damages occurred during high magnitude events.

The N–S direction of the eastern bank of the main old meander, as well as the eastern limit of the plateau, may reflect influences of the regional fault system illustrated in Figs. 2.6, 2.10 and 2.11.

Another problem that is considered in urban planning is the presence of paleobeds in old cities centers, Bucharest included.

The old meanders of Dâmboviţa river were associated with segments of regional or local fault systems, the N–S one being important in old Bucharest town, presently the city centre. Thick muddy sediments remained deposited in the shallow geological structure within the abandoned course segments and, together with the subsequent anthropogenic fillings, represented and still represent very poor locations for heavy buildings.

2.8 Conclusions

Active fault systems have influences in ground displacements during neotectonic vertical and horizontal movements or high magnitude earthquakes. In urban planning activities, besides a detailed seismic zonation, information on the areal distribution of active/capable faults and fractures crossing a city with heavily built districts is valuable in most cases.

Since a great number of specialists in civil engineering or geotechnical works are still reluctant to accept the presence of active or capable fault systems within the Bucharest area crossing Quaternary geological formations, data coming from geomorphology, rivers network, tectonics, neotectonics, geophysics and seismology were selected in this study, aiming at attesting through an integrated interpretation the presence within the city of such tectonic features.

Geomorphology and River Network At a regional scale, the sudden change in topographic heights along a N–S trending lineament between Bucharest and Ploiesti cities, in good agreement with neotectonic information on actual crustal vertical movements, illustrates the presence of a N–S regional fault representing the western boundary of a subsiding area (0.5 mm/y). It could represent an active transcrustal fault, the deepest seismic event recorded within Bucharest city at subcrustal depth being located on its southern prolongation. Such an N–S fault system was also interpreted in this study using the distribution of seismic epicenters activated by the March 4, 1977 high magnitude Vrancea earthquake.

The main directions of the rivers network in the area of Bucharest, which may be related to regional tectonics, are NW–SE: (Mostiştea, Colentina, Dâmboviţa, Sabar, Argeş), W–E (Câlniştea, Argeş, Ialomiţa) and NE–SW (tributaries of Ialomiţa, Mostiştea and Colentina).

Neotectonics Neotectonic information provided by a GPS regional velocity field illustrates in the south-eastern part of Romania a general trend of horizontal displacements from NW to SE. The central part of the Moesian Platform in Romania displays a transit from WNW to ESE (western part) toward NW to SE (eastern part), suggesting the presence of a NW–SE boundary between two sectors with different intensity of velocity field vectors. West of the Intramoesian Fault the velocity field is less homogeneous and less intense, the vectors trending W to E north of Bucharest, N to S within Bucharest and NNW to SSE at Măgurele, a few kilometers south-west of Bucharest.

Regional Fault Systems Bucharest is situated in the central part of the Moesian Platform, west of the transcrustal Intramoesian Fault, considered as a senestral strike-slip regional fault. The dominant crustal fault systems in the Moesian eastern compartment trend NW–SE, while the western compartment displays a more complex picture, with faults trending NW–SE, N–S, W–E and NE–SW. It was recently considered that in the Bucharest area the intersection of two major strike-slip systems are responsible for the active tectonics and high seismicity, important wrench zones occurring south of Bucharest.

The NE–SW lineaments observed in the Bouguer map built on mean gravity data are determined by perturbations of horizontal gradient type, due to 'Vrancean' faults affecting mainly the sedimentary cover. Bucharest is located between two such lineaments, the eastern one crossing the city's south-eastern outskirts.

Faults in the Bucharest Area Within Bucharest city limits the presence of faults affecting Pleistocene sediments was previously illustrated using borehole data, shallower vertical displacements of Quaternary formations along faults being interpreted in this study.

In the southern part of Bucharest the intermediate clay beds are lacking as compared to the northern part, the Mostiştea sands being in continuity with the Colentina gravels. This situation suggests vertical movements along a tectonic line, possibly along the old course of the Dâmboviţa river, determining an uplifted position of the southern sector and significant differences in the sedimentation regime.

The existing seismic reflection data recorded within Bucharest are of great significance for this study, shallow faults affecting reflectors located at cca 10 m depth being interpreted on a reflection seismic profile carried out in the eastern part of the city. Such faults can be considered as active or capable faults crossing Quaternary sedimentary deposits.

Seismicity Considering the Bucharest area seismicity, there are a number of seismic events whose hypocenters depths are generally located between 5 and 35 km (crustal earthquakes), their magnitude ranging between 2.0 and 3.2.

The earthquake epicenters scattered within the city area may be associated with NE–SW prolongations of Vrancea type active faults, with active segments of NW–SE faults related to the river courses (Colentina, Dâmboviţa, Argeş) or active/capable segments of the N–S fault system.

The Răcari-Cartojani seismic lineament of 'Vrancean' trending, represents an active segment of a regional NE–SW fault, limited by two NW–SE fault lines. The greatest magnitude of seismic events generated along this fault was $M_w = 3.1$, recorded on January 26, 2008.

Urban Planning and Local Tectonics The integrated analysis of all existing relevant data coming from geology, geomorphology, tectonics, geophysics and seismology represents a first stage of a study aiming at locating active/capable faults or fractures in the Bucharest city area and understanding their role in the spatial variability of seismic damaging effects on buildings, in view of a better future urban planning.

The March 1977 earthquake ($M = 7.2$) caused the collapse of tens of buildings, more than 200 being heavily affected. Most of the collapsed or damaged buildings were located in the central part of Bucharest, being already mentioned in the list with heavily damaged buildings after the 1940 earthquake. It illustrates similar consequences on buildings for different seismic events, enhancing the significance of site characteristics.

The effects of the 1977 earthquake in Bucharest showed great discrepancies between the distribution of the main damages and the seismic zonation. The concentration of the highest damages in the centre of Bucharest suggested other causes, besides construction errors, faults being considered to locally increasing the seismic intensity.

The detailed mapping of the destructive effects on buildings performed especially within the city center offered the possibility of interpreting fault lines that crosses Bucharest city centre on N–S and NNW–SSE directions. At this stage of the study they were traced using regional and local information as discussed above and by mapping elongated 'halos' of collapsed/damaged buildings and old concrete fences.

Considering the observations reported at regional scale the role of strike-slip faulting and wrench tectonics seems to be important in the Moesian Platform, in a regional sense, and within the Bucharest area, at a local scale.

Acknowledgments The authors are indebted to the reviewers for their valuable scientific comments, contributing this way to an improved final version of this paper.

References

Atanasiu I (1961) Earthquakes in Romania (in Romanian). Romanian Academy Publishing House, Bucharest

Bala A, Răileanu V, Zihan I, Ciugudean V, Grecu B (2006) Physical and dynamic properties of the shallow sedimentary rocks in the Bucharest metropolitan area. Romanian Reports in Physics 58(2):221–250, Bucharest

Bath M (1973) Introduction in seismology. Birkhauser Verlag, Basel, pp 395

Burton PW, Xu Y, Qin C, Tselentis G-A, Sokos E (2004) A catalogue of seismicity in Greece and the adjacent areas for the twentieth century. Tectonophysics 390:117–127

Constantinescu L, Marza VI (1980) A computer-compiled and computer-oriented catalogue of Romania's earthquakes during a millennium (AD 984–1979). Rev Roum Geophys 24:193–234

Constantinescu L, Enescu D (1985) Vrancea earthquakes (in Romanian). Romanian Academy Publishing House, Bucharest

Cornea I, Lazarescu V (1982) Geology and seismotectonics of Romania (in Romanian). In: Balan S, Cristescu V, Cornea I (eds) March 4, 1977 earthquake in Romania (in Romanian). Romanian Academy Publishing House, Bucharest

Enescu D, Cornea I, Misicu M (1982) The March 4, 1977 earthquake focal mechanism and associated directivity effects (in Romanian). In: Balan S, Cristescu V, Cornea I (eds) March 4, 1977 earthquake in Romania (in Romanian). Romanian Academy Publishing House, Bucharest

Georgescu EM (2007) Bucharest and earthquakes (in Romanian). Libra cultural foundation, Bucharest

Grenerczy G (2004) The motion of Adria and its effects on the Pannonian Basin. Reports on Geodesy 4(71):107–114, Warsaw

Grünthal G, Wahlström R (2003) An M_w based earthquake catalogue for central, northern and northwestern Europe using a hierarchy of magnitude conversions. J Seismol 7:507–531

Gutenberg B, si Richter CF (1956) Magnitude and energy of earthquakes. Ann Geofis 9:1–15

Hefty J (2004) Geokinematical modeling and strain analysis. Reports on Geodesy 4(71)39–42, Warsaw

Ioane D, Ion D (1992) Bouguer gravity anomaly map of Romania built on mean gravity values, scale 1: 1,000,000. Getech, Leeds

Ioane D, Stanciucu M, Chitea F, Diaconescu M (2010) Active fault systems and their significance for urban planning in Bucharest, Romania. Geophysical Research Abstracts, Vol. 12, EGU 2010, Vienna

Knopfler A, Schmitt G, Nuckelt, Marcu C (2009) Recent plate kinematics in Romania. Reports on Geodesy 1(86):67–76, Warsaw
Lăcătuşu R, Anastasiu N, Popescu M, Enciu P (2008) Geo-Atlas of Bucharest city. Est Falia Publishing House, Bucharest
Lupan M (1982) The damages on buildings in areas affected by earthquake. In: Balan S, Cristescu V, Cornea I (eds) March 4, 1977 earthquake in Romania (in Romanian). Romanian Academy Publishing House, Bucharest
Mândrescu N (1992) Data on seismic microzoning of Bucharest (in Romanian). Stud Cercet Geofizica 30:21–28, Bucharest
Mândrescu N, Radulian M, Mărmureanu G, Ionescu C (2008) Integrate research of the geological, geophysical and seismological data for local response evaluation in Bucharest urban area. Romanian Academy Publishing House, Bucharest
Mocanu V, Păunescu C, Andrei G, Munteanu L, Muntean A (2008) Relationship between the topography and tectonic processes in the Vrancea continental collision zone (in Romanian). MENER 2008—Environment 470–476, Bucharest
Munteanu L (2009) Kinematics of Vrancea region tectonic blocks using high accuracy satellite positioning (in Romanian). In: Marmureanu G (ed) On the management of earthquake disasters in Romania (in Romanian). Tehnopress Publishing House 482–546, Iasi
Nicolescu A, Roşca V (1992) The Bouguer gravity anomaly map. scale 1: 1,000,000. Institute of Geology and Geophysics, Bucharest
Oltean R (2005) Bucharest, the capital 1847–1854 (in Romanian). ART HISTORIA, Bucharest
Oncescu MC, si Bonjer KP (1997) A note on the depth recurrence and strain release of large intermediate Vrancea earthquakes. Tectonophysics 272:291–302
Oncescu MC, Marza VI, Rizescu M, Popa M (1999) The Romanian earthquake catalogue between 984–1997. In: Wenzel F, Lungu D, Novak O (eds) Vrancea Earthquakes: tectonics, Hazard and Risk Mitigation. Kluwer Academic Publishers, Dordrecht, pp 43–47
Papazachos BC, Papazachou C (1997) The earthquakes of Greece. Ziti Publications, pp 304
Paraschiv D (1979) Romanian oil and gas fields. Technical and economical studies. A—geological prospecting and exploration, 13, Institute of Geology and Geophysics, Bucharest
Polonic G (1998) The structure and the morphology of the crystalline basement in Romania. Monograph of Southern Carpathians (ed. Ioane D). Reports on Geodesy 7(37):127–131, Warsaw
Rabagia T, Tarapoanca M, Dinu C, Smith R (2000) Neotectonics of the Moesian platform—seismic implications. Romanian Geophysics. Vergiliu Publishing House, Bucharest, 284–285
Radu C (1979) Catalogue of strong earthquakes occurred on the Romanian territory. Part I—before 1901; Part II—1901–1979. In Cornea si C. Radu (eds) Seismological studies on the march 4, 1977 Earthquake, 723–752, Bucharest
Radu C, Polonic G (1982) Seismicity of the territory of Romania and of the Vrancea region (in Romanian). Romanian Academy Publishing House, 75–136. In: Balan S, Cristescu V, Cornea I (eds) The March 4, 1977 earthquake in Romania. Bucharest
Raileanu V, Dinu C, Radulian M, Popescu E, Diaconescu V, Bala A, Ardeleanu L, Popa M (2008) Crustal seismicity and associated fault systems in Romania (in Romanian). MENER 2008—Environment 470–476, Bucharest
Sandi H, Perlea V (1982) Civil engineering studies on the ground seismic movements (in Romanian). In: (Balan S, Cristescu V, Cornea I) March 4, 1977 earthquake in Romania (in Romanian). Romanian Academy Publishing House, Bucharest
Săndulescu M (1984) Geotectonics of Romania (in Romanian). Editura Tehnica, Bucuresti
Scherbaum F, Kuehn N, Zimmerman B (2010) Earthquakes focal mechanism from Wolfram demonstration. http://demonstration.wolfram.com
Trifu CI, Radulian M (1991) Depth-magnitude catalogue of Vrancea intermediate depth micro-earthquakes. Rev Roum Geophys 35:31–45
Visarion M, Săndulescu M, Gragoescu I, Drăghici M, Cornea I, Popescu M (1977) Map of recent crustal vertical movements in Romania, Institute of Geology and Geophysics, Bucharest
Visarion M (1998) Gravity anomalies on the Romanian territory. Monograph of Southern Carpathians (ed. Ioane D). Reports on Geodesy 7(37):133–138, Warsaw
(2011) Digital terrain model. http://srtm.csi.cgiar.org

Chapter 3
Capacities of Remote Sensing for Population Estimation in Urban Areas

Julia Kubanek, Eike-Marie Nolte, Hannes Taubenböck, Friedemann Wenzel and Martin Kappas

Abstract In the past few decades, devastating earthquakes have caused high social and economic losses in cities. Earthquakes cannot be avoided, but the devastating impacts, especially fatalities, can be minimized through pre-event emergency response planning and preparedness. The development of emergency plans strongly relies on up-to-date population and inventory data. However, existing techniques for population data generation do not meet the requirements of many of today's dynamic cities. In this context, the importance of remote sensing as a cutting-edge technology for data acquisition in urban areas is increasing. The present study analyzes the capacities and limitations of high resolution optical satellite imagery (IKONOS) for modeling population distribution in the district of Zeytinburnu in Istanbul, Turkey. The results show remote sensing to be an independent, up-to-date and area-wide data source. The use of remote sensing facilitates a mechanism to provide necessary quantitative information on urban morphology and population distribution in a fast and accurate way. The generated data do not have the quality of cadastral data sets but they meet the requirements of identifying bottlenecks, highly risky zones, etc. and can serve as a base for decision making.

J. Kubanek (✉)
Geodetic Institute (GIK), Karlsruhe Institute of Technology (KIT),
Englerstr. 7, 76131 Karlsruhe, Germany
e-mail: julia.kubanek@kit.edu

E.-M. Nolte
Universitätsrechenzentrum (URZ), Heidelberg University,
Im Neuenheimer Feld 293, 69120 Heidelberg, Germany

F. Wenzel
Geophysical Institute (GPI), Karlsruhe Institute of Technology (KIT),
Hertzstr. 16, 76187 Karlsruhe, Germany

M. Kappas
Institute of Geography, University of Göttingen,
Goldschmidtstr. 5, 37077 Göttingen, Germany

H. Taubenböck
Georisks and Civil Security, German Remote Sensing Data Center (DFD)
German Aerospace Center (DLR), 82234 Wessling, Germany

M. Bostenaru Dan et al. (eds.), *Earthquake Hazard Impact and Urban Planning*,
Environmental Hazards, DOI 10.1007/978-94-007-7981-5_3,
© Springer Science+Business Media Dordrecht 2014

Keywords Buildings (built-up area, floor numbers, inventory, occupancy, residential living space, usable space) · Earthquakes · Geographic Information System (GIS) · IKONOS · Istanbul · Population · Spatial distribution · Modeling · Remote sensing · Satellite imagery · Urban areas · Zeytinburnu

3.1 Introduction

Over the past few decades, the rapid growth of the world's population has led to dynamic and complex urbanization on a global scale. By 2050, it is predicted that 6 billion people—by then about 70 % of the world's population—will reside in urban areas (United Nations 2008). One example for the dynamic and uncontrolled development of urban areas is the city of Istanbul. Between 1950 and 1990, the population of Istanbul increased from 1,002,085 to 7,309,190 people. According to the 'Address Based Population Registration System', which was developed by the Turkish State Institute of Statistics, the population has reached 12,573,000 people in 2009. This means a growth of 70 % within the last 20 years (Dökmeci and Berköz 1994; TurkStat 2010).

Existing techniques for population data generation like traditional censuses do not match the dynamics of many of today's cities. Available inventory data— especially in developing countries—might be incorrect, outdated, generalized or not available at all. However, population data is an indispensable and important source of information, e.g. in disaster management. The effectiveness of emergency planning strongly depends on reliable information on city population and its dynamics. Especially the knowledge of the spatial distribution of the population may support rescue teams in the direct aftermath of a disaster. The problem of unreliable population information is important because megacities are, due to the high concentration of people, assets and economic relevance, often vulnerable to natural hazards (Mitchell 1999; Wenzel et al. 2007). The 1999 Izmit (Turkey) earthquake for example has illustrated the disastrous consequences of an earthquake striking a densely populated area. On August 17 1999, a magnitude 7.6 earthquake struck the area of Izmit in Turkey, causing about 18,000 fatalities and US$ 18 billion economic loss (EMI 2005; Erdik et al. 2004). The probability of a magnitude 7 earthquake striking Istanbul in the years between 2004 and 2034 ranges between 30 to 70 % (Parsons 2004). While being at the same time, a site of rapid development and a place of high risk due to natural hazards—especially earthquakes—the emergency planning systems of megacities like Istanbul require having up-to-date information on population and its dynamics.

To overcome problems like missing up-to-dateness and sparse demographic data, earth observation data have become an important source of information for urban areas. Technical advances in combination with the need for up-to-date information have led to a variety of automatic techniques for urban data extraction from high resolution satellite imagery (Chen et al. 2009; Rottensteiner et al. 2007; Sohn and Dowman 2007; Wurm et al. 2011). However, the apprehensible excitement and enthusiasm by the scientific community with the advent of high to very high resolution satellite imagery such as GeoEye (0.4 m), WorldView (0.5 m), QuickBird (0.6 m), and IKONOS (1 m), lead to the impression that increasing the spatial resolution of the images really leads to a better quality of automatically extracted data. But does the quality of the automatically obtained data really match the needs of emergency planning and disaster management? The accuracy of the generated data is often questionable; especially when comparing them to Geographic Information System (GIS)-based, cadastral data or manually digitized and visually inspected information. However, a rational discourse on the suitability of high resolution optical satellite images as a source of information for population modeling in urban areas is still missing.

The present study aims at constituting the first step towards a quantitative evaluation of the suitability of high resolution optical satellite imagery (IKONOS) for data generation relevant for population modeling in cities. Therefore, based on a comparison between population distribution modeled using IKONOS imagery and population distribution modeled using GIS and secondary sources, the limitations and advantages of IKONOS imagery for population modeling are discussed. We choose IKONOS imagery for the reason that a comprehensive study on urban data generation using this kind of satellite imagery had been conducted for Zeytinburnu by Taubenböck (2008). Additionally, the availability of different studies and data on Zeytinburnu facilitates the evaluation of the satellite based data generation approach.

3.2 Study Site Selection

For the methodology development, the district of Zeytinburnu in Istanbul, Turkey was selected as test site and is shown in Fig. 3.1. The city is situated in the earthquake prone Marmara region in the northwestern part of Turkey. In past decades, the city has developed into a megacity with similar characteristics to those megacities in the developing countries display (United Nations 2008). Among others, this includes high population concentration, social disparity, lack of management, uncontrolled and single-handled development in residential and industrial areas as well as an increasing irregular and illegal urbanization (Baycan-Levent 2003; EMI 2005; Görgülü 2002).

Located just 150 km northwest of the Kocaeli region that was strongly hit in 1999, Istanbul's earthquake risk is extremely high due to the combination of earthquake exposure and size of the city (Ambraseys and Finkel 1991; Sever et al. 2001). The rapid urban growth and overcrowding leads to improper land-use planning. As a result, informal and poor settlements are established in inappropriate areas, which in turn increase the earthquake risk (Cakir et al. 2008; Erdik and Durukal 2008). Increasing vulnerability emerges also from the aging and inadequate infrastructure, the use of inappropriate materials and construction techniques for buildings and lacking enforcement of building standards. Where norms and standards for minimizing the earthquake risk exist, their implementation is often complicated due to the high costs of retrofitting. The combination of these factors and the existing high earthquake hazard results in a high earthquake risk (Ansal et al. 2009; EMI 2005; Erdik and Durukal 2008).

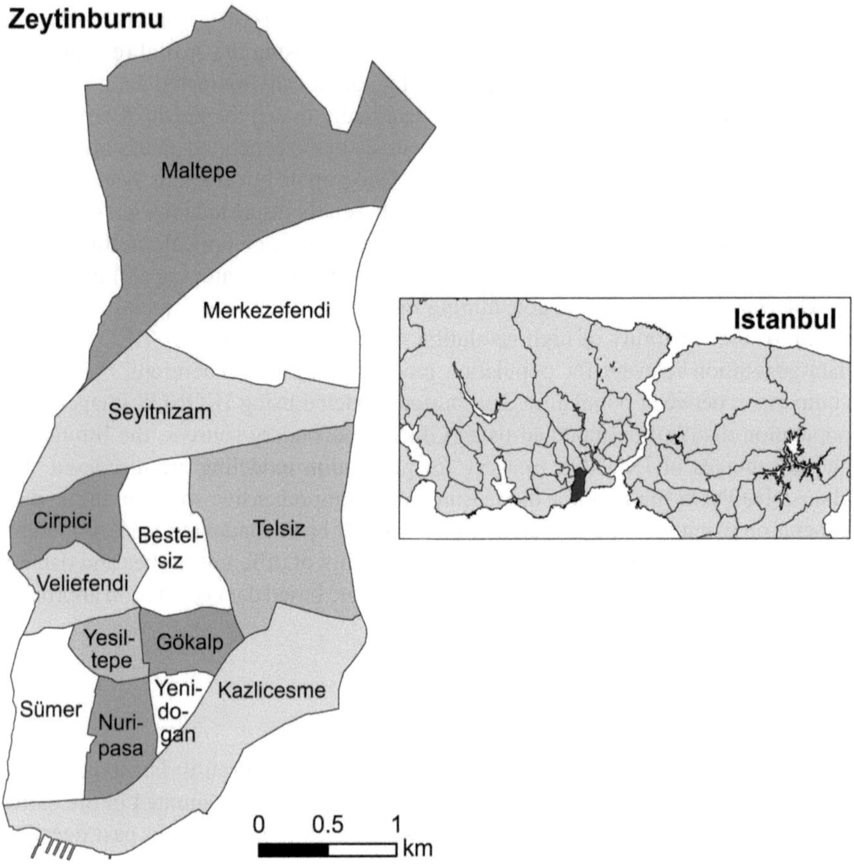

Fig. 3.1 Study area: the district of Zeytinburnu in Istanbul with its subdistricts

Beygo et al. (2006) describe the district of Zeytinburnu as "being a first focus of illegal settlements in Turkey and sheltering [a] major portion of the migration to Istanbul, it has lacked a proper urbanization process and thus, today suffers from a densely and unqualified development. This negative development is faced as a threat in the means of providing urban life quality as well as the future of the settlement according to natural disasters". In previous studies, the district of Zeytinburnu was identified as high risk area due to its unstable building assets. For the reason of its high vulnerability concerning earthquake hazards, Zeytinburnu was chosen as test site for the presented population modeling approach (see Fig. 3.1).

3.3 Remote Sensing in Urban Areas

To overcome problems like missing up-to-dateness and unsuitable spatial resolution of population data, the importance of remote sensing for information extraction in urban areas is increasing. A number of previous studies attempt to investigate the capabilities of satellite imagery for population estimation. Many of these studies use moderate resolution satellite imagery for deriving occupancy categories of urban areas. Harvey (2002) for example classified pixels of a Landsat TM (Landsat Thematic Mapper) image as residential or non-residential and distributed the population of each zone across its residential pixels for estimating the population of Ballarat and Geelong as two Australian provincial cities. Chen (1998) and (2002) linked remotely sensed data (Landsat TM) with areal census data with the aim to identify the correlations between pixel-based satellite data and zone-based census data. Using Landsat multi spectral scanner (MSS), Iisaka and Hegedus (1982) worked out a mathematical model to illustrate the relationship between the reflected electromagnetic energy from the surface and population distribution in sample areas. Further studies using moderate resolution satellite imagery (Landsat) for deriving urban occupancy categories were conducted by Lu et al. (2006) and Wu and Murray (2005, 2007). Utilizing SPOT (Satellite Pour l'Observation de la Terre) as another moderate resolution satellite, Lo (1995) developed different methods to extract population and dwelling unit data. Al-Garni (1995) developed four predictive mathematical models for population estimation in urban areas using SPOT imagery in combination with large scale aerial photographs. The models were tested for generating census-like data in urban areas of Saudi Arabia.

By this time, the progress in space technology has improved the image data quality in its spatial and temporal characteristics. Mesev (2005) for example used postal points to infer residential and commercial buildings from IKONOS imagery. Souza et al. (2003) analyzed the potential of IKONOS images for estimating the population of Sao Jose dos Campus (Brazil) using GIS technologies. Liu et al. (2006) explored the possible correlation between the population density and textures in IKONOS images for a study site in Santa Barbara County. Setiadi et al. (2010) combined a model generated based on conventional population data (census and household survey) with a 3D city model derived from high resolution IKONOS data, a Digital Surface Model (DSM) and a Digital Terrain Model (DTM). The intention of this study was to explore the potential of integrating remote sensing and social science for natural hazards research. Another study on image texture of very high resolution satellite imagery was conducted by Almeida et al. (2007) using QuickBird imagery to investigate the correlation between census population density and image texture.

The increasing resolution of satellite imageries also facilitates the extraction of the single building inventory, which can further be used as basis for population modeling. One study using IKONOS images for urban information extraction was conducted by Taubenböck et al. (2007, 2009, 2010). IKONOS images were used as the basis for classifying patterns of urban morphology and inferring socioeconomic parameters. The main intention of Taubenböck (2008) was to analyze the capabilities of remote sensing to assess earthquake risk and vulnerability in the megacity of Istanbul. The vulnerability assessment was conducted for the district of Zeytinburnu (Istanbul) including the development of building inventory categories, which are based on the following physical parameters: (1) building density, (2) building height, (3) building age, (4) size, (5) form, and (6) roof type. These parameters were then used to estimate the population density and to model its spatial distribution. The results of Taubenböck (2008) are used for the statistical comparison in this present study.

3.4 Inventory Generation for Population Modeling

The main objective of the present study is to analyze the capabilities and limitations of very high resolution optical satellite imagery for modeling population in urban areas.

The population modeling methodology proposed consists of two main parts: (1) inventory generation and (2) population modeling on single building level, whereas the inventory data developed in (1) serve as a basis for population modeling on single building level. The developed methodology follows the principle of top-down modeling. (A bottom-up approach using punctual information from a field survey has been conducted in Zeytinburnu as well. For further information see Taubenböck (2008)). In the first step of the present top-down approach, the district of Zeytinburnu is subdivided into its subdistricts. In a second step, the individual buildings are joined for each subdistrict from different data sources. Additionally, the total building footprint area on subdistrict level is calculated. In a third step, the number of floors and different occupancy categories are assigned to the individual buildings and from this, the usable building space is calculated on single building level. In combination with residential occupancy, the usable space of each building serves as an input for the population modeling procedure. The total residential living space for each subdistrict divided by the total subdistrict population results in the average residential living space per person. The total subdistrict population is provided by statistical data of the city (TurkStat 2010). Combining the average residential living space per person with the living space of each individual residential building, the number of people living in a single building is estimated.

The procedure for population estimation presented by Taubenböck (2008) involves the development of several intermediate data sets, which serve as an input for the following methodological steps. As the overall uncertainty and accuracy of the population estimates strongly depends on the level of accuracy of the input data, all intermediate results are quantitatively compared to the results from the GIS-based approach presented in this study. The schema of all intermediate steps consists of three steps: (1) GIS-data generation based on Taubenböck (2008) using IKONOS imagery (in the following also referred to as remote sensing (data), (2) GIS-data developed in this study (in the following also referred to as the manually developed (data), and (3) comparison of the results from step 1 and 2 and their analysis.

3.4.1 Development of Single Building Inventory

An object oriented approach is employed to automatically extract built-up polygons from the remote sensing data. The approach is subdivided into two major methodological steps: (1) segmentation and (2) classification. In the segmentation procedure, adjacent pixels with similar characteristics are merged to segments. The segments were improved using a methodology developed by Esch (2006). The result shows a segmentation level which contains small segments for small objects such as roofs and their shadows and large segments in homogenous areas like green spaces. This way, the approach attempts to present objects as an approximation of real world structures (Taubenböck and Roth 2007; Taubenböck 2008).

The second module is based on the optimized segmentation level developed in the segmentation in which the segments try to represent real world structures of urban areas. The general assumption of the second step is that urban characteristics such as buildings, streets, and green spaces generally feature the same spatial characteristics in every urban agglomeration all over the world. Considering the aspired transferability of the approach, it becomes necessary to use these urban characteristics for classification purposes. The classification process produces seven urban classes: (1) building, (2) street, (3) grassland, (4) tree, (5) bare soil, (6) water, and (7) shadow.

For identifying the suitability of the remote sensing data for inventory generation and so for population modeling, a GIS-based data set of the up-to-date single building inventory of Zeytinburnu is manually developed. The data set is clipped to the same administrative boundaries for comparison purposes (Kubanek 2011; Kubanek et al. 2010). In order to obtain a comprehensive dataset of all buildings, different data sources are used. As initial input data, the digital administrative boundaries of the district of Zeytinburnu and the city of Istanbul are available from the study 'A Disaster Prevention/Mitigation Basic Plan in Istanbul including Seismic Microzonation in the Republic of Turkey' (JICA study) conducted by the Japan International Cooperation Agency (JICA) and by the Istanbul Metropolitan Municipality (IMM) from 2001–2002 (JICA and IMM 2002a, b; Batuk et al. 2005). All generated GIS-layers (manually developed data as well as remote sensing data) are clipped to these administrative boundaries.

A digital building outline data set developed in the 'Zeytinburnu Pilot Projesi Kapsaminda Konut, Ticaret, Kücük Sanayi ve Sanayi Bölgelerinde very Derleme ve Degerlendirme Projesi' (Zeytinburnu Pilot Project) forms the basis for manually developing a GIS-based single building inventory for Zeytinburnu. The survey was conducted by the Turkish firm GENAR simultaneously to the 'Earthquake Masterplan for Istanbul' (Master Plan) from June to July 2004 and included questions on housing, trade, and industry. The main objective was to analyze the socio-economic structure of areas with a high earthquake risk (GENAR 2004a, b, c). This data set is primarily complemented by the single building outlines from the Master Plan, which was conducted by four Turkish universities with the goal of developing a holistic approach to reduce the earthquake risk in Istanbul (Baycan-Levent 2003; Erdik and Durukal 2008; Ergenc et al. 2005; Fernandez 2005; IEMP 2003).

It is important to note that the single building inventories from these two studies show significant deviations. To eliminate the ambiguities and to ensure the developed inventory data set only includes presently existing buildings, the two above mentioned data sets are compared and completed by Google Earth imagery (data from 2007–2008) as well as by the Interactive City Map of Istanbul (Istanbul Sehir Rehberi, IMM (2010), URL: http://sehirrehberi.ibb.gov.tr/map.aspx).

The main difference of the two developed datasets is that while the remote sensing data only contain polygons of built-up area, the manually developed building inventory actually contains the outlines of single buildings. Therefore, an overlay of both data sets and a direct, per building comparison is not feasible. Instead, the comparison is conducted on subdistrict level.

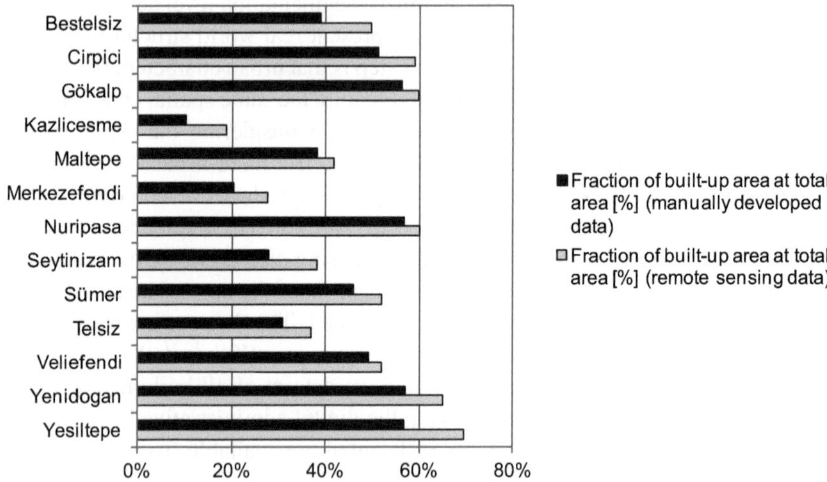

Fig. 3.2 Comparison of the manually developed and the remote sensing based built-up area in percent on subdistrict level

For the comparison between the manually developed built-up area and the remote sensing based built-up area, the ratio of the built-up area to the total area of each subdistrict is calculated and shown in Fig. 3.2 for the different subdistricts of Zeytinburnu. The results show that the built-up area is overestimated by an average of around 7 % in the remote sensing data compared to the manually developed data. A possible correlation of the grade of overestimation and other parameters is analyzed in the following sections.

3.4.2 Assessment of Different Floor Numbers

The number of floors is an indispensable parameter for estimating the usable building space and so the available residential living space of Zeytinburnu. The derivation of building heights and classification of 3D city models is a common research field in urban remote sensing. Multi-sensoral approaches using digital surface models in combination with optical satellite data have been presented (Wurm et al. 2011; Kocaman et al. 2006). For Zeytinburnu, a digital surface model was not available. Thus, an indirect approach correlating shadow lengths to buildings heights has been applied. Cheng and Thiel (1995), Hartl and Cheng (1995), Irvin and McKeown (1989), and Massalabi et al. (2004) derived building heights from shadow lengths. However, these methods have only been tested for isolated buildings and in not densely built urban areas. Thus, for the remote sensing data, we utilized an approach for estimating building heights while using shadow lengths in high agglomerated urban areas. The building height can be estimated by using the following parameters: (1) the position of the sun, (2) the sensor at the time of recording (Massalabi et al. 2004), (3) buildings with a totally visible shadow, and (4) the hypothesis, that neighboring buildings in physical homogeneous zones correspond to the calculated building heights from (3). The number of floors per building is subsequently calculated from a simple ratio of building height and assessed floor height.

The shadows are calculated based on the classification approach explained in 3.4.1; parameters (1) and (2) are stored in the metadata of the satellite imagery. The developed algorithm starts with single pixel scanning of the described classification. The direction of scanning depends to the azimuth of the sun.

In a first step, all pixels classified as 'building' are identified. The number of pixels classified as 'shadow' is counted afterwards in the scanning direction and the lengths of the shadow is calculated. Because a single building may have different shadow lengths at the same time, the procedure averages at the particular values calculating an uniform height for a building and also extrapolates values for buildings where the height could not be estimated for buildings containing minimal building heights.

The derivation of floor categories from building heights for the remote sensing data is based on the study 'Seismic Evaluation and Retrofit of Concrete Buildings Volume 1' which was conducted by the Applied Technology Council (ATC) of California in 1996. The building height classes correspond to a civil engineering differentiation, which is usually used to evaluate relevant building types due to their stability characteristics. This results in three building height classes. Buildings with a height up to 10.5 m are assigned to the first category with 1–3 floors. Structures with a height between 10.5 m and 22.5 m build the second category considering having 4–7 floors. The last category containing 8 and more floors includes all buildings higher than 22.5 m. Figure 3.3 shows the resulting floor categories of Zeytinburnu based on the remote sensing data. The left part of the figure shows the floor categories of the district of Zeytinburnu. On the right, the smaller figure depicts the details of the building inventory including floor categories.

For manually assessing the floor numbers, the up-to-date single building inventory developed in 3.4.1 is used as basis. Floor numbers are available from the Zeytinburnu Pilot Project, the Master Plan, and the JICA study. Due to significant deviations in these three data sources, only information from the Zeytinburnu Pilot Project was included. In the manually developed data, the number of floors range between 1 and 20 buildings, with no buildings having twelve or nineteen floors. Due to comparison purposes, this data set has to be reduced to the floor categories used in the remote sensing data. Figure 3.4 shows the manually developed floor categories. Like in Fig. 3.3, the left part contains the whole district while the smaller figure on the right shows a selected scene in more detail.

The majority of buildings in the manually developed data (60.16 %) as well as in the remote sensing data (76.25 %) have 4–7 floors, which becomes obvious from Figs. 3.3 and 3.4. The category 1–3 floors applies to a share of 38.68 % considering all buildings in the manually developed data set and to a share of 21.56 % in the remote sensing data. According to the manually developed floor numbers, only 0.68 % of all buildings in Zeytinburnu contain 8 or more floors whereas the remote sensing data result in a share of 2.19 % for this category. Only 0.48 % of the buildings belong to the unknown category which is only considered in the manually developed data.

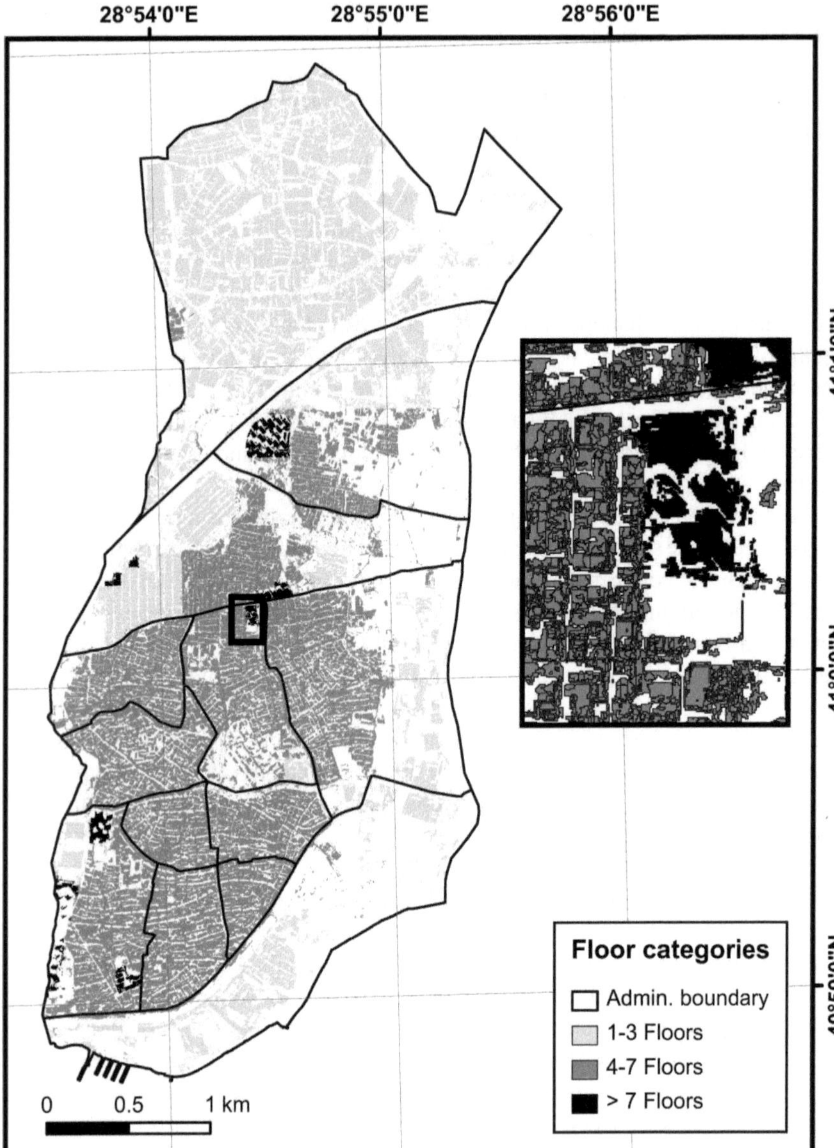

Fig. 3.3 Spatial distribution of floor categories in Zeytinburnu based on remote sensing data

The first category is underestimated in all subdistricts except of one by an average of 20.13 % in the remote sensing data. The second category is therefore overestimated in all subdistricts except of the same subdistrict by an average of 18.84 %. The number of buildings containing 8 and more floors is underestimated in some subdistricts and overestimated in other subdistricts. The mean deviation is with 2.05 % very low.

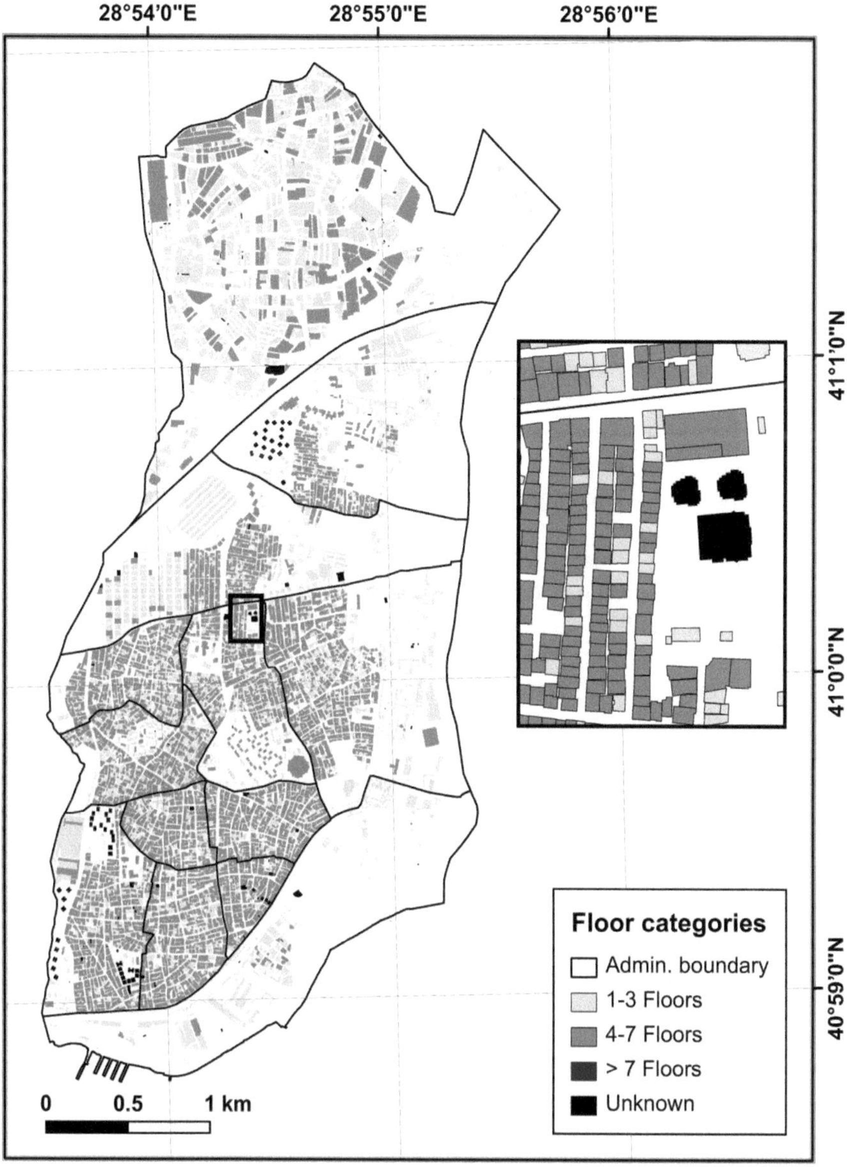

Fig. 3.4 Spatial distribution of manually developed floor categories in Zeytinburnu on single building level

3.4.3 Spatial Assessment of Occupancy Categories

The third step of the inventory generation contains the spatial distribution of occupancy categories in Zeytinburnu. For the remote sensing data, we assume that the physical urban morphology is strongly related to the functions of the buildings, allowing the estimation of land use for an urban area while using physical parameters like (1) land-cover, (2) building density, (3) building heights, and (4) roof or building types. According to this hypothesis, an industrial area in Istanbul is most likely defined by a flat roof and a large building footprint area. Additional influencing factors are comparatively low building heights and a homogenous texture of the considered area. In contrast, residential neighborhoods are characterized through gabled roofs, smaller building footprints as well as the typical red color of tiled roofs. Additional thematic classes such as hospitals and military areas have been added manually.

Based on these assumptions using building inventory extracted from the remote sensing data, a data set including the following occupancy categories for Zeytinburnu is developed: (1) commercial, (2) residential-mixed, (3) military area, and (4) hospital.

In a next step, the spatial distribution of occupancy categories is assessed based on the manually developed up-to-date single building inventory. The main data sources are the Zeytinburnu Pilot Project and the Master Plan. The output is a GIS-based data set including the following occupancy categories: (1) commercial, (2) residential, (3) industrial, (4) other, (5) depot, (6) service, (7) under construction, (8) unknown, (9) vacant, and (10) mixed.

For a consistent comparison, the available broad thematic spectrum of land use categories have to be grouped with respect to the indirect classifications using remote sensing data. The buildings with uniform occupancy are organized as follows: the 'commercial' class includes all buildings in which trade and commerce take place, like a supermarket or a car dealership. The category 'residential' comprises all residential buildings such as apartments or single-family buildings. It further includes the military quarters as well as boarding schools due to the assumption that these buildings are occupied comparable to residential buildings. All industrial facilities are grouped in the 'industrial' category, not distinguishing different sectors. The category 'depot' contains all kind of storage buildings; the 'service' category includes all buildings in which some kind of service is provided. The category 'other' includes a number of occupancies such as schools, nurseries, clubs, religious buildings, hospitals, or sports facilities. The 'unknown' category encompasses all buildings whose occupancy cannot be identified from the available data bases. In addition, a category is introduced to cover the 'vacant' buildings.

Besides the buildings with a single occupancy, numerous buildings are characterized through more than one function and are therefore merged into the category 'mixed'. This category includes for example an industrial facility that operates at the same time in manufacturing and trade. Another example for a mixed building is a multi-story building with a shop in the first floor and residential apartment on all other floors. This kind of mixed building occupancy is typical for Turkey.

Table 3.1 Deviations between manually developed and remote sensing based occupancy category ratios in Zeytinburnu on subdistrict level

Subdistrict	Residential-mixed [%]	Commercial [%]	Rest [%]
Bestelsiz	– 4.15	8.29	1.89
Cirpici	5.68	– 1.70	– 3.15
Gökalp	10.62	– 6.79	– 2.39
Kazlicesme	15.12	31.14	– 44.98
Maltepe	– 4.79	24.94	– 11.54
Merkezefendi	15.29	20.84	– 28.35
Nuripasa	8.27	– 5.30	– 2.24
Seytinizam	– 5.09	12.75	– 4.81
Sümer	16.72	– 8.88	– 7.27
Telsiz	4.70	13.11	– 17.06
Veliefendi	7.15	– 3.48	– 2.75
Yenidogan	12.27	– 10.53	– 1.06
Yesiltepe	4.75	– 2.31	– 2.04
Average	*6.66*	*5.54*	*– 9.67*

For a better comparison of the manually developed data with the remote sensing data, the occupancy categories of both data sets are limited to the following categories: (1) residential-mixed, (2) commercial, (3) rest, and in the manually developed data also (4) unknown. The category 'residential-mixed' as well as the 'commercial' category already exist in the remote sensing data. The 'rest' category encompasses 'military' areas as well as the 'hospitals'. For generating the same categories in the manually developed data, all buildings that are identified as 'residential' or 'mixed' are merged into the 'residential-mixed' category. In the 'commercial' category, all 'commercial' and 'industrial' buildings are combined. The 'rest' category contains all remaining occupancy categories except for 'unknown'. In the manually developed data, all buildings identified as (1) other, (2) depot, (3) service, (4) under construction, and (5) vacant are included. The category 'unknown' only exists in this data set and includes all buildings for which the occupancy could not be identified.

The majority of the built-up area in most subdistricts (in 11 out of 13) belongs to the residential-mixed occupancy category in both data sets. The residential-mixed occupancy category is overestimated in the remote sensing data in 10 subdistricts and underestimated in the remaining three ones (Table 3.1). The average deviation is 6.66 %. Here, the commercial category is overestimated in 6 subdistricts; the average deviation is 5.54 %.

3.4.4 *Calculation of Usable Building Space and Available Residential Living Space*

Based on the building footprint area and the floor categories, this section deals with the estimation of usable building space in Zeytinburnu, which is further combined with the occupancy information for estimating the available residential living space of the district.

The usable building space as well as the available residential living space is first assessed using the remote sensing data. The multiplication of the building footprint area calculated using GIS and the floor numbers results in the usable building space. For the reason that the remote sensing data only include floor categories instead of the exact number of floors, the usable building space is calculated using an averaging procedure. The total residential living space is further calculated based on the average usable building space in combination with the residential-mixed occupancy category. The categories of both data sets are based on a natural break distribution.

In the second step, the usable space of each individual building is calculated based on the manually developed data. Like in the remote sensing data, the classes are generated based on natural break procedure. For buildings with missing floor number, the usable space cannot be calculated. They are grouped in the category 'unknown'. The combination of residential occupancy and usable building space results in the total available residential living space. The classes used again depend on natural breaks.

With the remote sensing data, the usable building space is overestimated by a total of about 5,200 km^2 (excluding one subdistrict in which the usable building space is underestimated by about 1,000 km^2). The available residential living space is consequently overestimated by a total of 11,680 km^2.

3.5 Population Modeling on Single Building Level

This section includes the comparison of population distribution modeled based on the remote sensing data and population distribution modeled based on the manually developed data.

In this study, the night-time population modeling is based on the generalizing assumption that most people reside at home at night. In the first step, combining residential occupancy and usable building space gives the total available residential living space on subdistrict level as well as the living space of individual residential buildings. The 2000 Population Census provides information on population on subdistrict level. The division of the total residential living space per subdistrict by the total population per subdistrict results in the average residential living space per person in square meters on subdistrict level. The number of people living in a single building can be calculated while dividing the living space of each individual residential building by the average residential living space per person per subdistrict.

Based on the inventory attributes generated in section 3.4.1–3.4.4, the night-time population distribution is first modeled with the remote sensing data. The available residential living space is calculated based on the identified residential-mixed occupancy category. Modeling the night-time population only for the buildings identified as residential is not suitable here for the reason that the remote sensing data do not contain a single residential occupancy category. In contrast, there are two possibilities for calculating the available residential living space as basis for the population modeling according to the manually developed data. On the one hand, only buildings classified as residential are included whereas on the other hand we also used the residential-mixed occupancy category.

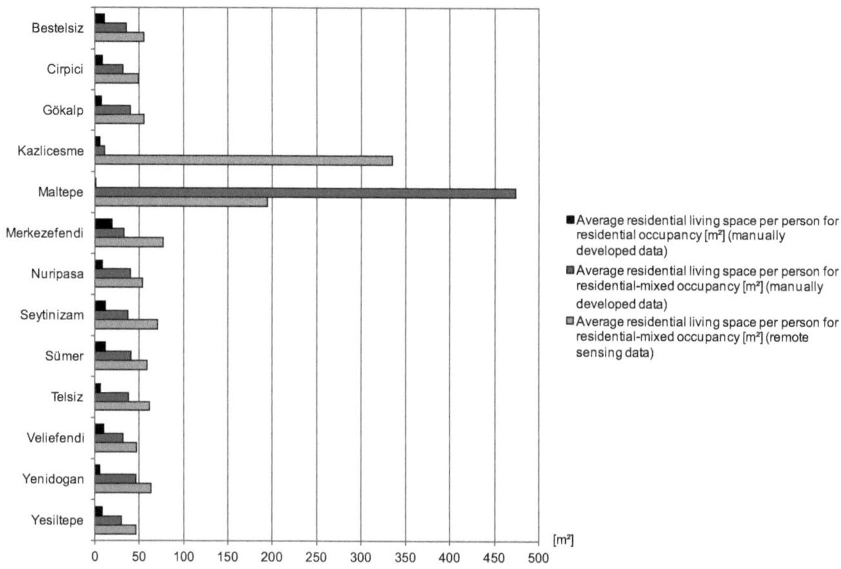

Fig. 3.5 Comparison of the average residential living space per inhabitant in square meters in Zeytinburnu on subdistrict level (1) based on the manually developed data using the residential occupancy and (2) using the residential-mixed occupancy and (3) based on the remote sensing data

The comparison of the night-time population distribution in Zeytinburnu constitutes the last step of this analysis. Altogether, the same places are identified as residential in both data sets containing residential buildings with night-time population. Figure 3.5 displays the comparison of the average residential living space per inhabitant in square meters in Zeytinburnu on subdistrict level (1) based on the manually developed data using residential occupancy and (2) using residential-mixed occupancy, and (3) based on the remote sensing data. In the latter one the average residential living space per inhabitant is overestimated in all subdistricts by an average of 80 m^2 when compared to the manually developed data using only residential buildings. The comparison of the average residential living space per inhabitant based on residential-mixed occupancy of the manually developed data and the remote sensing data shows similar results: In the latter one, the average residential living space per person is overestimated for all subdistricts, except of one, which is underestimated by about 280 m^2. Excluding this deviation the average overestimation is 46 m^2.

3.6 Results

It becomes obvious that the built-up area of each subdistrict is overestimated by approximately 7 % when using remote sensing data. This overestimation can be explained by the reason that while the manually developed building inventory actually contains single building outlines, the remote sensing based building inventory only shows polygons which represent built-up area. As buildings like other objects are identified from optical satellite images through their reflective signature, the observed deviations are mainly caused by similar spectral characteristics of land cover categories like built-up area and its surroundings such as streets or bare ground. The similarity strongly depends on the roofing material of the buildings. For example, concrete buildings with tiled roofs are easier to distinguish from tarmac than concrete roofs with the same greyish color.

Considering the floor categories, it becomes obvious that the remote sensing results contain similarities to the manually developed data. Altogether, the ratio of buildings containing 1–3 floors is underestimated for every subdistrict by an average of 17.12 % when using remote sensing data. The second category is overestimated for all subdistricts (except of one) to the same extent. The deviations concerning the third category (> 7 floors) between both approaches are with 1.51 % very small.

The deviations in the floor categories in the data sets can be explained through uncertainties in estimating building heights in densely built-up urban areas while using satellite images (in this case IKONOS). Defined by the building shadow, the position of the sun, and the characteristics of the sensor, the height of a building correlates with its shadow length. A scanning algorithm is employed for identifying building pixels and shadow pixels. By the reason that the shadow of a building is not displayed in its real dimension by lack of open space around a building, the high building density causes major problems in the scanning algorithm.

In the third step of this study, simplified occupancy categories on subdistrict level were compared and revealed the same deviations like the built-up area. Due to the very broad categorization scheme applied for the remote sensing data, almost no buildings are assigned to a 'rest' category and so this category is underestimated by the automatic extraction technique in every subdistrict by an average of 10 %.

The majority of the built-up area in most subdistricts (in 11 out of 13) belongs to the residential-mixed occupancy category in both data sets. The residential-mixed occupancy category is overestimated in the remote sensing data in 10 subdistricts and underestimated in the remaining three ones. The average deviation is 6.66 %. Using the remote sensing data, the commercial category is overestimated in six subdistricts; the average deviation is 5.54 %. Altogether, this comparison demonstrates the capabilities of high resolution satellite imagery to delineate occupancy categories in a simplified way.

The usable building space is constantly overestimated in the remote sensing data in every subdistrict except one. Excluding this underestimated subdistrict, the average overestimation is about 430 km^2. As a result, the total available residential living space is overestimated in the remote sensing data for every subdistrict, here by an average of 900 km^2 when compared to the manually developed data using only buildings classified as residential. Accordingly, the average area per inhabitant is overestimated in the remote sensing data for every subdistrict (except of one) when using residential-mixed occupancy. The overestimation is at an average of 80 m^2 when compared to the manually developed data using only residential occupancy and about 46 m^2 when using residential-mixed occupancy (see Fig. 3.5).

The overestimation of the average area per inhabitant has consequences for population modeling on single building level. Altogether, based on the remote sensing data, the population of every subdistrict is underestimated by an average of 500 inhabitants compared to the manually developed data using residential occupancy.

3.7 Discussion

The presented comparison proves that high resolution optical satellite images can be successfully employed for urban parameter extraction. However, some constraints were identified in this study. A main problem concerning the inventory generation procedure is that existing automated building extraction techniques have limitations in complex urban areas. Here the extraction of single buildings using existing scanning algorithms is not feasible yet and per-building modeling of population in the last step is not possible. The lack of a single building extraction methodology results in an overestimation of the built-up area especially in densely populated urban agglomerations due to difficulties in classifying different land cover categories caused by similar spectral characteristics of urban surfaces like roofs and tarmac. Like for the built-up area, IKONOS images can be used for generating information on floor categories when other data sources are not available. Although there are some complications when calculating the shadow lengths of the buildings for estimating building heights, extrapolating and averaging procedures provide acceptable results. The same statement can be made for the generation of occupancy categories while using high resolution satellite images. A main limitation concerning this step of the inventory generation is due to buildings featuring more than one occupancy. Especially in Istanbul, many buildings have shops in their first floors whereas residential living space can be found in all other floors. It is not possible to derive these mixed occupancy categories by image interpretation procedures, but the presented results demonstrate the capabilities of high resolution satellite imagery to delineate occupancy categories.

The high building density is responsible for the overestimation of the built-up area. Further, the buildings containing 1–3 floors are underestimated in the remote sensing data, and at the same time, the second category (4–7 floors) is overestimated for all subdistricts. Both attributes (1) built-up area and (2) number of floors are used for calculating the usable building space. As a result, the usable building space is overestimated and so is the available residential living space. The calculation of the average area per inhabitant is in turn based on the overestimated residential living space and is therefore also overestimated. This overestimation finally leads to an underestimation of the number of inhabitants of the different subdistricts.

3.8 Conclusion

The study focused on the quantitative assessment of the capabilities and limitations of optical high resolution satellite imagery (IKONOS) for population modeling in urban areas. It is important to note that high resolution optical satellite images constitute a rapidly accessible source of information with limited accuracy compared to cadastral information. However in cases where suitable cadastral data is lacking, remotely sensed information can serve as a first approximation. The findings of the present study confirm this statement for modeling population distribution on subdistrict level.

The overestimation of the built-up area due to the limitations of optical images constitutes a major source of error as it results in a constant overestimation of the residential area per subdistrict and consequently results in an overestimation of the average residential living space per person on subdistrict level. Considering an average overestimation of 7 %, the question arises whether this methodology is suitable to provide information for risk managers and urban planners. From the present study, it becomes obvious that modeling population distribution at a building level is feasible using automatically extracted building footprints only with a certain loss of accuracy. Satellite images prove to be a valuable source for supplementing existing building inventory data.

Another source of error is the estimation of the floor numbers from high resolution optical satellite images. As the floor number is estimated from the building height which is derived from the building shadow, the floor numbers can only be provided as categorical data. Inaccuracy also arises from the use of mixed occupancy categories. Buildings with different occupancies on different floors were classified as mixed. In order to calculate the residential living space in a building, the footprint area should be multiplied with the number of residential floors only and not with the total number of floors.

For obtaining a more precise distribution of the residential floors and thus of the population, further research may be conducted on the extraction of building-level parameters. A further methodological step may include the consideration of employment and commuting population to model the day-time population. The results may also be improved when using high resolution satellite imagery with a better resolution than IKONOS imagery (1 m) like GeoEye (0.4 m), WorldView (0.5 m), or QuickBird (0.6 m). The technical development in the field of hyperspectral images such as the EnMap mission are very promising in the sense that high resolution hyperspectral images will be available on a global scale within the near future.

Further research may also include the transferability of the presented method to other, much more dynamic megacities or cities featuring different social and cultural characteristics. The application of optical satellite imagery to cities like Mumbai or Delhi in India with different housing and living structures is a main challenge for future research. Particularly in slum areas it may be difficult to extract single buildings from satellite images and to model the population in such areas. Additional problems will occur when using remotely sensed imagery to model population of cities located in the tropics or subtropics. In these geographical areas, data generation by optical satellites fail due to clouds. Here, weather independent radar data are an alternative. However, the classification of urban areas using radar data shows lower thematic detail compared to classifications of high resolution optical satellite images (Dell'Acqua et al. 2009; Esch et al. 2010).

From an emergency planning perspective, the overestimation in the inventory generation step of the analysis is of minor importance. But the underestimation of people which may be affected during a disaster may have severe consequences for evacuation and supply operations. Although population modeling on single building level using remote sensing might be not sufficiently accurate for the coordination of rescue teams and fire brigade, the generated population data can provide a valuable insight to the distribution of potentially affected people at a district level in case of a disaster. Overall, remote sensing data are independent, up-to-date, and basically available from anywhere around the globe. Especially in developing countries, remote sensing data may be the only available data source for planning purposes.

Although the study site was selected because of the high earthquake hazard, the presented methodology is not limited to densely populated areas and earthquake risk. The knowledge of the spatial distribution of the inhabitants of an entire city including sparsely populated outskirts also plays a critical role for risk reduction and mitigation strategies for other hazards such as tsunamis, floods, or volcanic eruptions. Another important issue related to population modeling is the identification of crowded places and the assessment of the potential for human crowding. This information is essential for determining places at risk from terrorist attacks, for example. In addition, the health sector is interested in building level population data, for pandemic prevention or for identifying people at risk concerning the contamination e.g. of drinking water.

Acknowledgments The present study was developed in a cooperation of the Center for Disaster Management and Risk Reduction Technology (CEDIM) at the Karlsruhe Institute of Technology (KIT), the Cartography, GIS and Remote Sensing Section of the Institute of Geography at the University of Göttingen, Germany and the German Remote Sensing Data Center (DFD) at the German Aerospace Center (DLR).

We gratefully acknowledge the support of Mustafa Erdik, Tamer Soylu, and Dietmar Borst for providing comprehensive data on Istanbul. Special thanks go to James Daniell and the anonymous reviewers for their extensive professional comments that helped to improve the manuscript.

References

Al-Garni AM (1995) Mathematical predictive models for population estimation in urban areas using space products and GIS technology. Math Comput Model 22:95–107. doi:10.1016/0895-7177(95)00104-A

Almeida C, Souza I, Durand AC, Pinho C, Pereira M, Feitosa R (2007) Multilevel object-oriented classification of Quickbird images for urban population estimates. In: Samet H, Schneider M, Shahabi C (eds) Proceedings of the 15th International Symposium on Advances in Geographic Information Systems. New York. USA. doi:10.1145/1341012.1341029

Ambraseys N, Finkel C (1991) Long-term seismicity of Istanbul and of the Marmara Sea. Terra Nova 3:527–539. doi:10.1111/j.1365-3121.1991.tb00188.x

Ansal A, Akinci A, Cultrera G, Erdik M, Pessina V, Tönük G, Ameri G (2009) Loss estimation in Istanbul based on deterministic earthquake scenarios of the Marmara Sea region (Turkey). Soil Dyn Earthq Eng 29:699–709. doi:10.1016/j.soildyn.2008.07.006

Applied Technology Council (ATC) (1996) ATC-40. *Seismic evaluation and retrofit of concrete buildings*. Volume 1. Redwood City. California. USA. http://www.atcouncil.org/pdfs/atc40 toc.pdf. Accessed 15 Sep 2011

Batuk F, Sengezer B, Emem O (2005) The new zoning approach for earthquake risk assessment. In: Oosterom P van, Zlatanova S, Fendel E (eds) Geo-information for disaster management. Springer, Berlin-Heidelberg 1225–1237

Baycan-Levent T (2003) Globalization and development strategies for Istanbul: Regional policies and great urban transformation projects. 39th ISoCaRP Congress 2003. http://www.isocarp.net/Data/case_studies/359.pdf. Accessed 12 Oct 2010

Beygo C, Ozcevik O, Turk S, Akcakaya I, Sen K, Damen T (2006) Milestones of Zeytinburnu Municipality through sustainable urban regeneration: From vision to action. 42nd ISoCaRP Congress 2006. http://www.isocarp.net/Data/case_studies/777.pdf. Accessed 12 Oct 2010

Cakir G, Ün C, Baskent EZ, Köse S, Sivrikaya F, Keles S (2008) Evaluating urbanization, fragmentation and land use/land cover change pattern in Istanbul City, Turkey from 1971 to 2002. Land Degrad Dev 19:663–675. doi:10.1002/ldr.859

Chen K (1998) Correlations between census dwelling data and remotely sensed data. Proceedings: SIRC 98–10th annual colloquium of the spatial information research centre. Dunedin. New Zealand.

Chen K (2002) An approach to linking remotely sensed data and areal census data. Int J Remote Sens 23:37–48. doi:10.1080/01431160010014297

Chen K, Su W, Li J, Sun Z (2009) Hierarchical object oriented classification using very high imagery and LIDAR data over urban areas. Adv Space Res 43:1101–1110. doi:10.1016/j.asr.2008.11.008

Cheng F, Thiel KH (1995) Delimiting the building heights in a city from the shadow in a panchromatic SPOT-image—Part 1. Test of forty-two buildings. Int J Remote Sens 19:409–415. doi:10.1080/01431169508954409

Dell'Aqua F, Lisini G, Gamba P (2009). Experiences in optical and SAR imagery analysis for damage assessment in the Wuhan, May 2008 earthquake. Geoscience and Remote Sensing Symposium (IGARSS) (2009). IEEE International 4:37–40. doi:10.1109/IGARSS.2009.5417603

Dökmeci V, Berköz L (1994) Transformation of Istanbul from a monocentric to a polycentric city. Eur Plan Stud 2(2):193–205. doi:10.1080/09654319408720259

EMI (2005) Istanbul, Turkey. Disaster risk management profile. 3 CD city profile series—current working document. http://emi.pdc.org/cities/CP-Istanbul-09-05.pdf. Accessed 07 April 2010

Erdik M, Durukal E (2008) Earthquake risk and its mitigation in Istanbul. Nat Hazards 44:181–197. doi:10.1007/s11069-007-9110-9

Erdik M, Demircioglu M, Sesetyan K, Durukal E, Siyahi B (2004) Earthquake hazard in the Marmara Region, Turkey. Soil Dyn Earthq Eng 24:605–631. doi:10.1016/j.soildyn.2004.04.003

Ergenc M, Ilkisik O, Turk M (2005) Earthquake risk and mitigation studies in Istanbul. EGU General Assembly 2006. NH 9.07—natural hazards impact on urban areas and infrastructure. Poster Presentation. Vienna. Austria. http://www.euromedina.org/bibliotheque_fichiers/Doc_RisksIstanbul.pdf. Accessed 14 Sep 2011

Esch T (2006) Automatisierte Analyse von Siedlungsflächen auf der Basis höchstauflösender Radardaten. Dissertation, Bayerische Julius-Maximilians Universität Würzburg. Germany

Esch T, Thiel M, Schenk A, Roth A, Müller A, Dech A (2010) Delineation of urban footprints from TerraSAR-X data by analyzing speckle characteristics and intensity information. IEEE T Geosci Remote 48(2):905–916. doi:10.1109/TGRS.2009.2037144

Fernandez J (2005) Sound Practice No. 1. An earthquake Master Plan for Istanbul. 3 cd Sound Practice Series. Istanbul. Turkey. http://emi.pdc.org/soundpractices/Istanbul/SP1-Istan-Earthq-Master-Plan.pdf. Accessed 13 Sep 2011

GENAR (2004a) Zeytinburnu Pilot Projesi Kapsaminda Konut, Ticaret, Kücük Sanayi ve Sanayi Bölgelerinde veri Derleme ve Degerlendirme Projesi. Büyük Sanayi Raporu. Temmuz 2004. Istanbul. Turkey.

GENAR (2004b) Zeytinburnu Pilot Projesi Kapsaminda Konut, Ticaret, Kücük Sanayi ve Sanayi Bölgelerinde veri Derleme ve Degerlendirme Projesi. Kücük Sanayi Raporu. Temmuz 2004. Istanbul. Turkey.

GENAR (2004c) Zeytinburnu Pilot Projesi Kapsaminda Konut, Ticaret, Kücük Sanayi ve Sanayi Bölgelerinde veri Derleme ve Degerlendirme Projesi. Kücük Sanayi. Mahalle ve Bölge Raporlari. Temmuz 2004. Istanbul. Turkey.

Görgülü Z (2002) Istanbul. Die konsumierte Stadt. GeoPoint Nr. 11. Die sozialgeographische Hauspostille. Selbstverlag Lehrstuhl für Sozial- und Wirtschaftgeographie. Universität Augsburg. Germany

Hartl P, Cheng F (1995) Delimiting the building heights in a city from the shadow on a panchromatic SPOT-image: Part 2: Test of a complete city. Int J Remote Sens 16:2829–2842. doi:10.1080/01431169508954594

Harvey J (2002) Population estimation models based on individual TM pixels. Photogramm Eng Rem S 68(11):1181–1192

IEMP (2003) Earthquake Master Plan for Istanbul. Metropolitan Municipality of Istanbul. Planning and Construction Directorate. Geotechnical and Earthquake Investigation Department. Bogazici University, Istanbul Technical University, Middle East Technical University, Yildiz Technical University. 7. July 2003. http://www.dmc.metu.edu.tr/DMC/download.php?fname=./AnaSayfa/Dosyalar/Raporlar/IBB.IDMP.ENG-comp.pdf. Accessed 01 Sep 2011

Iisaka J, Hegedus E (1982) Population estimation from Landsat imagery. Remote Sens Environ 12:259–272. doi:10.1016/0034-4257(82)90039-6

Interactive City Map of Istanbul. Istanbul Sehir Rehberi, IMM. http://sehirrehberi.ibb.gov.tr/map.aspx. Accessed 10 Oct 2010

Irvin B, McKeown D (1989) Methods for exploiting the relationship between buildings and their shadows in aerial imagery. IEEE T Syst Man Cyb 19:1564–1575. doi:10.1109/21.44071

JICA and IMM (2002a) The Study on A Disaster Prevention/Mitigation Basic Plan in Istanbul including Seismic Microzonation in the Republic of Turkey. Final Report. Main Report. (JICA—Japan International Cooperation Agency; IMM—Istanbul Metropolitan Municipality)

JICA and IMM (2002b) The Study on A Disaster Prevention/Mitigation Basic Plan in Istanbul including Seismic Microzonation in the Republic of Turkey. Final Report Volume I. Summary. (JICA—Japan International Cooperation Agency; IMM—Istanbul Metropolitan Municipality)

Kocaman S, Zhang L, Gruen A, Poli D (2006) 3D city modeling from high-resolution satellite Images. Proceedings: ISPRS workshop on topographic mapping from space. Ankara. Turkey.

Kubanek J (2011) Comparison of GIS-based and high resolution satellite imagery population modelling. A case study for Istanbul. Ibidem-Verlag, Stuttgart.

Kubanek J, Kappas M, Nolte E, Wenzel F, Taubenböck H (2010) Modelling of population dynamics: GIS versus remote sensing—a case study for Istanbul. The International Emergency Management Society (TIEMS) 17th Annual Conference Proceedings. June 8th–11th. 2010. Beijing. China.

Liu X, Clarke K, Herold M (2006) Population density and image texture: a comparison study. Photogramm Eng Rem S 72(2):187–196. doi:0099-1112/06/7202–0187/$3.00/0

Lo C (1995) Automated population and dwelling unit estimation from high-resolution satellite images: a GIS approach. Int J Remote Sens 16(1):17–34. doi:10.1080/01431169508954369

Lu D, Weng Q, Li G (2006). Residential population estimation using a remote sensing derived impervious surface approach. Int J Remote Sens 27(16):3553–3570. doi:10.1080/01431160-600617202

Massalabi A, He D-C, Benie GB, Beaudry E (2004) Detecting information under and from shadow in panchromatic Ikonos images of the city of Sherbrooke. Geoscience and Remote Sensing Symposium (IGARSS), 2004. IEEE International. 3:2000–2003. doi:10.1109/IGARSS.2004.-1370740

Mesev V (2005) Identification and characterization of urban building patterns using IKONOS imagery and point-based postal data. Comput Environ Urban 29:541–557. doi:10.1016/j.compenvurbsys.2005.01.003

Mitchell J (1999) Mega cities and natural disasters: a comparative analysis. GeoJournal 49:137–142. doi:10.1023/A:1007024703844

Parsons T (2004) Recalculated probability of $M \geq 7$ earthquakes beneath the Sea of Marmara, Turkey. J Geophys Res 109:B05304. doi:10.1029/2003JB002667

Setiadi N, Taubenböck H, Raupp S, Birkmann J (2010) Integrating socio-economic data in spatial analysis: An exposure analysis method for planning urban risk mitigation. 15th International Conference on Urban Planning and Regional Development in the Information Society (REALCORP), Vienna, Austria. GeoMultimedia 2010. 367–374

Sever S, Erek E, Vanholder R, Akoglu E, Yavuz M, Ergin H, Tekce M, Korular D, Tülbek M, Keven K, Vlem B van, Lameire N (2001) The Marmara earthquake: epidemiological analysis of the victims with nephrological problems. Kidney Int 60:1114–1123. doi:10.1046/j.1523–1755.2001.0600031114.x

Sohn G, Dowman I (2007) Data fusion of high-resolution satellite imagery and LiDAR data for automatic building extraction. ISPRS J Photogramm 62:43–63. doi:10.1016/j.isprsjprs.2007.01.001

Souza I, Pereira M, Kurkdijan M (2003) Evaluation of high resolution satellite images for urban population estimation. Instituto National de Pesquisas Espacias. Aceito para apresentacao no 3° Simposio de Sensoriamento Remoto de Areas Urbanas, Istambul, 11–13 de junho, 2002– Turquia. Sao Jose dos Campos. Brazil. http://mtc-m12.sid.inpe.br/col/sid.inpe.br/marciana/2003/04.01.10.18/doc/publicacao.pdf. Accessed 10 Oct 2010

Rottensteiner F, Trinder J, Clode S, Kubik K (2007) Building detection by fusion of airborne laser scanner data and multi-spectral images: Performance evaluation and sensitivity analysis. ISPRS J Photogramm 62:135–149. doi:10.1016/j.isprsjprs.2007.03.001

Taubenböck H (2008) Vulnerabilitätsabschätzung der erdbebengefährdeten Megacity Istanbul mit Methoden der Fernerkundung. Dissertation, Bayerische Julius-Maximilians Universität Würzburg. Germany

Taubenböck H, Roth A (2007) A transferable and stable object oriented classification approach in various urban areas and various high resolution sensors. 2007 Urban Remote Sensing Joint Event. doi:10.1109/URS.2007.371828

Taubenböck H, Roth A, Dech S (2007) Linking structural urban characteristics derived from high resolution satellite data to population distribution. In: Rumor M, Coors V, Fendel E, Zlatanova S (eds) Urban Data Management. Annual 2007. Taylor & Francis Group, London. United Kingdom

Taubenböck H, Wurm M, Setiadi N, Gebert N, Roth A, Strunz G, Birkmann J, Dech S (2009) Integrating remote sensing and social science—the correlation of urban morphology with socioeconomic parameters. In: Urban remote sensing joint event. Shanghai. China. doi:10.1109/URS.2009.5137506

Taubenböck H, Esch T, Wurm M, Roth A, Dech S (2010) Object-based feature extraction using high spatial resolution satellite data of urban areas. J Spat Sci 55:117–133. doi:10.1080/-14498596.2010.487854

TurkStat (2010) Address based population registration system. Population census results, 2009. Turkish Statistical Institute, Prime Ministry, Republic of Turkey. Press Release. http://www.turkstat.gov.tr. Accessed 14 April 2010

United Nations (2008) World urbanization prospects: the 2007 revision. United Nations Department of Social and Economic Affairs/Population Division. New York. United States

Wenzel F, Bendimerad F, Sinha R (2007) Megacities—megarisks. Nat Hazards 42:481–491. doi:10.1007/s11069-006-9073-2

Wu C, Murray A (2005) A cokriging method for estimating population density in urban areas. Computers Environ Urban 29:558–579. doi:10.1016/j.compenvurbsys.2005.01.006

Wu C, Murray A (2007) Population estimation using Landsat enhanced thematic mapper imagery. Geogr Anal 39:26–43. doi:10.1111/j.1538-4632.2006.00694.xv

Wurm M, Taubenböck H, Schardt M, Esch T, Dech S (2011) Object-based image information fusion using multisensor earth observation data over urban areas. Int J Image Data Fusion 2:121–147. doi:10.1080/19479832.2010.543934

Chapter 4
The Vulnerability of Fragile Landscape and Complex Hazards

Areas with Potential Risks and Specific Types of Hazards and Vulnerabilities of the Natural, Anthropic and Cultural Landscape in Braila County, Romania

Cerasella Crăciun

Abstract This article is a component of the Study of the Protection, Management and Development of Landscape in Braila County. 2009–2011. Phase III. 2011, consisting of two parts: Identifying potential risky areas and determining the influence of risk on the county landscape. Identifying specific types of hazards and vulnerabilities relevant for territorial development. The study radiographed landscape in all its components (natural, anthropogenic and cultural), from a macro-area landscape, of a territorial county level. Our research methodology is the first one in landscape in Romania to be applied to case studies of Braila County, an area located on the Danube River, near the Danube Delta, an administrative area structure with significant heritage values. We highlighted areas of potential risk landscape and its interaction in fragile landscape typologies; e.g., specific hazards, and vulnerability of the natural environment and specific, complex anthropogenic landscape of human settlements (in terms of architectural, urban, agriculture and forestry), and specific cultural landscape hazards and vulnerabilities (traditions, customs and local beliefs, traditional costume, ancestral music and dance, ornaments and typology of popular architecture, traditional crafts, local personalities, anthropological components, etc). By identifying priorities for intervention, we created two versions of a conceptual schema proposal type at macro- and mezzo-levels of landscape in Braila county. In the end (Phase IV), the overlap reached a final version, which was detailed in the last phase of the study, with proposals focused on strategy and reference to each administrative unit of the county.

Based on a project work Study of the protection, management and development of landscape in Braila County. 2009–2011. Phase III. 2011—Identifying potential risky areas and determining the influence of risk on the county landscape, reproduced and translated with kind permission of the authorities for which it has been performed

C. Crăciun (✉)
Department of Urban and Landscape Design, "Ion Mincu" University of Architecture and Urbanism, Academiei 18-20, 010014 Bucharest, Romania
e-mail: cerasellacraciun@gmail.com

M. Bostenaru Dan et al. (eds.), *Earthquake Hazard Impact and Urban Planning*, Environmental Hazards, DOI 10.1007/978-94-007-7981-5_4,

© Springer Science+Business Media Dordrecht 2014

Keywords Fragile landscape · Valuable landscape · Unified green urban system · Braila county · Danube River · Urban organism · Anthropic heritage values · Periurban landscaping plan · Cultural landscape · Green-yellow-blue Belts · Elongated finger green poles · Concentric green belts · Vulnerability of landscape

4.1 Introduction

The study regarding landscape protection, management and design in Braila County (Phase III. 2011) represents an effort of the local public administration in Romania, oriented towards improving the quality of life for the citizens of Braila county, aimed at the rehabilitation and capitalization of valuable landscape, as well as the integration in a current and unified green urban system of the planted public spaces in accordance with the principles of sustainable development and with the Romanian laws in protection of Natural Areas[1,2]. Braila County is located in the South-East of Romania, on the river Danube and in the vicinity of the *Little Island of Braila Natural Park*, which are two major landscape structuring elements in Romania. The county must take on the leading role in maintaining the microclimate and the health of the urban organism (Craciun 2008a), of the natural as well as urban environment, as it is a part of an area with important natural and anthropic heritage values, but also in regards to the dynamics of the geographical environment (Andronache 2009). The aims that were set out for the present research arise out of the unique character of the landscape at the level of the County of Braila, which is especially important not only in a local and territorial context but also at national level. The special character of Braila's representative ambience must be protected and conserved. But it is also known to be an important point from a touristic and functional point of view; it must support intense visitation of county residents, people from other parts of Romania, and international visitors and tourists. The conclusions of the study will be integrated into the territorial development strategy (AMTRANS[3]) within the framework of the

[1] LEGEA PRIVIND REGIMUL ARIILOR NATURALE PROTEJATE, CONSERVAREA HABITATELOR NATURALE, A FLOREI SI FAUNEI SALBATICE. LEGEA nr. 462 din 18 iulie, pentru aprobarea Ordonantei de urgenta a Guvernului nr. 236/2000) (LAWONTHE GENERAL CONDITIONS OF PROTECTED NATURAL AREAS AND ON THE CONSERVATION OF NATURAL HABITATS AND OF WILD FLORA AND FAUNA. LAW nr. 462 of the 18th of July approving Government Emergency Ordinance nr. 236/2000)

[2] OUG 57/2007 privind regimul ariilor naturale protejate, conservarea habitatelor naturale, a florei şi faunei sălbatice (GEO 57/2007 on the general conditions of protected natural areas and on the conservation of natural habitats and of wild flora and fauna)

[3] AMTRANS—Model Conceptual Şi Metodologic Pentru Stabilirea Sistemului De Relaţii Specifice Planificării Strategice Regionale În România Din Perspectiva Dezvoltării Durabile A Zonelor Funcţionale/Metropolitane Şi A Aglomeraţiilor Urbane. 2002–2003 (Conceptual And Methodological Model For Establishing The Specific Regional And Strategic Planning Relations For Metropolitan/Functional Areas And Urban Agglomerations In Romania From The Perspective Of SustainableDevelopment), research study—National Programme for Research—Development—Innovation, PhD Architect Cerasella Craciun, PhD Architect Gabriel Pascariu, Research, Design, Expertise and Consulting Centre (RDECC) of the "Ion Mincu" University of Architecture and Urbanism (IMUAU) Bucharest

County's Spatial Planning and of the Periurban Landscaping Plan, as well as into the regulations provided at local level through the General Urban Plan and through the Local Urban Regulation of the Municipality of Braila.

4.2 Specific Vulnerability

4.2.1 Specific Vulnerabilities of the Environment and of the Landscape

The specific vulnerabilities of the environment and of the landscape (ECOLOC[4]) are: the lack of specific infrastructure that would protect the environment and ensure an effective protection status for the valuable elements of flora and fauna, as well as the presence of anthropic activities within protected areas, in the areas of weakened dams due to erosion, which could lead to their collapse and to the flooding of the areas left unprotected. Also, we can distinguish in the county a lack of protection areas for the gas supply system, the destruction of the water improvement system during the communist period or the lack of irrigations, which is absolutely necessary due to the arid climate in the area (Govindasamy et al. 2003, pp. 391–404), which leads to a tendency to reduce the cultural diversity that once existed there. Another problem is the policy focusing on the embankment of the Danube in order to reduce the risk of floods and to provide land for agriculture, policy which led during the communist period to a dramatic decrease of fisheries, and most of all since the overlapping of Natura 2000 sites and fisheries leads to diminished stocks of various fish species and the vulnerability of habitats, their fragmentation, is under the pressure of anthropic insertions such as building new roads with uninterrupted traffic flow—motorways and expressways, but also land improvement works that bring about the transformation of natural ecosystems into agricultural ecosystems.

We can add other hazards to this list, such as earthquakes of magnitude 7.1 or 8.1 (with an average period of recovery from earthquakes of 50 years), land sliding in the areas with weak, degraded soils, floods due to the increase of the level of the Danube, waterways and soil pollution with waste that affects the natural environment, embankment by changing the landscape which increases the risk of shoreline erosion. Also, another dysfunction is the advanced degree of wear and tear of municipal networks (90–100 %), of sector distribution canals which serve landscaped areas, irrigated by semi-stationary pumping thermal units, as well as the pollution through drainage systems (that carry pollutants to the emissary).

---

[4] ECOLOC—Reconfigurarea starii actuale a spatiilor rezidentiale din mediile urbane din Romania sub efectul schimbarilor de mediu. (Reconfiguring the current state of urban residential areas in Romania under the effect of climate changes). 2008–2011. National Centre for Programme Management (NCMP), Programme 4—Partnerships in priority areas (research study—Project Manager IMUAU—PhD Architect Cerasella Craciun, Research, Design, Expertise and Consulting Centre (RDECC) of the "Ion Mincu" University of Architecture and Urbanism Bucharest

4.2.2 *Specific Vulnerabilities of the Architectural and Forest Landscape*

The specific vulnerabilities of the agricultural and forest landscape are due to the loss of natural habitat for the local flora and fauna that occurs in Braila through: excessive deforestation in forest areas and protective forest areas (we include here both those in the period after 1989 as well as those before 1989), draining wetlands to gain proper areas for agriculture and horticulture, the appearance of the desertification phenomenon, the degradation of soils through the use of defective agricultural techniques (see Order 223/28th of May 2002 on the approval of Methodology on the drawing up of pedological and agrochemical studies of the National and Municipal Monitoring System for Agricultural Lands[5]), the collection of natural specimens that led to the removal of fauna in the areas of contact with the anthropic element.

All of these aspects generated losses at the level of the number of individuals as well as the reduction of the surface these species occupy.

Furthermore, the decrease in the number of animals belonging to endangered species, the lack of crop rotation which lowers the adaptability of the species to environmental conditions and requires a high degree of human intervention, as well as the focus on certain types of crops, with decreased interest in horticulture and livestock have been observed. Another vulnerability is the agricultural production that uses genetically modified seeds, but also the decrease in the number of people involved in agricultural, horticultural and livestock production processes and the fact that the environmental degradation is worsened through an uncontrolled system of water improvements, through excessive mineralization of forest and agricultural areas located at the periphery of urban settlements, through increasing the built-up area due to investment pressure and through the destruction of green areas and forests located in the vicinity of specific oil and crude oil networks; the destruction of the identity of the cultural landscape associated with agricultural and forestry practices, as well as the creation of vulnerable areas due to major communication ways (motorways) and municipal networks.

Along with climate changes that bring about an increase in soil temperature of over 60 degrees Celsius in summer, the type of technologies used in local agriculture increase the risk of fire because, despite bans to burn crop fields (to remove most part of the weed seeds and to achieve an improvement in soil structure), this practice is still quite common in the rural area of small farmers. Additionally, the severe drought of recent years, climate changes generated by deforestations, the pollution of the water table through specific agricultural production activities and the invasion of insect pests can be added here (Hijmans et al. 2005, pp. 1965–1978).

By being located in the same area, genetically modified plants can pollinate typical species, leading to the growth of products that contain different genetic material from

[5] Ordinul 223/28 mai 2002 privind aprobarea Metodologiei întocmirii studiilor pedologice şi agrochimice, a Sistemului naţional şi judeţean de monitorizare sol-teren pentru agricultură (Order 223/28th of May 2002 on the approval of Methodology on the drawing up of pedological and agrochemical studies of the National and Municipal Monitoring System for Agricultural Lands)

the original material of the mother plant and these plants are not labelled as genetically modified organisms. Also, there arises the problem of using the seeds obtained from such cultures as sowing material which would lead to the possible fertilization with a new type of pollen from genetically modified plants. Since this phenomenon can not be controlled, the only possibility is to refuse crops of genetically modified plants.

4.3 Complex Hazards

As complex hazards in Braila County, in terms of natural landscape, we note the loss of agricultural production capacity (human + environment/soil + type of crop + technology). As a result of anthropic activities a series of factors with several implications were triggered, such as the intervention on habitats through deforestation, agricultural crops, livestock activities (excessive grazing, designated to be carried out on small surfaces, a fact that makes it impossible to sustain even only local needs and thus leads to grazing being done in areas that have other purposes—agriculture, forestry, generating a degradation of these invaded areas), excessive irrigation aimed at obtaining high agricultural production. Other factors with complex hazards implications are the use of improper agricultural techniques and technologies, meant to increase production capacity, but which are poorly correlated with the needs and capabilities of the environment; soil degradation; climate changes resulting from the intervention on the environment.

All of these have resulted in the loss of agricultural production capacity of a percentage of lands on which such activities were carried out. Added to these are: the environmental aggression exerted on the natural environment, with repercussions on the cultural and anthropic environment, the aggression exerted on the specific forest and agricultural landscape in its complexity (both as natural and as cultural landscape), the massive deforestations having effects on the territory, resulting in desertification, in changes of the environmental conditions of human settlements, in generating the green house effect, in the loss of important habitats for flora and fauna etc.

4.4 Risks

The risks regarding the increasing degree of fragility of the landscape in Braila County are: the fires caused by drought or obsolete ways of farming; the floods that require ecological reconstruction (Stoiculescu 2008), especially in crops following changes in the land for benefits in agriculture (erosion due to deforestation in areas with uneven terrain); desertification made worse by faulty agricultural practices; soil quality deterioration due to excessive additions (fertilizers, compost); soil salinization through irrigation works carried out improperly; water quality changes; affecting the human genome by introducing genetically modified plants; loss of habitat for local species; gradual loss and destruction of the traditions associated with the

Cultural Landscape; possible turbulence brought about by wind parks that can affect the migration of birds or that could accentuate the desertification phenomenon.

European data also allow highlighting changes in land use and occupation. For this, according to the methodology provided by Petriş or (2008), during the first period at level I of the classification, which indicates the occupation of land, the changes work out to approximately 30 ha (29 ha—the flooding of agricultural land and 1 ha—the transformation of an agricultural area into a built area), and at level III, which indicates land use, the affected areas total about 1423 ha. The analysis of these data show that the changes inflicted on built land (urban restructuring) represent 0.1 % of all the changes, the processes of restructuring in agriculture (land abandonment) represent 85.2 % while the remaining 14.7 % is due to afforestation works (see the images below). In the second period, the changes amount to 2.5 ha, 0.9 ha representing changes in land coverage and 1.6 ha representing changes in land use, and the changes are due to urbanization (0.7 ha—30 %), the development of agriculture (1.4 ha—54.4 %) and deforestation (4.0 ha—15.6 %).

4.5 Other Vulnerabilities

In regards to the Areas vulnerable to extreme weather events, statistics indicate a general increase in the county of Braila—in other words, climate forecasts indicate only an increase in temperature; there are no regions for which to expect a cooling that can be compared with the current situation regarding temperatures. Also, climate forecasts indicate a general decrease in rainfall; the region for which an increase compared to the current rainfall situation is expected represents a reduced area.

The analysis of the results of the self-monitoring performed by economic agents and communicated to the Romanian Water Agency in Braila regarding the quality of the discharged wastewater found that compared to the maximum concentrations allowed by the existing norms and regulatory acts there were excesses at the level of the indicators for extractible substances, suspension materials, BOD5, chlorides, COD-Cr, total N, sulphates, ammonia, sulphites, biodegradable synthetic detergents, SO_4^{2-}, total Fe, ammonia nitrogen, pH, fixed residue, Zn^{2+}; etc.

On the specific issue regarding the Anthropic/Complexly Built Landscape of Human Settlements, the following vulnerabilities have been observed: the rapid urbanization (population growth and migration to urban areas) in the settlements in close proximity to the Municipality of Braila similar with the situation of Bucharest city[6], the lack of green areas and protection zones in the central and western area of the County of Braila, the lack of a middle segment, i.e. medium sized towns

[6] DRTTFSV—"Definirea regimului tehnic al terenurilor cu funcţiune de spaţiu verde, în scopul protejării şi ameliorării condiţiilor de mediu în Municipiul Bucureşti" (Defining the technical conditions of lands that function as green areas in order to protect and improve environmental conditions in the city of Bucharest). 2005–2006. Research study—PhD Architect Cerasella Craciun, PhD Architect Florin Machedon, Research, Design, Expertise and Consulting Centre (RDECC) of the "Ion Mincu" University of Architecture and Urbanism (IMUAU) Bucharest

(20,000–50,000 inhabitants) within the structure of the city network of the County of Braila, the decline of youthful population and the increase of elderly population will be felt on the labour market as well.

4.6 Relationship Between Anthropic and Complex Hazards

Anthropic hazards are present through air pollution (dust, pollutants) in the industrial towns of the county, in areas with an active risky environment due to the technological risks arising from fires, explosions in the technological field (Insuratei, Stancuta, Ulmu, Bordei Verde, Ianca) and due to the Galati steel plant located near Braila county. We can add here the areas that present a tendency towards desertification when interacting with the anthropic/built landscape which is due to large variations in temperature between winter and summer (Gumelniţa) and to the cases of displaced population during communism (relocation of settlements—Maxineni) or increased deforestation.

Regarding the combination of hazards—complex hazards of the anthropic landscape, they refer to the following: the expansion of the built up area in towns which is due to external pressures and the approval of urban planning documents (detailed urban plans) in villages which is due to pressure from foreign investors. Also, the County of Braila faces the risk of explosions due to the presence of the nearby Galati steel plant, as well as the technological risk for towns whose location is in oil fields, but also vulnerabilities which include the naval transport of dangerous, toxic chemicals as well as of nuclear materials.

From the point of view of the Cultural Landscape, hazards are natural (non-anthropic: floods, desertification and anthropic: population displacement, deforestation), and the vulnerabilities are given by the change in architectural style to adapt to the new territory. Winds, the lack of forests (Braila is the county with the smallest percentage of forests in all Romania), open areas, crop changes and the lack of "green" protection areas lead to dramatic changes in landscape and to its transformation in dismal sites which often times lack landscape value.

The elements that constitute risk lead to major losses with respect to population density and structure, to the quality of the built up area, lead to the loss of identity for the landscape, the loss of local traditions and customs, contributing to the irreversible aggression of the landscape which requires actions that have a quick landscape impact and whose result must not have to do with hazard and with the different interests in the territory.

4.7 Intervention Priorities

The intervention priorities for landscape development, protection, conservation and restoration of nature in the County of Braila and the improvement of the quality of life at the level of the environmental issues regarding complex ecosystems in the sense of sustainable development are the following:

- Identifying representative ecosystems for the Romanian and European biogeographical landscape that are relatively intact, in a state close to the natural state, and including them in the list of protected natural areas (listed in the LAW ON THE GENERAL CONDITIONS OF PROTECTED NATURAL AREAS AND ON THE CONSERVATION OF NATURAL HABITATS AND OF WILD FLORA AND FAUNA—Law nr. 462 of 18th of July, approving Government Emergency Ordinance nr. 236/2000[1]);
- Expanding protected natural areas by creating buffer zones or sustainable development zones in the case of those areas threatened by anthropic pressure;
- Expanding the forest area;
- Developing greening actions in priority areas, especially those within or near protected natural areas;
- Developing programmes to reduce pollution and to stimulate economic agents who implement pollution control or monitoring (at the source) measures;
- Closing down non-compliant landfills;
- Making the waste collection and treatment system more efficient;
- Measures aimed at controlling the changes in land use, especially in regards to the expansion of settlements, mainly in protected natural areas;
- Measures aimed at controlling soil erosion due to flooding such as planting or afforestation of desert areas;
- Developing management plans for protected natural areas that do not benefit from these tools;
- Identifying custodians for protected natural areas that are not currently managed;
- Eliminating the sources of air, soil and water pollution;
- Improving the waste management system in some localities and closing down those that are non-compliant;
- Limiting changes in land use, especially controlling urbanization, deforestation and crop abandonment;
- Measures to adapt to climate changes and to reduce climate changes;
- Identifying bodies that could manage certain protected areas and developing management plans;
- Promoting actions aimed at greening affected areas targeting the settlements adjacent to the river Danube, to the river Buzău, Siret and the localities whose administrative territory includes lakes/ponds;
- Protecting and preserving the variety of terrestrial and aquatic ecosystems (specific meadow forests, grasslands, ponds and lakes, channels with alluvial banks), the ecosystems that are typical of the steppe biogeographical region (which is currently found on the creeps formed by subsidence, in the spaces between agricultural parcels, on the roadside, in temporarily not cultivated areas);
- Protecting through legislative actions the two endemic species—Campanula rotundifolia L., ssp. Romanica Savulescu Hayeck (bell flower) and Achillea coarctata Poir (the yarrow with yellow flowers) that only grow on Popina Blasova archaeological mound—a witness to Hercynic erosion (natural monument), a vestige of a mountain range of which only the Mountains of Măcin stand today;

- Protecting specific zoocoenoses, the most complex zoocenoses being typical of forests (mixed forests) and permanent ponds;
- Preserving the faunal richness (90 species) considered community interest (GEO 57/2007 on the general conditions of protected natural areas and on the conservation of natural habitats and of wild flora and fauna) for which special conservation areas and special avifaunistic protection areas must be implemented; in this context, one represents a priority species (the mink—Mustela lutreola) and 49 species are under strict protection regime. Of the total of 305 inventoried vertebrate species at the level of the county, 160 require strict protection, of which 121 species are of community interest and 39 are of national interest.
- Protecting the species of fish, amphibians and birds—of the 67 inventoried fish species, 4 species are vulnerable and 4 are endangered, one species—the sea trout (Salmo trutta labrax) being critically endangered. 2 species of amphibians of the 7 inventoried at the level of the county are vulnerable, 3 are almost threatened and the crested newt of Dobrogea (Triturus dobrogicus), whose presence was observed in the Danube floodplain, in areas that are part of the sites of Community importance ROSCI0006 Little Island of Braila Natural Park and ROSCI0012 Măcin is endangered. Among the bird species, 22 are vulnerable, 13 species are endangered and 3 species are critically endangered (Lesser White-fronted Goose, White Tailed Eagle, Greater Spotted Eagle).
- Preserving and conserving the diversity of ecosystems and habitats
- Oligotrophic to mesotrophic still waters with vegetation of Littorelletea uniflorae and/or Isoëto-Nanojuncetea
- Rivers with muddy banks with vegetation including Chenopodion rubri and Bidention
- Molinia grasslands on calcareous, peaty or argillic soils (Molinion caeruleae)
- Groups of border vegetation with hygrophilous tall grass from plains, to mountains and alpine surroundings.
- Alluvial grassland made of Cnidion dubii
- Low altitude grasslands (Alopecurus pratensis, Sanguisorba officinalis)
- Riparian mixed forests with Quercus robur, Ulmus laevis, Fraxinus excelsior or Fraxinus angustifolia, along the great rivers (Ulmenion minoris)
- Groves with Salix alba and Populus alba
- Riparian galleries and shrubbery (Nerio-Tamaricetea and Securinegion tinctoriae)
- Strongly oligotrophic and mesotrophic waters with benthic vegetation of Chara species
- Ponto-Sarmatian steppes
- Communities with Salicornia and other annual species that colonize wetlands and sandy lands
- Pannonian and Ponto-Sarmatian saline grasslands and marshes

The final proposal of the study on fragile landscape and its vulnerabilities resulted in two versions of conceptual scheme type proposals for the development of landscape at macro- and mezzo- level in the County of Braila.

4.8 Strategic Options

4.8.1 First Strategic Option

The first strategic option of restoration of nature proposes setting up a series of green penetrations, in the shape of elongated finger-like green poles (DRTTFSV[6]), having as main focal point the Municipality of Braila that has great historical significance (see Paltanea 1994) and the periurban area such as the Little Island of Braila Natural Park. These main green corridors located in the NE-SW direction, penetrate the administrative territory of the County through plantations and forests that connect the focal area that represents a green pole around the city and the Little Island of Braila Natural Park with the other corridors whose purpose is also restoring nature: the area of the Southern bank of the river Prut, located in the north of the administrative boundary of the county, the adjacent area of the river Buzau in the NW direction, the adjacent area of the railway infrastructure corridor CF 700 and DN 23, in the NE–SW/Braila–Traian–Ianca–Faurei, the DN 21-E 584 area in the NE–S direction/Braila, Unirea, Viziru, Insuratei, Baraganul, the area located to the West of Cremenea Channel and to the East of Vilciu/Macin Channel, as a green protective area for the national heritage zone represented by the Insula Mica a Brailei National Park, the area located to the east of the Old Danbe (Dunarea Veche) channel, near the Great Pond of Braila as green protective area.

In the N–S and NW directions, collateral protection areas and green-yellow-blue protective belts are set forth, whose aim is to protect, preserve and restore the wetlands and the valuable farmlands (Fig. 4.1).

4.8.2 Second Strategic Option

The second strategic option proposes setting up green concentric belts in the shape of waves located gradually from the SW toward the NE, in the city of Braila and its periurban area as a main development point in the territory. The green waves will be laid out in NS and NW directions, where green and blue wave-like corridors are proposed, that will connect the following areas: the area adjacent to the river Calmatui, between the localities of Stancuta, Zavoia, Ciresu and Jugureanu, the area adjacent to the railway that connects the localities Barganul, Dudesti, Ciresu, Faurei, Galbenu, the central area of the county, in the direction of the localities Frecatei, Marasu/Zaton, Stancuta, Tufesti, Viziru, Bordei Verde, Ianca, Sutesti, Racovita, the area located near the localities: Voinesti, Sihleanu, Gemenele, Traian, Tichilesti, the area of the green belt located near the Municipality of Braila and its periurban area (Fig. 4.2).

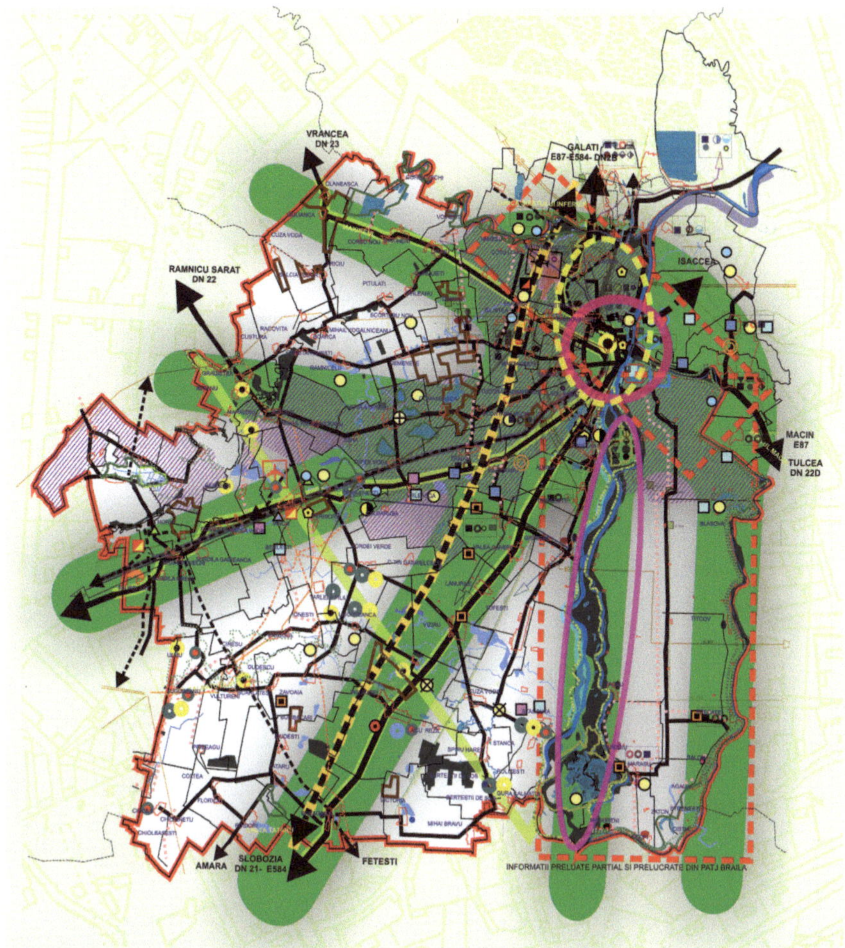

Fig. 4.1 Strategic Option 1 Scheme—*Green* elongated poles in the *shape of fingers* from the Study of the protection, management and development of landscape in Braila County, Phase IV, 2001—Progress option; Complex Project Coordinator: PhD. Reader Architect Cerasella Craciun, Research, Design, Expertise and Consulting Centre (RDECC) of the "Ion Mincu" University of Architecture and Urbanism Bucharest[7]

---

[7]STUDIU PRIVIND PROTECTIA, MANAGEMENTUL SI AMENAJAREA PEISAJULUI DIN JUDETUL BRAILA, 2009–2011; Etapa III. 2011—Identificarea zonelor cu potential de risc si influenta riscului asupra peisajului judetean. Identificare tipurilor specifice de hazard si vulnerabilitati cu relevanta asupra proceselor de dezvoltare teritoriala; Etapa IV. 2011—Varianta de Progres: Propuneri de solutii practice privind renaturarea teritoriului judetean pentru limitarea efectelor adverse microclimatice si punerea in valoare a potentialului economic si turistic; Etapa V. 2011—Varianta finala. (Study of the protection, management and development of landscape in Braila County. 2009–2011. Phase III. 2011—Identifying potential risky areas and determining the influence of risk

Fig. 4.2 Strategic Option 2 Scheme—*Green* concentric *wave-like* belts (from the Study of the protection, management and development of landscape in Braila County, Phase IV, 2001—Progress option; Complex Project Coordinator: PhD. Reader Architect Cerasella Craciun, Research, Design, Expertise and Consulting Centre (RDECC) of the "Ion Mincu" University of Architecture and Urbanism Bucharest

---

on the county landscape. Identifying specific types of hazards and vulnerabilities relevant for territorial development; Phase IV. 2011—Progress version: Proposals for practical solutions regarding the restoration of nature on county territory with the aim of limiting adverse microclimate effects and of highlighting the economic and touristic potential of the area; Phase V. 2011—Final Version.) Beneficiary: Braila County Council; Complex Project Coordinator: PhD Architect Cerasella Craciun, Research, Design, Expertise and Consulting Centre (CCPEC) of the "Ion Mincu" University of Architecture and Urbanism (IMUAU) Bucharest

Fig. 4.3 Strategic Option—Final Proposal Scheme (from the Study of the protection, management and development of landscape in Braila County, Phase V, 2001—Final version; Complex Project Coordinator: PhD. Reader Architect Cerasella Craciun, Research, Design, Expertise and Consulting Centre (RDECC) of the "Ion Mincu" University of Architecture and Urbanism Bucharest

4.8.3 Common Features

Both proposals and both strategic versions will take into consideration the following components:

- Connections within a green territorial system between the existing and the proposed green areas, by setting up a "green mesh-like web" whose aim is to preserve, restore and protect the ecological balance of the county and form a link with the green areas of the nearby counties.

- Integrating the periurban area as well as the peripheral area of the city of Braila in a green belt whose aim is to provide protection for the city
- Setting up tree windbreaks in the direction of dominant winds to the NE (through partnerships with neighbouring towns located in these directions)
- Preserving, restoring and protecting all the existent wetlands as well as restoring other wetlands in other areas with potential
- Integrating intensely planted "shadow roads" in the county territory, on secondary communication ways
- Developing planted areas on the boundaries of agricultural parcels and inside large plots of land in order to create green shelter-type retreats for local flora and fauna, having the ecological role of maintaining the microclimate (GEO 57/2007)
- Integrating specific green protection areas for the technical city networks and communication ways
- Diversifying agricultural land through green-yellow-blue belts and different types of crops
- Landscaping national roads (European national road 2B, 21, 21B, 22, 22B, 23) as well as important entries, accesses and road and railway communication networks in localities
- Expanding forest areas and increasing forest potential
- Creating "green buffer zones" and sustainable development zones, by expanding protected areas in the case of the areas that are under anthropic pressure
- Controlling soil erosion due to flooding by planting on and afforesting desert areas.

Finally, the two strategic versions will overlap to achieve a network of green spaces that will serve to reduce the malfunctions and dangers which Braila County faces regarding the vulnerability of fragile landscape (Fig. 4.3).

References

Andronache I (2009) Dinamica mediului geografic in Balta Brailei (The Dynamics of the Geographical Environment in Balta Brailei). PhD Thesis. University of Bucharest, Faculty of Geography, Bucharest

Crăciun C (2008a) Metabolismul urban. O abordare Neconventionala a Organismului Urban (The Urban Metabolism. An Unconventional Approach to the Urban Organism). "Ion Mincu" University Publishing House, Bucharest

Crăciun C (2008b) Conturarea unui model de analiza si diagnosticare de tip neconvenţional (Outlining an unconventional analysis and diagnostics model), in Preocupări recente în planificarea spaţială. Spre Confluenta tendinţelor europene cu priorităţi naţionale (Recent pursuits in the field of spatial planning. Towards the confluence of European trends and national priorities). "Ion Mincu" University Publishing House, Bucharest

Govindasamy B, Duffy PB, Coquard J (2003) High-resolution simulations of global climate, part 2: effects of increased greenhouse cases. Climate Dynamics 21:391–404

Hijmans RJ, Cameron SE, Parra JL, Jones PG, Jarvis A (2005) Very high resolution interpolated climate surfaces for global land areas, Int J Climatol 25:1965–1978

Petriş or AI (2008) Levels of biological diversity: a spatial approach to assessment methods. Romanian Rev Regional Studies 4(1):41–62
Paltanea P (1994) Istoria orasului Braila de la origini pana in 1918 (The history of the city of Braila from the beginning until 1918). Braila. Porto-Franco
Stoiculescu CD (2008) Reconstructia ecologica a zonei inundabile a Dunarii Romanesti/Ecological Reconstruction of the Easily Flooded Region of the Romanian Danube. Bilingual Edition. Green Steps SRL. Bucharest
*ELC—EUROPEAN LANDSCAPE CONVENTION. 2000. Florence, ratified in Romania through Law 451/2002

[illegible]

Part II
Earthquake Risk Assessment

Chapter 5
Scenarios for Local Seismic Effects of Tulcea (Romania) Crustal Earthquakes—Preliminary Approach of the Seismic Risk Characterization for Tulcea City

Ştefan Florin Bălan, Dumitru Ioane, Carmen Cioflan, Ionelia Panea, Bogdan Apostol, Zina Malita, Florina Chitea and Mirela Adriana Anghelache

Abstract This paper is a multidisciplinary presentation of the seismogenic area situated in the North-Dobrogea Orogen (Tulcea). This zone is characterized by significant crustal seismic activity, with crustal earthquakes of magnitude Ms $\geq$ 5.0 on the Richter scale. Geological and geophysical data for the area are presented. The seismicity of the region is presented by making use of the latest catalogues, exemplified with maps and a 3D figure. Focal mechanisms with their parameters for several earthquakes are analyzed together with the observation data (provided by different seismological stations). A brief presentation of the main geological features, which are characteristic of the tectonic units that build up the North-Dobrogean Orogene, outlines the diverse petrographic constitution of the various structural levels. In order to discuss the local seismic site effects two scenarios are considered, both of which take into account the characteristics of the seismogenic area. The first one considers the city exposed to an earthquake (superficial) from the E Vrancea zone and the second one considers the city exposed to a seismic event with magnitude $M_w = 5.1$ from Sf. Gheorghe fault. The earthquake epicentres are located in very active seismic areas. The important features taken into account are the nonlinear behavior of the upper soil strata, the effect of the bedrock elasticity, and different shear modulus and damping of the linear-equivalent-type system. Additionally, several local amplification functions are presented.

S. F. Bălan (✉) · C. Cioflan · B. Apostol · Z. Malita
Department of Engineering Seismology, National Institute of Research-Development for Earth Physics, 12 Călugăreni St., Măgurele, Romania
e-mail: sbalan@infp.ro

D. Ioane · I. Panea · F. Chitea
Faculty of Geology and Geophysics, Department of Geophysics, University of Bucharest, 6 Traian Vuia str., 020956 Bucharest, Romania
e-mail: dumitru.ioane@g.unibuc.ro

M. A. Anghelache
Institute of Geodynamics of Romanian Academy "Sabba S. Stefanescu", Bucharest, Romania

M. Bostenaru Dan et al. (eds.), *Earthquake Hazard Impact and Urban Planning*, Environmental Hazards, DOI 10.1007/978-94-007-7981-5_5,

© Springer Science+Business Media Dordrecht 2014

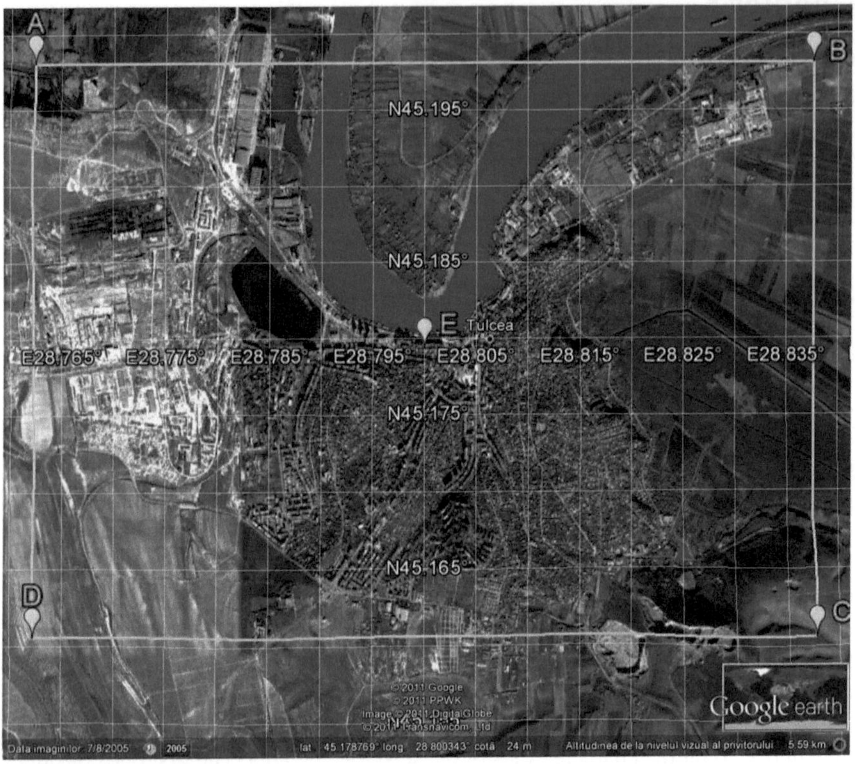

Fig. 5.1 Map of the study area

Keywords Seismogenic area · Seismic hazard · Faults · Epicenter distribution · Focal mechanisms · Local seismic effects · Nonlinear behavior of soils

5.1 Introduction

This paper refers to the well known seismogenic area situated in the North-Dobrogea Orogen (Tulcea), the zone is characterized by a significant crustal seismic activity. Crustal earthquakes with magnitude Ms $\geq$ 5.0 on the Richter scale are generated in this area. Considering the shallow depth of the seismic shocks ($\sim$ 1–30 km) and their existence in the proximity of an important city, a major seismic risk is induced in this densely populated territory. Therefore, there is a necessity for study of the seismic hazard, which is based on the information regarding the modeling of the seismic source and of the seismic response of the lithosphere in the propagation process of the seismic waves generated by these crustal earthquakes. The seismicity is monitored by the Macin and Tudor Vladimirescu seismological stations, both of them are located in the Tulcea area.

Fig. 5.1 shows the city of Tulcea and its surroundings.

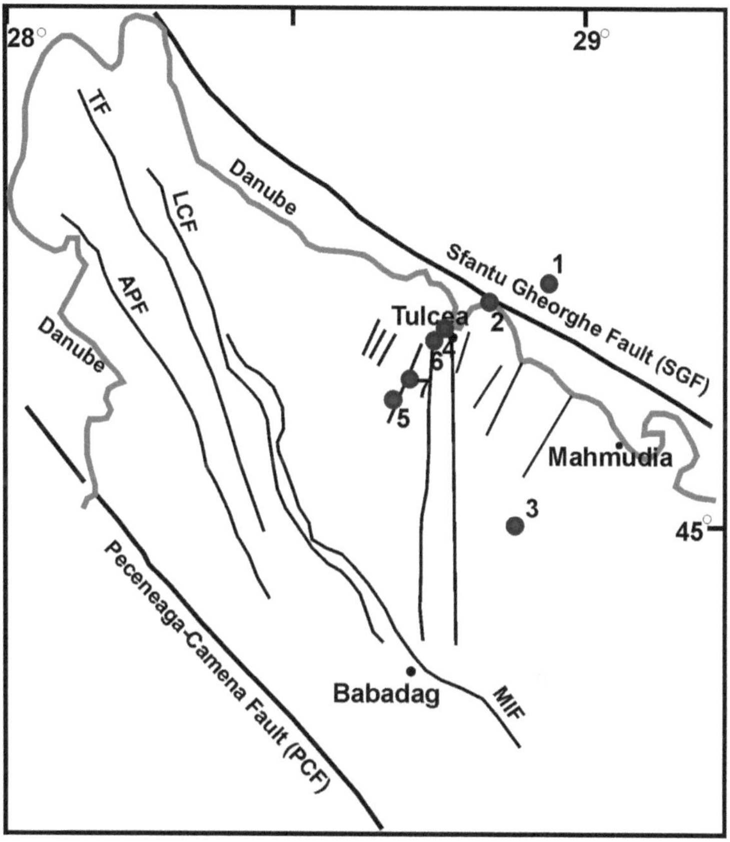

Fig. 5.2 Tectonic sketch of the North Dobrogean Orogen; *LCF*Luncavita–Consul Fault, *MIF*Meidanchioi–Iulia Fault, *APF*Ac Punar Fault, *TF*Taita Fault (Seghedi 1998, 2008); *Black dots*(1–7): epicenters of the Tulcea 1981 earthquake (Oncescu et al. 1989)

5.2 Geological and Tectonic Setting

The city of Tulcea is located in the North Dobrogean Orogen, an early Alpine intracratonic orogenic belt bounded by major tectonic faults: Sfantu Gheorghe Fault (SGF) to the north-east and the Peceneaga-Camena Fault (PCF) to the south-west (Fig. 5.2). The North Dobrogean Orogen was divided into three main tectonic units: Macin, Niculitel and Tulcea (Sandulescu 1984).

The basement of the North Dobrogean Orogen represents a part of the northern European Hercynides displaced during Permo-Triassic time by strike-slip movements related to the Pangaea super continent break-up. It experienced rifting and subsequently thrusting during the Triassic-Jurassic inversion (Seghedi 1998).

The metamorphic rocks of North Dobrogea belong to three main groups: Orliga (Precambrian, medium-grade metamorphism), Megina (Precambrian, medium-grade metamorphism) and Boclugea (Proterozoic-early Cambrian, low-grade metamorphism).

The sedimentary cover starts with Palaeozoic deposits, which in the northern part of the Tulcea zone consists of deep marine Silurian-Devonian sequences, considered to represent a fragment of an old accretionary prism. The Triassic-Jurassic sedimentary cycle include carbonate platform sequences and deeper water formations. The late Cretaceous calcareous to detrital sediments were deposited over a Kimmerian deformed basement, creating a major geological unconformity (Seghedi 1998).

The main tectonic units of the North Dobrogean Orogen overthrust each other along thrust-faults from south-west toward north-east, a suite of Alpine tectonic events (Sandulescu 1984). The front of the Macin tectonic unit is represented by the NW–SE trending Luncavita-Consul Fault (LCF), a thrusting lineament between the Macin and Tulcea zones. East of LCF and almost parallel with this fault is located the Meidanchioi-Iulia Fault (MIF), considered to be associated here with the emplacement of basic and acid magmatic rocks (Mutihac and Ionesi 1974).

The sigmoidal aspect of the main tectonic units within the North Dobrogean Orogen, oblique with respect to the major bounding regional crustal faults, may be considered as an effect of wrench tectonics processes. The main structure is represented by a large synclinorium trending NW–SE, including several faulted syncline and anticline folds, displaying thrusting features from SW to NE.

The Uzum Bair anticline, with metamorphic rocks outcropping in the Redi hill, presents an asymmetrical shape, with a steep north-eastern flank.

The Tulcea–Mahmudia anticline follows closely the Sfantu Gheorghe Danube river branch, trending NW-SE and displaying metamorphic and Palaeozoic sediments in areas located in the vicinity of these towns.

The basement of the Tulcea city area consists of low-grade metamorphic rocks outcropping in the Redi hill: metagraywacke, sericito-cloritous schists with intercalations of basic effusive rocks and quartzites.

The sedimentary deposits are represented by Silurian (quartzites and limestones), Devonian (sandstones and argillites), Triassic (conglomerates, sandstones, argillites, limestones, marls) and Quaternary (loess) (Mutihac and Ionesi 1974). In the city area the basement is covered by Triassic sediments: Werfenian conglomerates and Anisian limestones (Mirauta and Panin 1979).

Fault lines in the area of Tulcea city are difficult to be located by means of geological mapping since most of the geological structure is concealed by Quaternary deposits. However, a recent compilation of geological and tectonic information for the North Dobrogean Orogen depicts three main fault systems (Fig. 5.2; Seghedi 2008):

A. The NW–SE fault system, represented mainly by the two regional faults, Pecineaga-Camena and Sfantu Gheorghe. It is also represented by important faults located especially in the Macin zone: Ac Punar, Taita, Luncavita-Consul and Meidanchioi-Iulia;
B. The N–S fault system, represented by faults developed between Tulcea and Babadag;
C. The NE–SW fault system, developed mainly in the Tulcea zone.

Since both the N–S and the NE–SW fault systems displace the axes of the NW–SE trending faulted anticlines, the NW–SE fault systems is the oldest.

The Monument outcrop, located at the city eastern limit, shows a normal fault affecting the basement (Mutihac and Ionesi 1974) or Devonian sediments (Mirauta et al. 1967; Mirauta and Panin 1979)and the Werfenian (Lower Triassic) conglomerates, its northern compartment deepening northward, toward the Danube. Detailed geological mapping showed that the WNW–ESE normal fault is crossed by a NE–SW trending one (Mirauta and Panin 1979), the latter being interpreted in this paper as a strike-slip fault.

5.3 Geophysical Data

The interpretation of Bouguer gravity data for North Dobrogea was performed (Botezatu et al. 1961) soon after the regional gravity mapping depicted significant geophysical information, such as the NW–SE trending low gravity within the Tulcea zone and the parallel high gravity, developed between Tulcea and Mahmudia. Tulcea city is situated at the north-western end of this high gravity, a fact that suggests that the uplifted high density metamorphic rocks and Palaeozoic formations are rapidly deepening west of Tulcea, along N–S and NE–SW faults. The high intensity lineaments of gravity horizontal gradient, located at both flanks of the Tulcea-Mahmudia high gravity represent the effect of important fault systems trending NW–SE, the northern one being the Sfantu Gheorghe Fault.

Since there are no highly magnetic rocks beneath the Tulcea city area, the magnetic and aeromagnetic data are not particularly useful in locating tectonic features, such as fault lines or fractures. However, the splitting of a large high anomaly of vertical component of the magnetic field suggests the presence of NE–SW fault systems, developing south-west of Tulcea (Airinei et al. 1983).

A regional refraction seismic crustal study, entitled Vrancea 2001, offered important data related to the deep structure of the North Dobrogean Orogen and its relationship with the Moesian Platform (Hauser et al. 2007). The seismic profile started south of Tulcea and was directed toward WNW to Braila, and then crossing the Eastern Carpathians and the Transylvanian Basin, to Aiud.

The geophysical model presented in Fig. 5.3, based on seismic velocity discontinuities, represent the eastern part of the resulted crustal model, situated between Tulcea and Focsani. Considering only the Tulcea city area, the eastward decrease of seismic velocities from 5.6–5.8 to 5.0–5.2 km/s may be related to rapid modifications in the "shallow" (1–2 km depth) geological structure, from metamorphic rocks and Palaeozoic formations south-west of Tulcea, to Middle and Upper Triassic sedimentary deposits between Tulcea and Mahmudia. Going deeper within the crust, the upper part of the consolidated crust does not show velocity variations until 13–14 km depth (suggesting the absence of significant petrographic variations), where a velocity discontinuity was detected (from 6.0 to 6.3 km/s). Two other velocity discontinuities were obtained after the seismic data processing, at 26 km depth (from 6.4 to 6.7 km/s) and at 43 km depth (from 7.1 to 8.0 km/s).

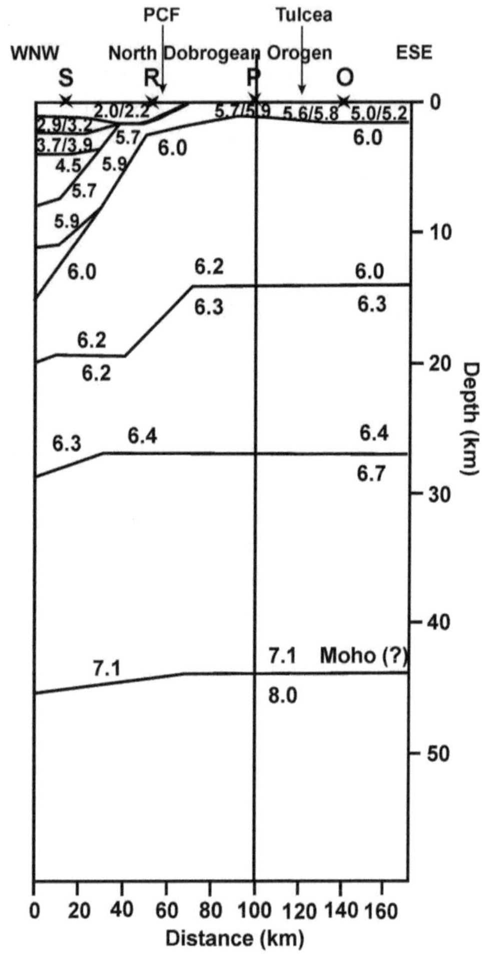

Fig. 5.3 Crustal model based on seismic velocity discontinuities: P wave velocities in km/s (Slightly modified after Hauser et al. 2007)

As it can be observed on the geophysical model, the whole crustal structure rapidly modifies west of the Peceneaga-Camena Fault (PCF), the contact with the Moesian Platform.

5.4 Seismicity and Earthquakes

The seismicity of the Romanian territory was first monitored during the end of the 19th century and the first half of the 20th century, the earthquakes intensity being mainly estimated by considering the damages of buildings and localities infrastructure (Atanasiu 1961).

At a regional scale, the high seismicity line Tulcea–Galati, trending NW–SE, was found to be activated by "Moldavian" earthquakes occurring in Vrancea, by "pre-Balkan" earthquakes occurring in Bulgaria south of Danube river and by "Kimmerian" earthquakes occurring in Dobrogea, along the Cogealac–Topolog seismic line.

The intensity of seismic events occurred along the Tulcea-Galati line was evaluated to range between 4 and 6, local amplification at Tulcea city area reaching 6+for a pre-Balkan earthquake (October 1902) and 7 for a Vrancea earthquake (November 1940).

Besides the seismicity of Tulcea determined by remote located earthquakes, a local seismicity associated with active faults was also observed (January 1872) or recorded (November 1981).

The earthquake that occurred in November 1981 in Tulcea area, recorded in eight stations of the Romanian seismic network, is of special interest for the city seismicity due to its characteristics: multishock event and quite high magnitude (M = 4.6) for this region. The location of the first shock and the subsequent six shocks is presented in Fig. 5.1, the hypocenters being situated up to 10 km depth. A NE–SW migration of seismic activity was noticed along 23 km during 34 h. The shallow location of the main shock (h = 4 km) was related to the relatively strong effects of this earthquake in the city of Tulcea. The fault plane solution computed on the basis of P-wave first motion suggested that the main shock was generated on a NE–SW dextral strike-slip fault (Oncescu et al. 1989). This shock probably occurred at the intersection between the regional NW–SE Sfantu Gheorghe Fault and the NE–SW fault system, geologically mapped in the Tulcea city area. The NE–SW distribution of epicenters for the shocks number 2, 4, 5, 6 and 7 show that the main active fault belonging to the latter system is crossing the town. The epicenter location of shock number 3, in an area situated in the vicinity of Mahmudia town, indicates that other NE–SW faults may be considered as active faults in this region.

5.4.1 The Seismicity of the Region

As already indicated, I. Atanasiu (1949) mentioned the seismic focal region of Tulcea, where the 26 January 1872 earthquake had occurred (the epicentral intensity of that earthquake was III). In the same paper, the Macin–Cerna and Tulcea–Isaccea seismic lineaments were additionally mentioned. Previously, Ioan G. Popescu (1938) provided the locations of 11 earthquakes which had occurred in the Galati–Sulina zone, during the time interval 1871–1929. All that information was included in the seismic catalogues devised by A. Florinescu (1958) and Constantinescu and Marza (1980).

The catalogue prepared by us made use of various Romanian sources (the INCDFP–ROMPLUS catalogue; the Constantinescu, Marza 1980 catalogue)and of foreign ones (the Mediterranean Seismological Centre of Strasbourg, France; The International Centre of Edinburgh, Great Britain; The National Seismological Centre

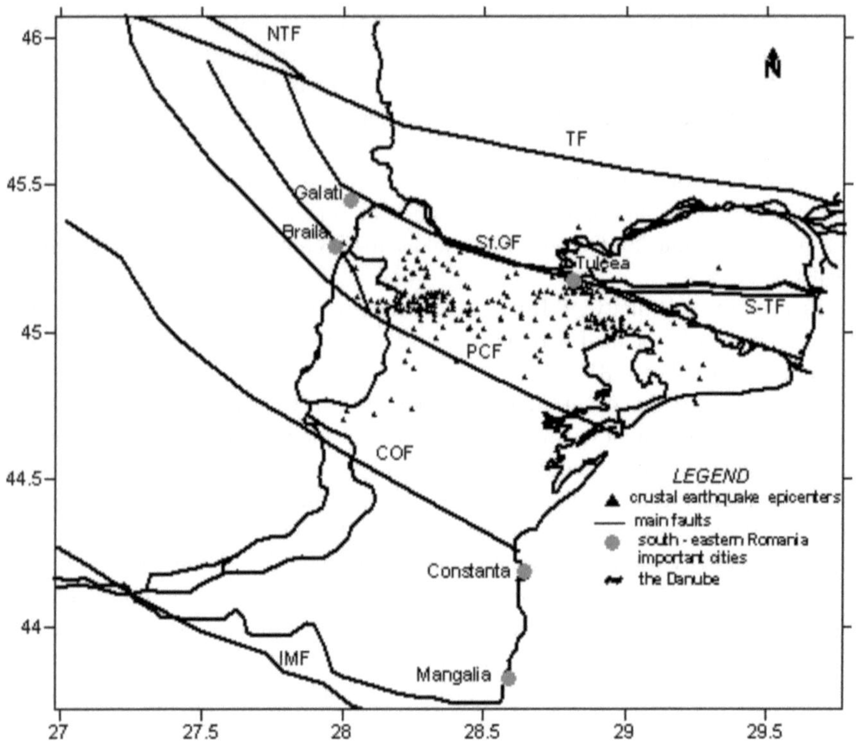

Fig. 5.4 The epicenter distribution of the earthquakes which occurred during the time interval: 1844–2006. Romania's tectonic background with main faults: *IMF*Intramoesian fault, *COF*Capidava–Ovidiu fault, *PCF*Peceneaga–Camena fault,Sulina–Tarkhankut fault, *Sf. GF* Saint Gheorghe fault,*TF*Trotuş fault,*NTF*Noua falie Trotus (New Trotus fault)

of Denver, Colorado, USA). As a result, 263 crustal earthquakes have been identified for the time interval 1844–2009, 58 of which had magnitudes (Ms, Mw, md, mb, MD, ML) above 3. Among them, there are worth mentioning the shocks which occurred on 3 July 1864 (MS = 4.8), 27 April 1865 (MS = 5.6), 27 June 1890 (MS = 4.6), 24 July 1927 (MS = 4.9), 13 November 1981 (MW = 5.1) and 3 October 2004 (mb = 4.8).

In Fig. 5.4. there is illustrated the distribution of the epicenters of the earthquakes which occurred during the specified time interval (1844–2006).

It is worth noticing the seismic activity concentration in the central zone of the analyzed area (south of Tulcea) and in the western region, south of Macin (in the Turcoaia–Peceneaga zone). The earthquakes focal depths ranged between 0 and 33 km. Out of the earthquakes with $M > 3$ (53), 18 earthquakes occurred at the surface, i.e. within the sedimentary lithosphere; the deepest earthquakes (33 and 31 km) occurred on 20 August 1995 ($m_b = 3.7$) and 3 October 2004 ($m_b = 4.8$), their hypocenters being located in the lower part of the crystalline lithosphere (within the "basalt" layer). In addition, we have to also consider the 19 January 2001 shock, which occurred at 33 km depth. Two of these earthquakes (the ones having occurred in 1995 and 2004) were located west of Tulcea (SE of Niculitel and SW of Macin), while the third one was located east of Tulcea (in the Nufaru–Victoria region).

Except for the 27 April 1865 earthquake (M = 5.1), crustal earthquakes of low magnitude (3.1–3.9) are characteristic to the seismic activity of this orogenic region, and only a few other events exceeded a magnitude of 4 (specifically, those of 1864, 1878, 1890, 1906, 1927, 2004). The hypocenters were situated mainly in the upper part of the crystalline crust, few of them in the medium crust (in the 20–24 km depth range), while those already mentioned previously were located deeper, in the 31–33 km range. Crust thickness estimates for this zone indicate 42–48 km. The depth distribution of the hypocenters of these earthquakes is indicated in Fig. 5.5.

5.4.2 *Focal Mechanisms*

Among the earthquakes having occurred in the region, the 13 November 1981 seismic event was studied by Oncescu et al. (1989). The main shock ($M_W = 5.1$) was followed by 6 aftershocks of lower magnitudes (2.9–3.3). The investigators determined that the seismic focal depths ranged between 0 km and 9 km and that the strike of the focal plane (for the main shock) was N120°E (ENE–WSW). The data-records analysis indicated that the strike of the compression axis (P) was WSW–ENE, while that of the tension axis (T) was NNW–SSE.

There were subsequently analyzed the focal parameters of several earthquakes for which the observation data (provided by different seismological stations) complied with the quality requirements. It was thus possible to determine for another 17 earthquakes, which had occurred during the time interval 1980–2005, the elements which defined the dynamic process in the focal region (Table 5.2). The Richter magnitudes of those events ranged 2.7–5.1.

The concerned earthquakes were located east of Tulcea (9), SE of Niculitel (4), SW of Macin (2) and NE of Greci (2). Their mechanisms were of thrust (10) and of strike-slip (7) faults (Table 5.1).

Figure 5.6 illustrates the focal mechanism solutions, in terms of compartments undergoing compression (white) and tension (black). One may notice that the compression axes strike prevalently NW–SE; there occur, however, also E–W or NE–SW directions. Figs. 5.7, 5.8 and 5.9 illustrate these focal mechanism solutions for all the 17 crustal earthquakes under consideration.

In Tulcea area, the Hercynian crystalline basement is overlain by Triassic formations consisting of conglomerates, limestone deposits and, at the top, sandstones and shales. A series of diabase and porphyry bodies intrude this sedimentary complex, especially in the Niculitel–Consul area. Those formations are largely equivalent to the Jurassic gypsum and anhydrites existing in Cernavoda area (in Ilie Barza quarry), where there were measured seismic velocities (V_P) of 3600–3800 m/s (at 125–130 m depth) (Radulescu et al. 1982).

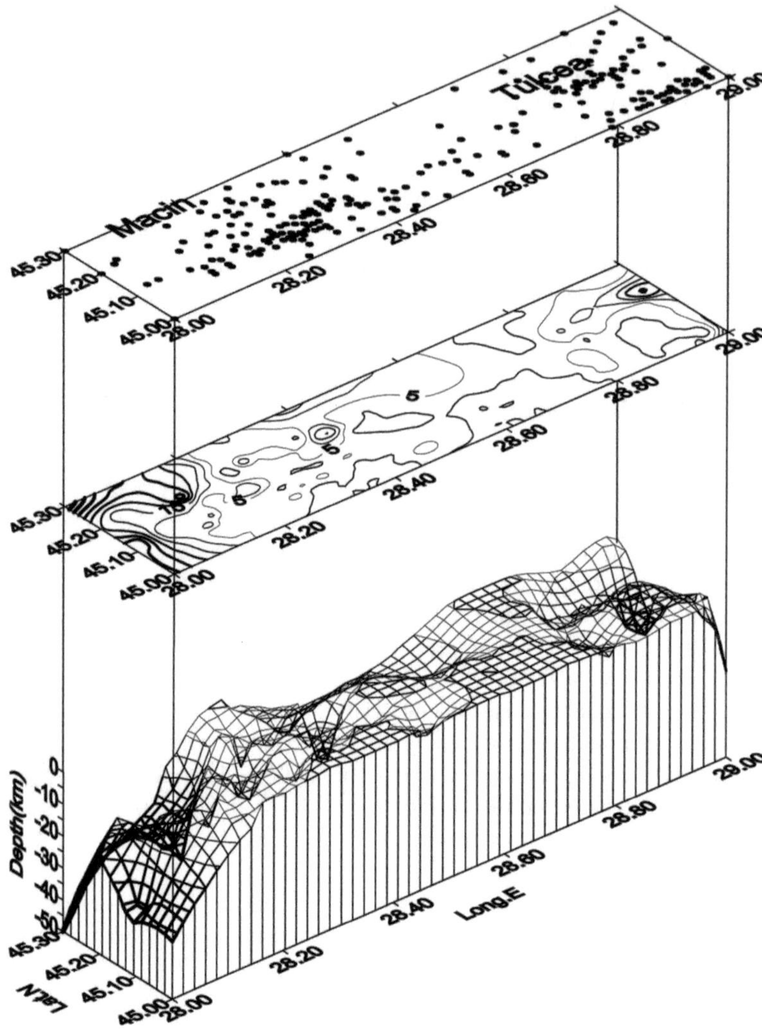

Fig. 5.5 The epicenter distribution, the hypocenter envelope depth-contour map and 3D image for Tulcea region

At the surface, there is a relatively continuous layer consisting of loess deposits with V_P velocities of 300–1000 m/s and densities of 1.57–1.63 g/cm^3.

Quantitative data concerning the seismic waves propagation velocities across the respective metamorphic and sedimentary complexes, corroborated with information referring to rocks with similar lithologies and geological ages (especially in terms of V_S—the transverse wave velocity—and of specific weight values) resulted in devising lithospheric profiles (including thicknesses, seismic velocities and specific weights of the concerned crustal domains) for the regions of Turcoaia (south of Macin) and

Table 5.1 The focal mechanisms catalogue of 17 earthquakes from Tulcea zone

Date	hh:mm:ss	Lat. (°N)	Lon. (°E)	h (km)	m_b	M_w	M_D	M_L	Fault type	
									Plane 1	*Plane 2*
11.09.1980	23:24:25	45.32	28.03	20.4		4.2			TS	TF
13.11.1981	09:07:13	45.17	29	15		5.1			TF	TF
09.05.1991	15:21:36	45.04	28.81	10					TF	TS
03.06.1992	20:21:43	45.2	28.9	10		3.1		3.9	TF	TF
24.11.1994	10:24:39	45.15	28.41	10			3.3		TS	TF
13.06.1997	09:23:59	45.02	28.89	5					TF	TF
21.08.1997	08:49:42	45.12	28.36	10			3.1		TF	TF
06.08.1998	11:03:45	45.12	28.39	5			3.4		TS	TF
08.05.1999	11:06:18	45.12	28.39	5			3.4		TF	TF
19.01.2001	15:12:47	45.12	28.03	33					TF	TS
03.10.2004	09:02:07	45.2	28.97	31	4.8				TF	TF
08.11.2004	08:55:55	45.22	28.24	0				3.1	TF	TF
24.11.2004	18:36:55	45.16	28.79	3				3.8	TF	TF
18.01.2005	17:15:30	45.14	28.88	0				3.8	TF	TF
24.01.2005	11:34:54	45.03	28.94	0				2.7	TS	TS
18.04.2005	12:29:21	45.18	28.94	5				3.5	TS	TS
23.04.2005	05:18:34	45.22	28.24	0				3.3	TF	TF

Note:
Stress regime defined in the following manner:
TF: Thrust faulting
TS: Thrust faulting with strike-slip component
SS: Strike-slip faulting

Tulcea, which are the most significant seismic activity areas. The physical data were extended into the lithosphere down to depths of 24 km (in Turcoaia zone) and of 18 km (in Tulcea zone), those levels corresponding to the estimated location of the Conrad discontinuity (the reference surface between the upper and the lower crustal seismic layers). Velocity values and specific weight values were also ascribed to the domain which was positioned beneath this seismic horizon and which hosted the deepest earthquakes of the area.

To conclude, it is worth underscoring the fact that except for the shallow layer consisting of Quaternary loess (a rock of small seismic velocities and densities), the Proterozoic metamorphic complexes and the Paleozoic and Mesozoic sedimentary ones consist of hard rocks, subject to high seismic velocities.

5.4.3 *Litosphere Profiles*

The brief presentation of the main geological features, which are characteristic to the tectonic units that build up the North-Dobrogean Orogene, outlined the diverse petrographic constitution of the various structural levels. The Quaternary loess is underlain by Paleozoic formations (in the western part, in the Macin unit) and by Triassic formations (in the eastern part, in the Tulcea unit). They are all intruded by alkaline (diabases, rhyolites) and acid (granites) igneous rocks, which are well developed especially in the Macin and Niculitel zones, where they may reach remarkable

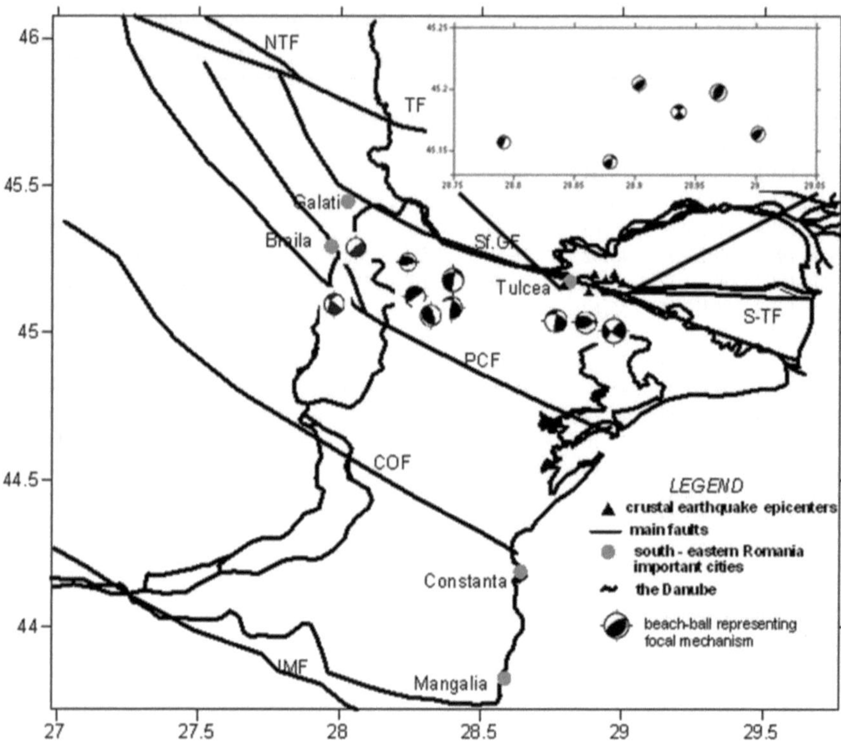

Fig. 5.6 Map with the focal mechanisms of several earthquakes from Tulcea zone

thicknesses. The Paleozoic sediments (of Devonian and Permo-Carboniferous age) overlie a Hercynian basement consisting of crystalline schists (of Proterozoic age).

The lower section of the Pre-Silurian metamorphic formations consists of amphibolites, amphibolites schists, meta-gabbros, meta-diorites, quartz-sericite schists and schists with muscovite and biotite, while the upper section consists of epimetamorphic sedimentogenic deposits (phyllites, quartzite). The indicated crystalline formations have relatively high seismic wave velocities and high values of their elastic parameters. Specifically, as a result of seismic studies conducted in Central Dobrogea, values of V_P(the longitudinal waves velocity) in the 4500–4700 m/s range were determined in the green schists complex of Celea Mare (6 km SW of Harşova) (Rădulescu et al. 1982), the latter formations being rather similar to the ones mentioned above. Higher values were indicated by seismic logs performed in certain drill holes that had intercepted the crystalline basement in Tulcea area; specifically, 5420 m/s were recorded at 3700 m depth, and 6100 m/s at 1800 m depth (Damian 1974).

The Paleozoic formations of Macin area consist of quartzites, amphibolites, micaschists, shales and conglomerates; in addition, there have to be mentioned the granites of Macin, Iacobdeal and Atmagea. Seismic measurements performed in a

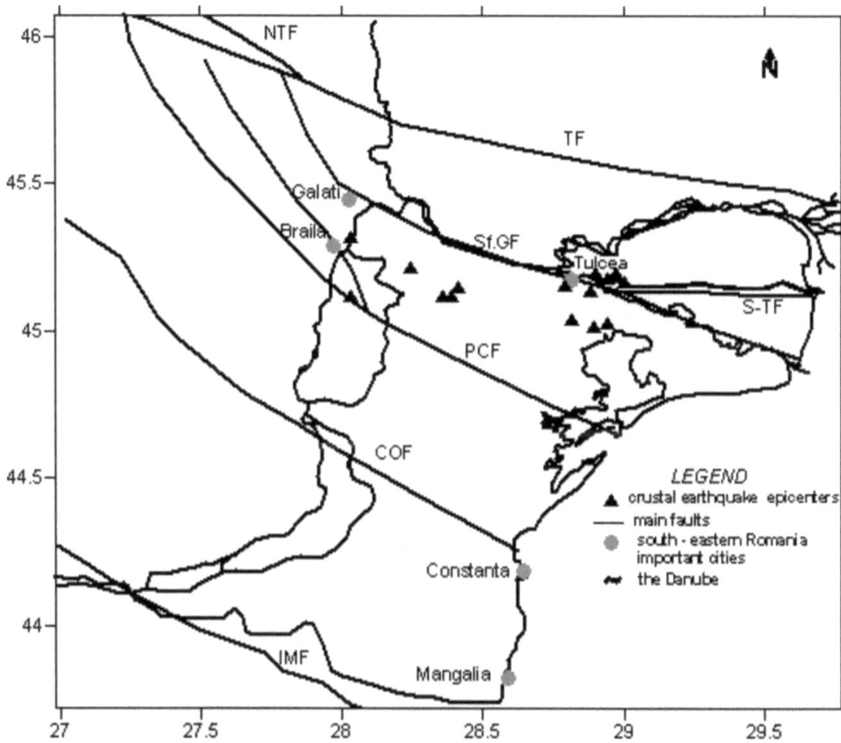

Fig. 5.7 The epicenters of the earthquakes for which focal mechanisms could be determined

drill hole near the seismological station Carcaliu (south of Măcin) indicated in the Paleozoic rocks complex (Silurian) $V_P$velocities of 3500 m/s (Răileanu et al. 1998).

5.5 Local Seismic Effects in Tulcea City Area

Two recorded crustal events, typical for this region, were used to study local seismic effects in the Tulcea city area, in order to obtain the local amplification between bedrock and surface of the seismic signal.

The characteristics of the chosen two events are: the first event produced on 27.04.1986 (**ev1**), magnitude $M_W = 5$; long[deg] 26.95, lat[deg] 45.48; source depth h = 40 km; strike receiver 226°, fault dip 44° and rake 78°, and the second event occurred on 13.11.81 (gh1), $M_W = 5.1$; long[deg] 29; lat[deg] 45.17; source depth h = 15 km; strike receiver 117°, fault dip 57° and rake 65°. These data were used to seismic signal generation and modeling, to propose seismic scenarios, to obtain a local amplification function in a chosen point, in order to perform our aim, a preliminary approach to study local seismic effects in the area. A single point was used because we had a complete geotechnical profile only for this area.

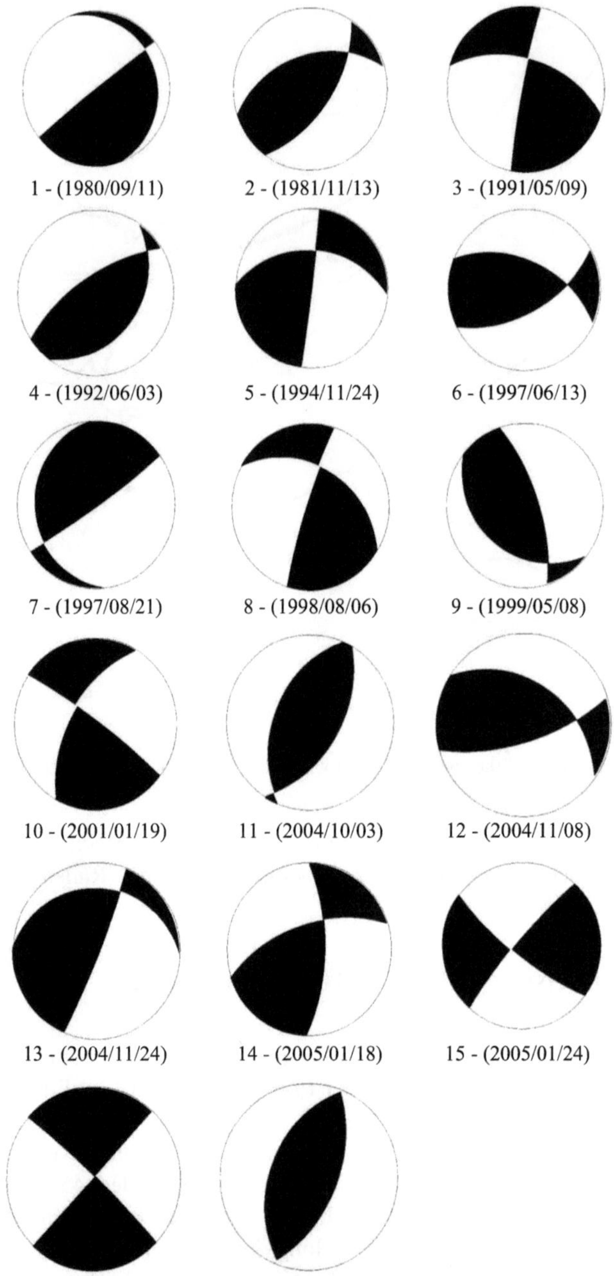

Fig. 5.8 The focal mechanisms of several characteristic earthquakes of Tulcea area

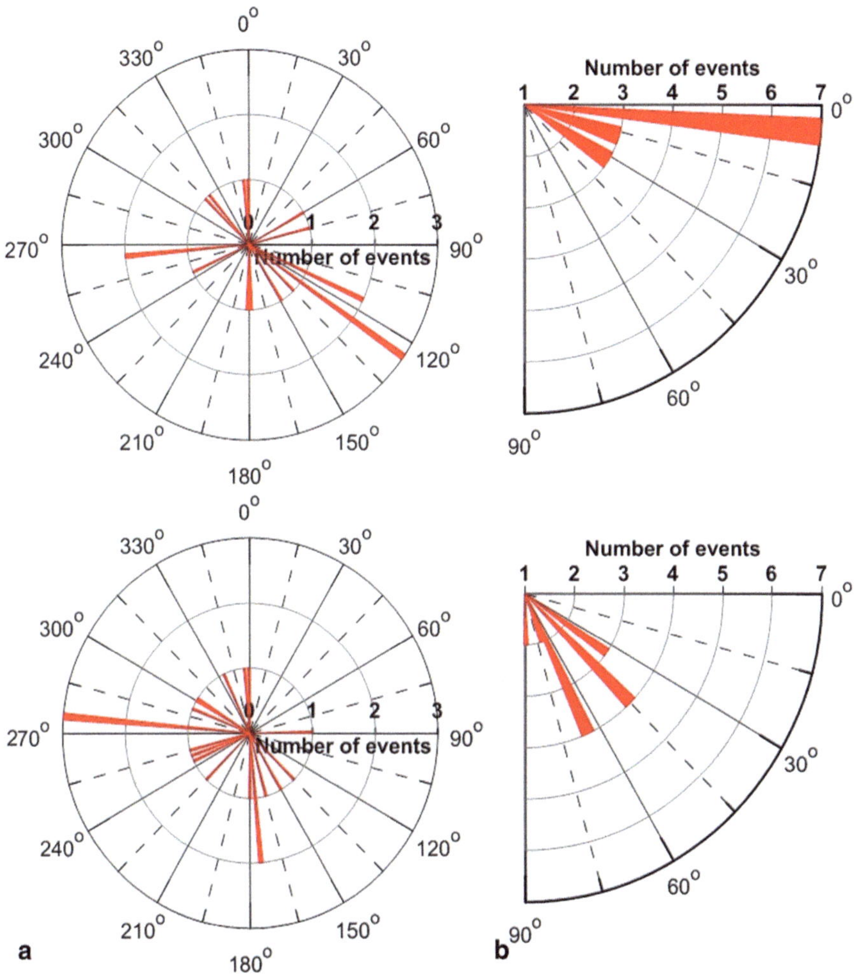

Fig. 5.9 Azimuth **a** and plunge **b** of the *P* and *T* axes

The Tulcea profile shown in Table 5.2 has been used in the computational procedure:

The seismic signal generation on the source-bedrock path was performed using seismological characteristics of the sources in the interest area. The simulation of the synthetic signals used the above described characteristics of the real events. This way the synthetic signals are transported to the bedrock. Then the simulated seismic signal was transferred to the surface using SHAKE 2000 software, the modern version of SHAKE91, which is able to compute the seismic response anywhere in the considered area.

The important features taken into account were the nonlinear behavior of the upper soil strata, the effect of the bedrock elasticity, and different shear modulus

Table 5.2 Tulcea profile

Thickness [m]	Material	V_P[m/s]	V_S[m/s]
0–5	Loess	300	173
5–15	Loess	400	231
15–30	Loess	1000	578
	Bedrock	4000	2312

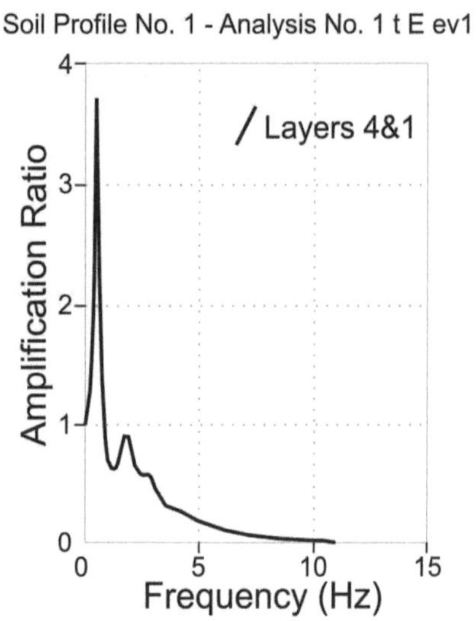

Fig. 5.10 Seismic amplification function

and damping of the linear-equivalent-type system. Earthquakes induce nonlinear deformations into the soil strata, which result in important variations of shear modulus and damping ratios of each strata function of deformations. In this analysis there were used data obtained from tests in the resonant column Drnevich, with samples from the location.

In the Figs. 5.10–5.13 there are presented the seismic amplification functions between bedrock and surface in site E (see Fig. 5.1). In Figs. 5.10 and 5.11 there are presented the seismic amplification functions due to the transversal and radial component of the seismic input from the source ev1, and in Figs. 5.12 and 5.13 there are presented the seismic amplification functions due to the transversal and radial component of the seismic input from the source gh1.

5.6 Conclusions

Three main fault systems were mapped within the Tulcea city area, either by geological, geophysical or seismological observations: NW–SE, N–S and NE–SW.

The NW–SE system seems to be the oldest, while the NE–SW system is the youngest one, being expressed in geological structures or geomorphology.

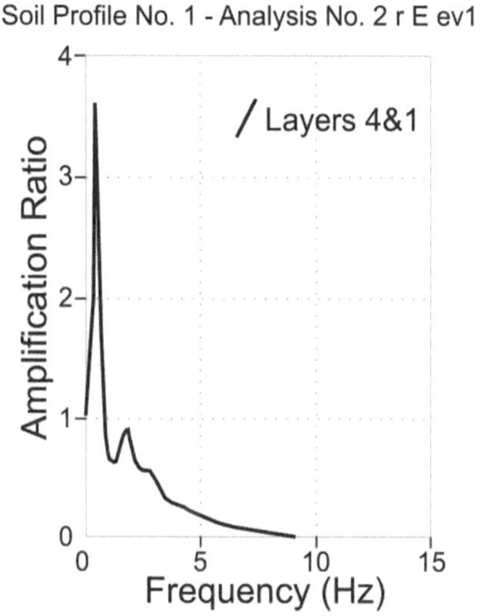

Fig. 5.11 Seismic amplification function

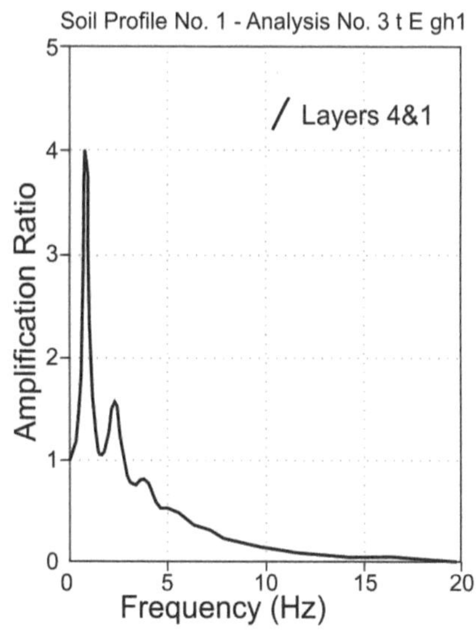

Fig. 5.12 Seismic amplification function

Tulcea city is located at a junction of these fault systems, a site of high seismic hazard since all of them may be considered as active fault systems at both regional and local scales.

The North Dobrogea earthquakes emphasize local monokinetic seismic sources, (Cornea and Lazarescu 1980) of low energy ($M \leq 5$), initiated within the crust, from which the isoseists are primarily elongated NW and NE in some places. This suggests their direct contact with fractures, especially Hercynian, not yet welded, and less with the Alpine ones, but the position of the epicenters corresponds usually to the intersection of these two generations of faults.

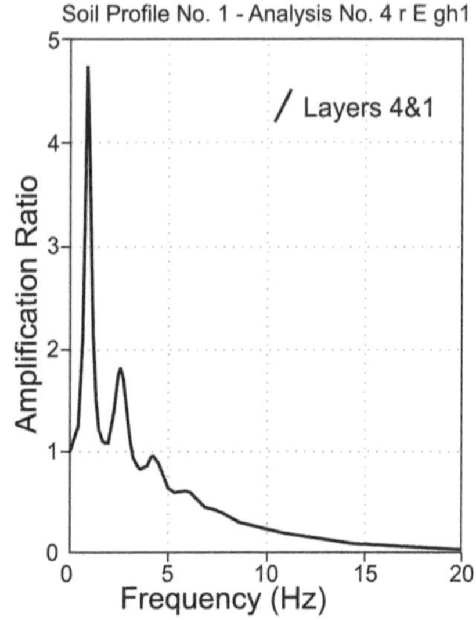

Fig. 5.13 Seismic amplification function

The Tulcea area is made of a mosaic of tectonic blocks. Some of them are isometric, while others are elongated, of large dimensions, which resulted by joining a system of NW-SE and WNW-ESE oriented fractures with another transversal one, oriented NE-SW. These faults are active, alongside of which a series of epicenters are ranged.

The mechanisms within the focal point of the earthquakes have shown the fact that the stress accumulated in this area is of the compressive type, the mechanism solutions indicating two types of faults: reverse faults and strike-slip faults.

In this paper we have approached the assessing of the perspective seismic risk of potential earthquakes, starting from the geologic and geophysical information upon the terrestrial crust.

The main appreciation is that the seismic risk grows where the most criteria sum up, indicating the dangerous zones.

From Figs. 5.10–5.13 it could be observed that the amplification ratio between the bedrock and surface for both sources, ev1 and gh1, attains expected values. Thus, the loess-type soil present in the superficial crust of the area induces typical behavior for wave propagation through this seismic region, as it was also shown in the paper (Georgescu et al. 1982).

These observations show us the importance of taking into account the crustal seismic activity, as these earthquakes can produce damage to certain buildings in the area. For populated areas like Tulcea, knowing local conditions is very important, for a better understanding of the regional seismic hazard and mitigation of seismic risk.

Acknowledgments This work has been done in the framework of the Projects: PN 09-03 04, PN 09-03 06, PN-II-RU-TE-2011-3-0072 and PN-II-PT-PCCA-2011-3.2-1448.

References

Airinei S, Stoenescu S, Velcescu G, Romanescu D, Visarion M, Radan S, Roth M, Besutiu L, Besutiu G (1983) The anomaly map of vertical component of the magnetic field (ΔZ_a). Institute of Geology and Geophysics, Bucharest

Atanasiu I (1961) Earthquakes in Romania (in Romanian). Editura Academiei R.S.R., Bucharest, p 194

Botezatu R, Bacioiu T, Andrei J (1961) The relationship between the gravity anomalies and the geological structure of North Dobrogea and the adjacent regions of South Moldavia and East Muntenia (in Romanian). Geological Studies and Research 1:33–56

Constantinescu L, Mârza VI (1980) A computer-compiled and computer-oriented catalogue of Romania's earthquakes during millennium (984–1979). Rev Roum Géol Géophys Géogr Sér Géophys 24(2)193–234

Cornea I, Lazarescu V (1980) Tectonica si evolutia geodinamica a teritoriului Romaniei. Sn, Bucharest

Damian A (1974) Viteze de interval în fundamentul unor unitati structurale si semnificatia lor geologică. BTS—IPGG IV(4):23–33

Georgescu S, Costea M, Sandru O, Ticoca P, Tabirna G (1982) Aspecte seismologice si inigneresti ale cutremurului de la Tulcea din 13 noiembrie 1981. Studii si cerceteri INCERC 1–68, 5–6

Hauser F, Raileanu V, Fielitz W, Dinu C, Landes M, Bala A, Prodehl C (2007) Seismic crustal structure between the Transylvanian Basin and the Black Sea, Romania. Tectonophysics 430: 1–25

Mirauta O, Mutihac V, Bandrabur T, Dragulescu A (1967) Tulcea geological map scale 1: 200,000. Geological Institute, Bucharest

Mirauta E, Panin N (1979) Tulcea geological map scale 1: 50,000. Institute of Geology and Geophysics, Bucharest

Mutihac V, Ionesi L (1974) Geology of Romania (in Romanian). Editura Tehnica, Bucharest

Oncescu M, Bazacliu O, Popescu E (1989) The Tulcea earthquake of November 13, 1981. Rev Roum Géol Géophys Géogr Sér Géophys 33:23–26

Radulescu F, Raducanu M, Raileanu V, Dragu S, Pompilian A, Sova V (1982) Parametrii elastici ai unor masive calcaroase din Dobrogea. St cerc fiz 34(2)141–154

Raileanu V, Diaconescu C, Mateciuc D, Diaconescu M (1998) Velocity crustal models under the Romanian telemetered seismological observatories. Rom Rep Physics 50(1–2):123–141

Schnabel PB, Lysmer J, Seed HB, SHAKE91 (1972) A computer program for conducting equivalent linear seismic response analyses of horizontally layered soil deposits. University of California, Davis

Seghedi A (1998) The Romanian carpathian foreland. In: Ioane D (ed) Monograph of Southern Carpathians. Reports on Geodesy, 7, vol 37. IGWiAG PW, Warsaw, pp 21–48

Seghedi A (2008) The geological and structural frame of the Black Sea neighboring terrains focusing on its north-western limit (in Romanian). In: Oaie G (ed) Natural hazard: tsunami events in the Black Sea. Bucharest, pp 11–26

Chapter 6
High-Detail Damage Pattern in Towns Hit by Earthquakes of the Past: An Approach to Evaluate the Reliability of the Historical Sources

Fabrizio Terenzio Gizzi, Lucia Tilio, Nicola Masini, Beniamino Murgante, Maria Rosaria Potenza and Cinzia Zotta

Abstract Historical data relating to damage in urban areas are useful to evaluate and mitigate seismic risk. However, documentary data need to be considered in relation to their quality. The paper describes a mathematical approach to assess reliability of historical sources referring to damage effects caused by earthquakes in the past. The method applied is based on Rough Set Theory. Its application has been explained by analyzing a huge amount of archive data reported by documents concerning earthquake effects in eight towns of Basilicata (southern Italy), heavily damaged by the 1930 Irpinia earthquake. Mathematical analysis has been applied to evaluate whether technical reports citing the effects on buildings were affected by voluntary or involuntary "manipulation" by compilers. This has been possible by comparing damage descriptions reported in technical reports to damage inferred by the analysis of independent technical-economic-administrative data. Results seem to suggest that the sentences on damage were subjected only to limited and involuntary "falsification" that, however, caused negligible changes in the spatial distribution of damage.

Keywords Historical earthquakes · Seismic damage · Rough Set Theory · Computer science · Southern Italy

6.1 Introduction

Recent and historical earthquakes teach that severity of ground shaking is significantly influenced by site features. This is due to the effects of surface geology or topography on ground motion, which may cause different damage to neighboring

F. T. Gizzi (✉) · N. Masini · M. R. Potenza · C. Zotta
CNR-IBAM (Institute of Archaeological and Monumental Heritage),
C.da S.Loja, 85050 Tito Scalo (PZ), Italy
e-mail: f.gizzi@ibam.cnr.it

L. Tilio · B. Murgante
School of Engineering, University of Basilicata,
Via dell'Ateneo Lucano 10, 85100 Potenza, Italy

M. Bostenaru Dan et al. (eds.), *Earthquake Hazard Impact and Urban Planning*,
Environmental Hazards, DOI 10.1007/978-94-007-7981-5_6,
© Springer Science+Business Media Dordrecht 2014

buildings having comparable features (Chavez-Garcia et al. 1997; Mucciarelli and Monachesi 1988) (Sanchez-Sesma 1985; Tertulliani 2000).

Knowing which urban areas are subjected to potential increase in effects for future events is a crucial task to mitigate seismic risk. Usually, identifying such “sensitive” areas, in situ geological and geophysical investigations are performed. However, an important help can be provided from an in-depth analysis of the effects of earthquakes that have already occurred in a certain site. Indeed, historical sources often refer to different seismic effects, especially sources documenting quakes occurred in Italy, the “cradle” of historical seismology.

Usefulness of historical documents concerning consequences of past earthquakes is due to the circumstance that they refer about areas where high levels of damage were already recorded in previous earthquakes, showing the result of a test to which the building-soil system was exposed. From this point of view, the analysis of damage experienced in past events represents a lesson from which obtaining useful data to identify areas to undergo detailed geological-geophysical investigations (Gizzi 2006).

Analysing the literature it emerges that several historical cities, especially Italian ones, underwent analyses to delineate damage “scenarios”, such as Rome (Tertulliani and Riguzzi 1995), Syracuse (Boschi et al. 1995), Florence (Guidoboni and Ferrari 1995), Catania (Boschi and Guidoboni 2001), and Palermo (Guidoboni et al. 2003). By some of these studies it has been possible to better constraint *in situ* investigations (e.g.: Guidoboni et al. 2003).

Ascertaining that the analysis of the consequences of past events can contribute to put into the field proper strategies to mitigate seismic risk, a crucial phase to consider when analysing historical documents, we are concerned that the identification of the quality of written sources, i.e. to evaluate if the descriptions reported in original documents are affected by voluntary or involuntary “errors”, made by those in charge to survey the damage. By this approach, it will be possible to avoid a misrepresentation analysis of the effects in certain urban areas. This is particularly important when the analysis of effects is performed with high spatial resolution, i.e. building-by-building or housing-by-housing.

The traditional and irreplaceable approach to the evaluation of reliability of written sources is strictly connected to the critical criterion proper of historical seismology, which allows one to assess the quality of the descriptions considering the social-economic-political context in which the natural event occurred (e.g. Albini et al. 2004; Guidoboni 2000 and bibliography therein).

Starting from these preliminary remarks, the work illustrates a possible complementary criterion to assess reliability of historical sources.

The approach proposed is that of Rough Set Theory. This theory, developed in early 1980’s by Zdzisław Pawlak (Pawlak 1982, 1997, 1998a, 1999), allows to tackle imprecise and uncertain knowledge, providing efficient algorithms for finding hidden patterns in data, and approximately classifying knowledge using available information (Komorowski et al. 1999a; Pawlak 2002; Soetanto and Van Geenhuizen 2007). It has found application in other fields of geological risks, as in the analysis of landslide susceptibility (Gorsevski and Jankowski 2008).

The approach to macroseismic data has been tested on a sample area of Basilicata Region (southern Italy), where a strong earthquake caused damage in several towns on 23 July 1930.

For this event a rich technical documentation (thousands of dossiers) is available, useful to delineate damage patterns in each town of the test area (Gizzi and Masini 2004, 2006). All data, having parametric and textual features, have been implemented in a GIS: the database contains information for eight of the most damaged towns.

After a preliminary test, concerning macroseismic data analysed only for the town of Rapolla (Gizzi et al. 2010a), and after a first extension concerning five of the nine earthquake affected towns (Gizzi et al. 2010b), a further investigation is proposed in this paper, concerning almost all (8 of 9) most affected towns of Basilicata, in order to refine the reliability of the information.

The paper is organized in four sections. The first one summarizes the main aspects of Rough Set Theory, useful to understand the followed approach; the second part discusses the study cases, with the description of the database; the third section presents an overview of the analysis; the fourth one discusses the results.

6.2 Methodology: An Essential Framework

Rough Set Theory is based on the observation that every object in the universe is always associated to some information (Krawiec et al. 1998), and objects characterized by the same information cannot be distinguished, they are indiscernible. Rough Set Theory allows us to tackle imprecise, vague and uncertain knowledge (Pawlak 1997, 1998a), classifying objects in an indiscernible way, defining indiscernibility relations. In other words, Rough Set Theory is a methodology for data analysis, with the main objective to find hidden patterns in data, searching for dependencies, or, in other words, pursuing “cause-effect” relations (Pawlak 1998b).

To better understand the main concepts of Rough Set Theory, it is necessary to consider some examples in the field of this research.

If we consider two buildings, let us call them x_i and x_j, for which, after an earthquake, same damage characteristics are recorded, then x_i and x_j are indiscernible; they belong to the indiscernibility relation. Formally, this is described as:

$$b(x_i) = b(x_j), \rightarrow [x_i]_{Ind(B)}, \left[x_j\right]_{Ind_{(B)}}.$$

x_i and x_j, constitute an atom, also called an elementary set, that represents the smallest block of information knowledge about data.

In a formal way, the two buildings are two elements of a set U, called Universe, and that is finite and not empty, $U = \{x_1, \ldots x_m\}$,while information about damage are elements of a set A, that is finite and not empty, and characterized by a value function: $\forall a \in A$ there is a set of values $\mathrm{V_a}$, called the domain of A, $\exists f_a : U \rightarrow V_a$. The indiscernibility relation has been built considering information about damage as a subset of A: $B \subset A$. U and A are an Informative System: $IS = \{U, A\}$. When a decision attribute d influences classification, then it can be defined a Decision System $DS = \{U, A \cup d\}$.

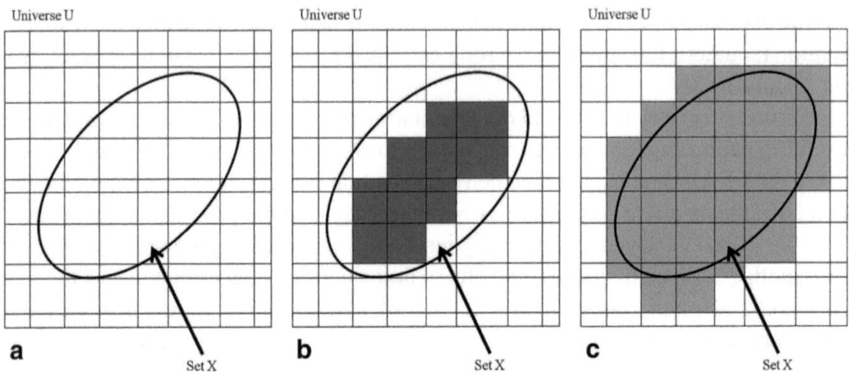

Fig. 6.1 *Lower* and *upper* approximation

A rough set approach allows to face the typical problem of defining a membership of equivalence classes to be a defined set $X \subset U$: indeed, considering available information, it may result that some equivalence classes surely belong to X, some others do not surely belong, but some others potentially belong to X. Figure 6.1 shows this situation:

In Fig. 6.1a the original condition is shown, with the set X in the Universe U; in Fig. 6.1b the simplest situation is represented: equivalence classes are included into X, so that they surely belong to X; the union of all equivalence classes for which this condition is verified is called lower approximation, and is indicated as $\underline{BX}$. Formally, lower approximation is defined as follows:

$$\underline{BX} = \left\{x_i \in U : [x_i]_{Ind(B)} \subset X\right\}.$$

In Fig. 6.1c some equivalence classes are not completely included into X, nor are they completely outside. Their intersection with X is not empty, and the union of all equivalence classes for which this condition is verified is called upper approximation, and is indicated as BX. Formally, upper approximation is defined as follows:

$$BX = \left\{x_i \in U : [x_i]_{Ind(B)} \cap X \neq 0\right\}.$$

The difference between upper and lower approximation is the Boundary Region of X. When a Boundary Region exists—i.e. when BX−$\underline{BX} \neq 0$—then X is a Rough Set, and it can be further classified depending on lower and upper approximation characteristics (Walczak and Massart 1999).

Cardinality of Lower and Upper approximation gives information about accuracy and quality of classification: in particular, accuracy is defined as

$$\mu_b(X) = \frac{card(\underline{BX})}{card(\overline{BX})}$$

and can vary in the range [0,1]. When accuracy is < 1, then X is indefinable. Quality of classification, instead, is defined as follows:

$$\gamma_b(X) = \frac{\sum_{i=1}^{m} card(\underline{BX_i})}{card(U)}.$$

It gives a measure of the number of correctly classified objects referred to all objects in the system (ProSoft 1999).

Definition of the indiscernibility relation, moreover, is useful to identify dependences between attributes; this means that if Ind(A) = Ind(A-a_i), then a_i is superfluous, it depends on other attributes and its contribution is not relevant for knowledge in Information System; then, it can be eliminated without information loss. If Ind(A)$\neq$Ind(A$-a_i$), instead, then a_i is necessary and it cannot be eliminated or omitted. As a consequence, it is possible to identify the minimal subset of condition attributes discerning all objects discernible by the original information table that is called *reduct* (Soetanto and Van Geenhuizen 2007). The common part of all reducts, i.e. intersection, is the core.

Finally, Rough Set Analysis allows to identify patterns and to extract relations, identifying cause-effect relation. Identified patterns are represented through a decisional rule set, induced from a decision table, where rules are expressed in the "*if...then*" form (Komorowski et al. 1999b), and are used to explain phenomena and processes related to data and which can be interpreted as descriptions of cause-effect relations (Greco et al. 2001). Objects are assigned to a decision class if they satisfy the conditions of an identified rule; rule strength is determined by number of objects satisfying that condition; at the same time, this number of objects also gives a measure of uncertainty into decision class assignment (Gorsevski and Jankowski 2008).

Rules can be *exact* or *approximate*, depending on whether consequences are univocal or not. This means that a rule is exact when in the assignment (that is, in the "then" part of the rule) the decision attribute has only one possible value, and the whole rule is only sustained by elements of lower approximation; the assignment concerns a decision attribute, that is not assignment decision concerning an attribute. In this research, currently, only exact rules are taken into account, as better explained in the following paragraph.

6.3 Study Case

A Rough Set Approach (RSA) has been used to assess the reliability of historical macroseismic data, testing the huge database concerning the Irpinia Earthquake, which on 23 July 1930 damaged several towns in Basilicata (southern Italy) (Fig. 6.2).

Information collected by the Office of Civil Engineers allowed them to build a rich database, useful to delineate housing-by-housing damage pattern in each town,

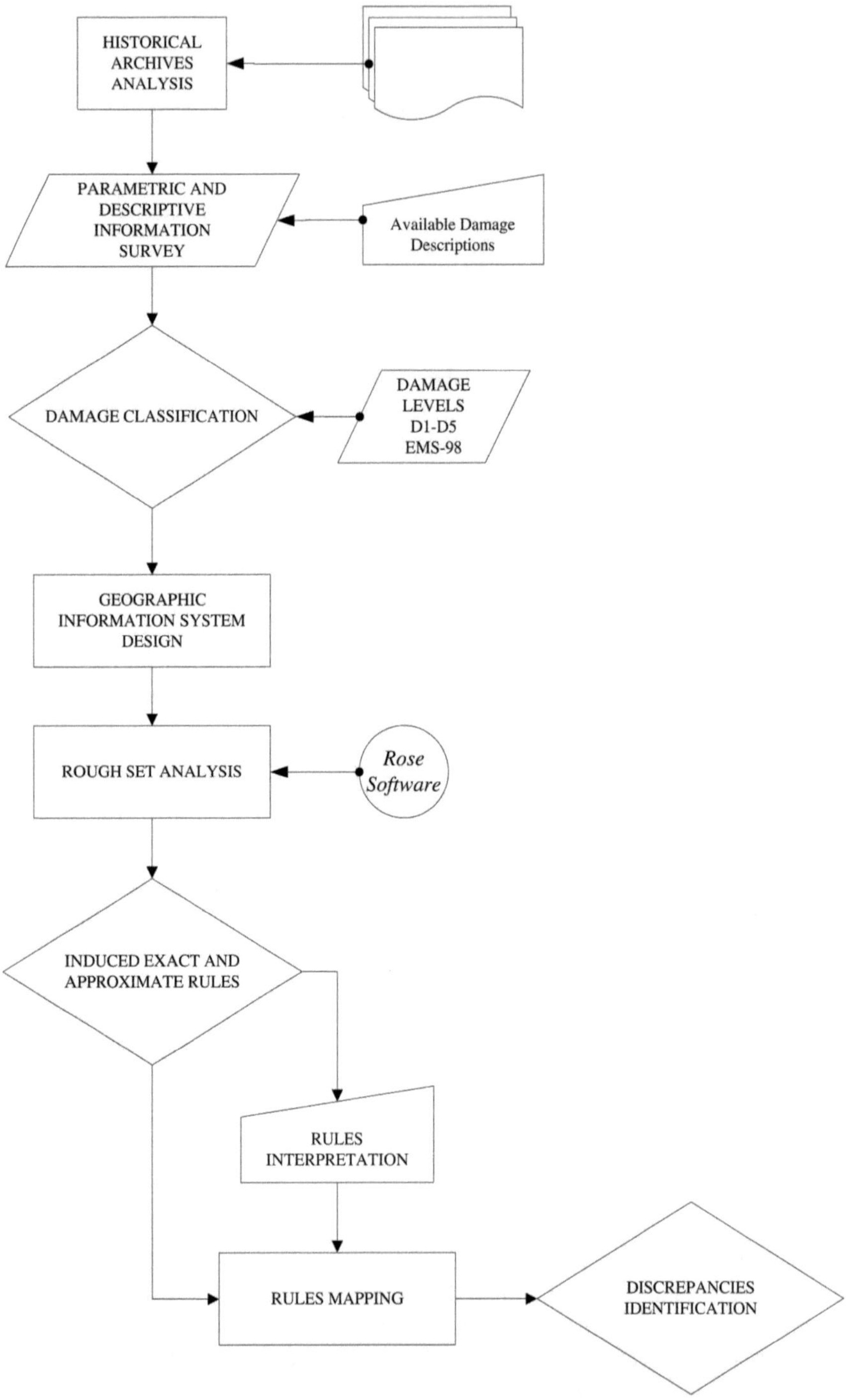

Fig. 6.2 Flow chart of the methodology adopted for analysing macroseismic information

Fig. 6.3 The geographical context of research: centres of Basilicata (southern Italy) damaged by 1930 Irpinia Earthquake

analysed through RSA in order to identify the degree of reliability of written sources evaluating, for instance, whether technical reports were affected by voluntary or involuntary "errors" by the compilers.

As shown in Fig. 6.3, historical data concern Vulture (north-west section of Basilicata) area and in particular eight of nine centres seriously damaged (with a macroseismic intensity ranging from VIII to IX degree in Mercalli-Cancani-Sieberg scale) by the natural event.

As mentioned before, data concerning the earthquake have been recorded in a geographic database. Information include the following aspects: post-seismic repairing procedures; technical-economic-administrative aspects such as repairing costs, date of start and end of repairing works. Moreover, a field of the database includes damage level deriving from the parameterisation of the descriptive information in EMS-98 scale (for details about the features of the information see Gizzi and Masini 2004, 2006). As regards the parameterisation, it must be emphasized that technical features of damage sentences are allowed to remove or reduce problems deriving from subjective choices and judgments based on personal experiences during the conversion of the qualitative descriptions into quantitative synthetic parameters (Ferrari et al. 1995, Monachesi and Moroni 1995) proper of the EMS-98 scale (Grünthal 1998). This is confirmed, as will be discussed below, from the results of the analysis by means of the Rough Set Approach.

Criteria followed to attribute the damage to the units according to the EMS-98 scale are shown in Table 6.1 (for details see also Gizzi and Masini 2006).

The availability of different kinds of data allows performance of cross-correlated analysis (for example between damage and costs, duration and typology of repairing works), to evaluate information reliability through the Rough Set Approach. The aim was to highlight possible "mistakes" in the technical reports, implying an underestimation or overestimation of the damage level during the procedure of parameterization. Table 6.2 indicates and describes attributes considered during data analysis.

Analyses have been carried out with ROSE software (Rough Set Data Explorer)[1], and results, in terms of rules, have been mapped on a GIS.

The authors assumed different rules among towns because the typology and completeness of available descriptions changed from town to town. This is due to the circumstance, for example, that the groups of technicians entrusted to survey buildings were different for each site.

With this in mind, it has been necessary to consider rules fitting the different features of written sources.

Generally, data tables have been prepared ready to be run with ROSE, but in some cases, a completion has been necessary, automatically carried out by the software itself, that assigns a value to missing data, using the most frequent value for the given attribute (ProSoft 1999). In order to limit this automatic completion, and to avoid some possible effects on quality of classification, not all properties have been included in the analysis. In particular, in the cases of uncertainty in database completion, units have not been considered.

Table 6.3 shows a summary of the results. Information in each column of the table refers to concepts of Rough Set that have been described in the second paragraph.

Remembering that the quality of classification refers to the number of objects (properties in this case) correctly classified in relation to all objects, Table 6.3 shows that Rionero, Barile, and Ruvo del Monte are the worst ones in performance: atoms are few, quality of classification is really low. Atella and Ripacandida show a higher number of data in the core, that is the minimum set of attributes necessary to discern all information, due to the fact that they are characterised by a higher number of attributes (and in particular those relating to type of interventions).

However, for each town, rules have been induced. Table 6.4 shows a synthesis of the number of rules for each town.

Rules are useful to identify patterns and to extract relations, identifying cause-effect relations; patterns, as explained in the second paragraph, are represented through a decisional rule set, where rules are expressed in the "*if . . . then*" form, and can be exact or approximate. Housings are assigned to a decision class if they satisfy the condition of an identified rule; each rule can concern one or more units (there is not a "one to one" relationship, but "one to many"). For example, we can consider the Rule 1 obtained for Melfi.

[1] ROSE (ProSoft 1999) is a modular software system implementing basic elements of Rough Set Theory and rule discovery techniques, developed at the Laboratory of Intelligent Decision Support Systems of the Institute of Computing Science in Poznan.

Table 6.1 Criteria followed to attribute the damage to the units according to the EMS-98 scale

	General damage grade	Structural damage grade	Structural element												
			Wall						Vault					Floor	Roof
			Structural element damage												
			slig	dam	hea	o.pl.	coll	slig	dam	hea	coll	dam	coll	dam	coll
D1	Slight	Negligible *(min/max)*	x												
			x					x							
D2	Moderate	Slight (min/max)	x	x											
			x	x				x	x	x		x		x	
D3	Substantial to heavy	Moderate (min/max)	x	x	x										
			x	x	x			x	x	x		x		x	
D4	Very heavy	Heavy: out of the plane (min/max)	x	x	x	x									
			x	x	x	x		x	x	x		x		x	
		Heavy: partial collapse (min)					x			x			x		x
D5	Destruction	Very heavy: total collapse					x			x			x		x

Table 6.2 Database description

Field	Description	Range
Destroyed	The attribute reveals if the property was destroyed after the earthquake	No–Yes (0–1)
	The attribute indicates the presence or not of scripts and documentation about building, revealing a link to damage level: generally scripts are strongly related to a consistent damage level	No–Yes (0–1)
Revoked aid	The attribute indicates the revocation of financial aid to rebuilding; it reveals a link to damage level	No–Yes (0–1)
Estimated costs	This attribute classifies the estimated cost of rebuilding works, referred to the currency at the earthquake time	0 (EC < 0£); 1 (0£ < EC 2000£); 2 (EC > 2000£)
	Survey Date is classified considering three period: surveys lead before 1930 earthquake, surveys lead in the next three months, and surveys lead later	0 (SD < 23.07.30); 1 (23.07.30 SD 23.10.30); 2 (SD > 23.10.30)
Financed Survey datecosts	This attributes indicates the finance cost of works, significant if compared to estimated costs	–
Financed-estimated costs	This attribute indicates if the difference between financed costs and estimated costs is positive or negative	0 (F-E < 0); 1 (F-E > 0)
Starting works date	This attribute indicates the starting work date, classified considering three period: works lead before 1930 earthquake, works lead in the next four months, and works lead later	0 (SD<23.07.30); 1 (23.07.30 SD 23.11.30); 2 (SD>23.11.30)
Works length	This attribute indicates the works duration, difference between ending and starting working date, expressed in number of days	0 (WL=0); 1 (WL 60 days); 2 (WL>60 days)
Owner income	Owner Annual Income, referred to the currency at the earthquake time	"0 (EC<0£); 1 (0£< EC 5000£); 2 (EC>5000£)."
Interventions	These fields describe the intervention type, indicating, for each one, the presence or not of intervention and its main characteristics.	
DD	This attribute indicates that after the earthquake the property was declared not reconstructable	No–Yes (0–1)
EMS_DAMAGE	Damage Level according to the EMS-98 scale	D1, D2, D3, D4, D5

$$
\begin{aligned}
&\textit{if}\ \{SCRIPTS = YES\}\ \textit{and}\ \{REVOKEDAID = YES\}\\
&\textit{and}\ \{DD = NO\}\ \textit{and}\ \{ESTIMATEDCOSTS = 1\}\\
&\textit{then}\ \{EMS\_DAMAGE = D1\}
\end{aligned}
$$

Table 6.3 Synthetic results of analysis

	Quality of classification	# of Atoms	# of Reducts	# of attributes in Core
Atella	0,7219	39	2	8
Barile	0,15	71	1	10
Melfi	0,2765	332	1	10
Rapolla	0,5839	228	1	15
Rionero	0,1436	236	1	10
Ripacandida	0,4965	222	1	15
Ruvo del Monte	0,3866	40	1	10
San Fele	0,9205	125	3	9

Table 6.4 Synthesis of induced rules

	# of Rules	# of Exact rules	# of Approximated rules
Atella	22	15	7
Barile	43	24	19
Melfi	198	123	75
Rapolla	106	90	16
Rionero	145	75	70
Ripacandida	101	75	26
Ruvo del Monte	30	19	11
San Fele	51	46	5

In this example, there are four conditions before assignment: units that respect those conditions are assigned to decision class of EMS_DAMAGE = D1. This is the typical structure of an exact rule. Rough Set Analysis, however, supplies also approximate rules, characterised by a not unique statement in the assignment, as the following rule 106, obtained for Rapolla:

$$
\begin{aligned}
&if \left\{ \begin{array}{c} \{DESTROYED = NO\}\ and\ \{REVOKEDAID = YES\} \\ and\ \{INTERVENTIONS[INTCOPERTU] = ND\} \\ and\ \{INTERVENTIONS[SCUCICUCI] = ND\} \\ and\ \{DD = YES\}\ and\ \{OWNERINCOME = 0\} \end{array} \right\} \\
&then \left\{ \begin{array}{c} \{EMS\_DAMAGE = D1\}\ or\ \{EMS\_DAMAGE = D2\} \\ or\ \{EMS\_DAMAGE = D3\}\ or\ \{EMS\_DAMAGE = D4\} \end{array} \right\}.
\end{aligned}
$$

6.4 Results Interpretation

Induced rules have then been mapped on GIS and grouped according to the damage classification (in EMS-98 scale), before going to the following phase, consisting of an expert (human) assessment of the coherence of *exact* rules. At this stage of the research, approximate rules have not been considered due to the objective difficulties of performing analyses of coherence by human judgment; anyway, exact rules are more significant, because they are obtained considering only dwellings well classified (that are those belonging to lower approximation).

Table 6.5 Discrepancy in rules

	# of Exact Rules	Cases of overestimating (%)		Cases of underestimating (%)		Incidence of under-overestimating on total exact rules (%)	
Atella	15	2	13,33	0	0,00	2	13,33
Barile	24	0	0,00	0	0,00	0	0,00
Melfi	123	14	11,38	2	1,63	16	13,01
Rapolla	90	2	2,22	1	1,11	3	3,33
Rionero	75	8	10,67	0	0,00	8	10,67
Ripacandida	75	4	5,33	4	5,33	8	10,67
Ruvo del Monte	19	2	10,53	1	5,26	3	15,79
San Fele	46	1	2,17	3	6,52	4	8,70

Table 6.6 Discrepancy affecting D2 and D3

	# of Exact rules	Discrepancy in case of D2 (%)		Discrepancy in case of D3 (%)		Percentage of discrepancy for D2 or D3
Atella	15	0	0	2	100	100
Barile	24	0	0	0	0	0
Melfi	123	2	12,5	12	75	87,5
Rapolla	90	1	33,3	2	66,67	100
Rionero	75	1	12,5	6	75	87,5
Ripacandida	75	3	37,5	4	50	87,5
Ruvo del Monte	19	1	33,3	1	33,33	100
San Fele	46	0	0	2	100	0

As mentioned before, once the rules have been fixed, their consistence with the damage class attributed according to historical descriptions has been evaluated.

The identification of inconsistent rules has been possible thanks to the thorough knowledge of the features of institutional response to natural events (for details about this aspect see Gizzi and Masini 2004). With these constraints it has been possible to select some cases of discrepancy between the damage class attributed only by the descriptions of damage effects and the *corpus* of independent data.

As expected, the damage evaluated by descriptions has been *underestimated* or *overestimated* in some cases. Generally, this has been noticed principally for intermediate damage levels, such as D2 and D3 as Tables 6.5 and 6.6 show. Cases of overestimation are more frequent than those of underestimation.

On the whole, the analysis of rules founded by Rose software allowed to identification of only 33 overestimated events of damage and 11 underestimated ones.

In the following Table 6.7 a confusion matrix is presented; the matrix has been built for each town, and shows the number of properties affected by discrepancy for each damage level of database (in the first column), as shown also in maps presented in Figs. 6.4, 6.5, 6.6, 6.7, 6.8, 6.9 and 6.10.

Table 6.7 Confusion matrix showing the number of units affected by a discrepancy in damage level attribution for (a): Atella; (b): Barile; (c): Melfi; (d): Rapolla; (e): Rionero; (f): Ripacandida; (g): Ruvo del Monte; (h): San Fele; (•) indicates a case of overestimation: e.g. housings with damage level D3, but interpretation by RSA considers D2; (▲) indicates a case of underestimation: e.g. units had D2 but interpretation by RSA considers D3

(a)	D1	D2	D3	D4	D5
D1					
D2					
D3		7(•)			
D4					
D5					

(b)	D1	D2	D3	D4	D5
D1					
D2					
D3					
D4					
D5					

(c)	D1	D2	D3	D4	D5
D1					
D2			2(▲)		
D3		45(•)			
D4			4(•)		
D5					

(d)	D1	D2	D3	D4	D5
D1					
D2			1(▲)		
D3		16(•)			
D4					
D5					

(e)	D1	D2	D3	D4	D5
D1					
D2	1(•)				
D3		13(•)			
D4			1(•)		
D5					

(f)	D1	D2	D3	D4	D5
D1		1(▲)			
D2			3(▲)		
D3		13(•)			
D4					
D5					

(g)	D1	D2	D3	D4	D5
D1					
D2			2(▲)		
D3		3(•)			
D4			1(•)		
D5					

(h)	D1	D2	D3	D4	D5
D1					
D2			6(▲)		
D3					
D4			1(•)		
D5					

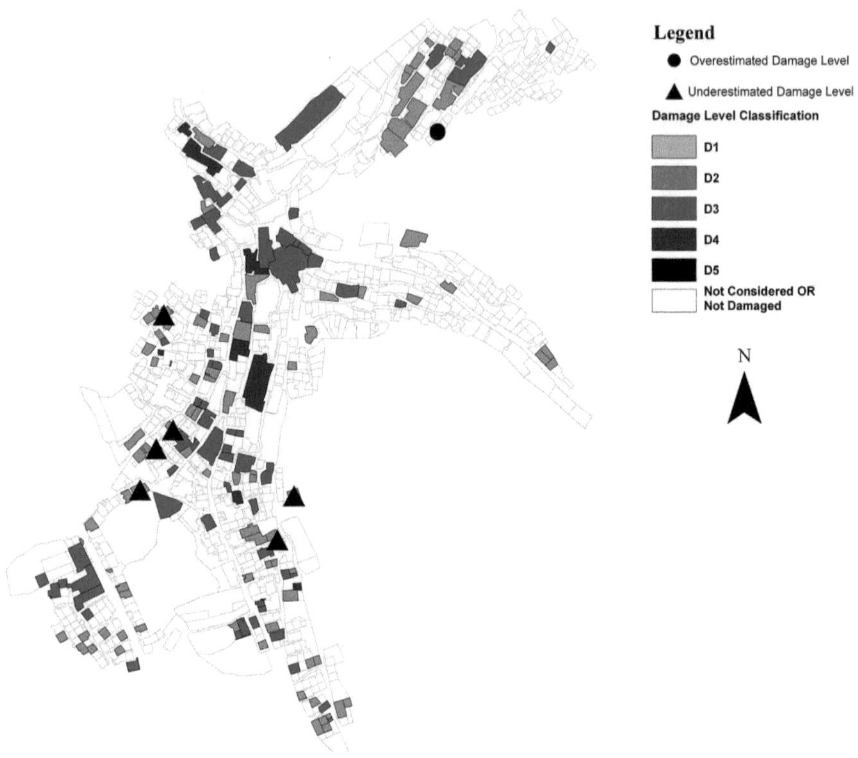

Fig. 6.4 The EMS-98 damage as assessed by analysing historical sources. Symbols (*circles* or *triangles*) showing units for which the analysis of rules suggests overestimation or underestimation of damage are overimposed on the map for the town of San Fele

The analysis of the rules led to exclude the possibility that the discrepancy was due to troubles encountered in extracting quantitative parameters (damage class in EMS-98 scale) from textual-qualitative information. This was because of the "incorrect" rules concerning only buildings for which the damage records did not lead to any ambiguity of interpretation.

The number of properties affected by overestimation or underestimation is limited; moreover, from a spatial point of view, differences in damage distribution are not clustered and, therefore, no significant changes in damage pattern can be observed (Figs. 6.4, 6.5, 6.6, 6.7, 6.8, 6.9 and 6.10). Another aspect that confirms the good quality of the archive data is the observation that differences in damage effects concern all sectors of the towns, characterised by different social and building features (both rich and poor owners, mansions and poor housings, well preserved and lacking in maintenance buildings, etc.). Therefore, on average, a voluntary "manipulation" of seismic effects observed by technicians during the phase of building survey (e.g. due to favouritisms requested by owners) does not seem evident. Rather, differences may be imputable to involuntary factors, among which are:

Fig. 6.5 Some of the rules with clear discrepancy in damage level attribution for the town of Atella

Fig. 6.6 Some of the rules with clear discrepancy in damage level attribution for the town of Melfi

Fig. 6.7 Some of the rules with clear discrepancy in damage level attribution for the town of Rapolla

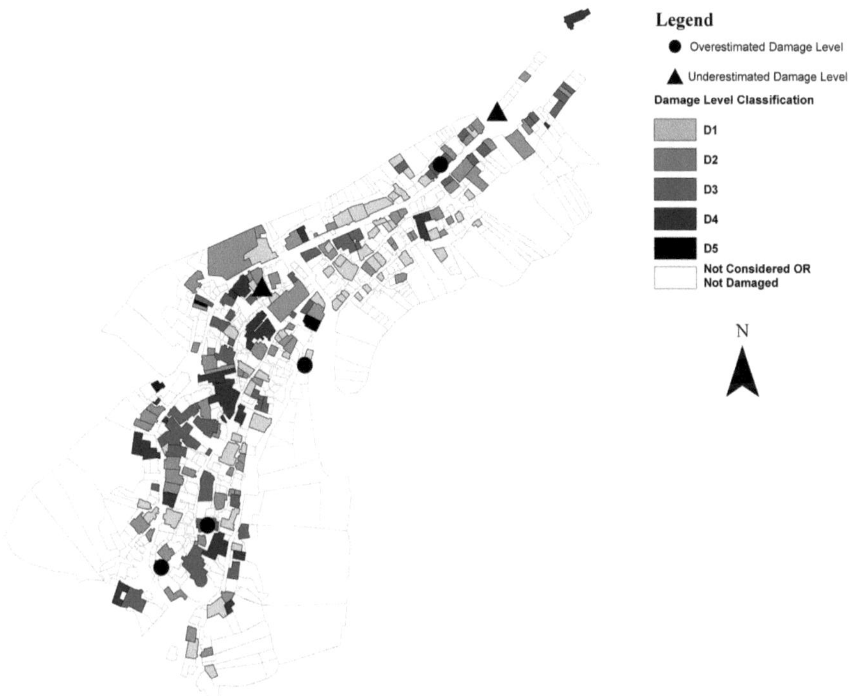

Fig. 6.8 Some of the rules with clear discrepancy in damage level attribution for the town of Ruvo del Monte

- Rough surveys of damage conditions (e.g.: damage attribution was limited only to a survey on the outside or to some rooms);
- Different vocational training of technicians entrusted to survey affected housings;
- Repairing works not strictly connected to the seismic event but included in damage description;
- Vague or inexistent rules about criteria to perform damage surveys in the earthquake aftermath;
- Occurrence of aftershocks. Descriptions in some cases do not report the increase of the effects caused by the aftershocks. Conversely, this issue is considered in the corpus of parametric-textual information on which rules have been evaluated.

High discrepancy affecting D2 and D3 may be due to the fact that intermediate levels of effects were harder to assess by technicians as respect to the lower and higher damages that show typologies of consequences on buildings easier to detect.

Summarizing, the use of cross-correlated data concerning the administrative-technical-economic aspects has confirmed the overall reliability of historical information used to delineate damage "scenarios" and, at the same time, has allowed one to select probable causes of discrepancy in damage attribution.

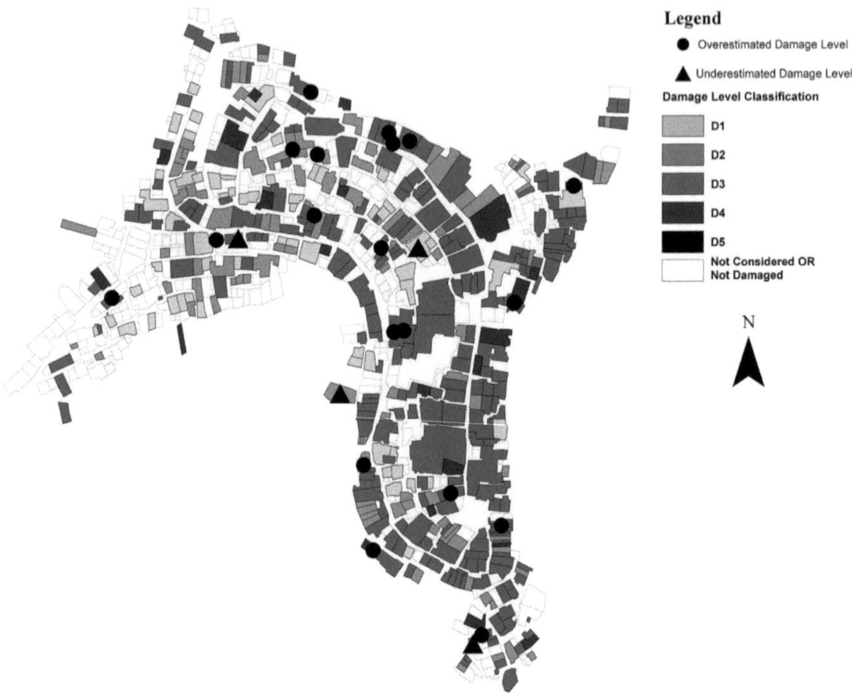

Fig. 6.9 Some of the rules with clear discrepancy in damage level attribution for the town of Ripacandida

6.5 Discussion

The analysis of effects caused by historical earthquakes can help in performing strategies of seismic risk mitigation because it is useful to identify areas prone to an increase of damage for particular geological-geomorphological features of the site.

However, a crucial aspect to consider when analysing historical documents is to identify their quality, i.e. to evaluate if descriptions reported in written documents are affected by voluntary or involuntary "falsification". The traditional approach to interpret and to evaluate reliability of information from the past involves the methodology proper of historical seismology. However, computational techniques can be especially of help in evaluating quality of information to analyse huge amounts of data.

The paper discusses the use of the Rough Set Approach to evaluate information quality concerning the consequences caused by an earthquake in the past. The event analysed is that of 23 July 1930, that strongly hit a wide area of southern Italy, including eight towns in Basilicata. For these towns a rich technical documentation was consulted and analysed.

Results allowed to show some insights about the assessment of information. On the whole, data do not appear to be affected by voluntary "falsification". Moreover, the limited involuntary "falsification" of damage descriptions is not clustered so that changes in spatial distribution of damage are negligible.

Fig. 6.10 Some of the rules with clear discrepancy in damage level attribution for the town of Rionero in Vulture

The encouraging application of this computer-aided analysis shows a glimpse of possibly more robust interpretations to run using other intelligent methods, such as Visual Analytics, multiform Bivariate Matrix, Self-Organising Map (SOM), and

Parallel Coordinates Plot (PCP), in order to compare RSA results. However, it must be emphasised that, as for the traditional approach to evaluation of quality of historical documents, also analysis by computational procedures involves an "expert judgement" about the proper use of information.

References

Albini P, García AV, Musson RMW, Stucchi M (eds) (2004) Investigating the records of past earthquakes. 21st course of the int. school of geophysics, Erice (Italy), 1–7 July 2002. Ann Geoph 47(2–3):335–911

Boschi E, Guidoboni E (2001) Catania, terremoti e lave dal mondo antico alla fine del Novecento. Editrice Compositori, Bologna

Boschi E, Guidoboni E, Mariotti D (1995) Seismic effect of the strongest historical earthquakes in the Syracuse area. Annali di Geofisica XXXVIII:223–253

Chavez-Garcia FJ, Field RM, Edward HF, Hatzfeld D (1997) Topographic site effects; a comparison of two non reference methods. Bull Seism Soc Am 87(6):1667–1673

Ferrari G, Gasperini P, Guidoboni E (1995) Macroseismic intensity evaluation with the "Fuzzy Set Logic". Annali di Geofisica XXXVIII(5–6):811–826

Gizzi FT (2006) To what degree can historical seismicity records assist in seismic microzonation? Eng Geol (87):1–12

Gizzi FT, Masini N (2004) Damage scenario of the earthquake on 23 July 1930 in Melfi: the contribution of the technical documentation. Ann Geophy 47(5):1641–1665

Gizzi FT, Masini N (2006) Historical damage pattern and differential seismic effects in a town with ground cavities: a case study from Southern Italy. Eng Geol 88:41–58

Gizzi FT, Masini N, Potenza MR, Zotta C, Tilio L, Danese M, Murgante B (2010a) Assessing macroseismic data reliability through rough set theory: the case of Rapolla (Basilicata, Southern Italy), Lecture Notes in Computer Science vol 6016, pp 320–330. Springer, Berlin. doi:10.1007/978-3-642-12156-2_25

Gizzi FT, Masini N, Potenza MR, Zotta C, Tilio L, Danese M, Murgante B (2010b) Assessing macroseismic data reliability through Rough Set Theory: application on Vulture area (Basilicata, Southern Italy). In: Phillips-Wren G, Jain LC, Nakamatsu K, Howlett RJ (eds) Advances in Intelligent Decision Technologies, Smart Innovation, Systems and Technologies Vol 4, VII, pp 279–288, Springer, Berlin. ISSN:2190-3018, doi: 10.1007/978-3-642-14616-9_27

Gorsevski PV, Jankowski P (2008) Discerning landslide susceptibility using rough sets. Environ Urban Syst 32:53–65

Greco S, Matarazzo B, Slowinski R (2001) Rough set theory for multicriteria decision analysis, Eur J Oper Res 129:1–47

Grünthal G (ed) (1998) European Macroseismic Scale 1998. Cahiers du Centre Européen de Géodynamique et de Séismologie, Luxembourg

Guidoboni E (2000) Method of investigation, typology and taxonomy of the basic data: navigating between seismic effects and historical contexts, In: Boschi E, Guidoboni E, Ferrari G, Mariotti D, Valensise G and Gasperini P (2000) Catalogue of Strong Italian Earthquakes from 461 B.C. to 1997, Introductory texts and CD-ROM, Version 3 of the Catalogo dei Forti Terremoti in Italia. Annali di Geofisica 43 (4)

Guidoboni E, Ferrari G (1995) Historical cities and earthquakes: Florence during the last nine centuries and evaluations of seismic hazard, In: Boschi E et al (eds) Earthquakes in the past: multidisciplinary approaches. Annali di Geofisica XXXVIII(5–6):617–647

Guidoboni E, Mariotti D, Giammarino MS, Rovelli A (2003) Identification of amplified damage zones in Palermo, Sicily (Italy), during the earthquakes of the last three centuries. Bull Seism Soc Am 93(4):1649–1669

Komorowski J, Pawlak Z, Polkowski L, Skowron A (1999a) Rough sets: a tutorial. In: Pal, Skowron (ed) Rough fuzzy hybridization—a new trend in decision making, S.K., Springer, pp 3–98
Komorowski J, Pawlal Z, Polkowski L, Skowron A (1999b) B6. A rough set perspective on data and knowledge. The handbook of data mining and knowledge discovery. Klosgen, Zytkow (eds) Oxford University Press
Krawiec K, Slowinski R, Vanderpooten D (1998) Learning of decision rules from similarity based rough approximations. In: Polkowski L, Skowron A (eds) Rough sets in knowledge discovery, vol 2. Physica, Heidelberg, pp 37–54
Monachesi G, Moroni A (1995) From historical earthquake records to intensity value: some results of a study. Annali di Geofisica XXXVIII(5–6):827–836
Mucciarelli M, Monachesi G (1988) A quick survey of local amplification and their correlation with damage observed during the Umbro-Marchesan (Italy) earthquake of September 26, 1997. J Earthq Eng 2(2):325–337
Pawlak Z (1982) Rough sets. Int J Comp Inform Sci 11:341–356
Pawlak Z (1997) Rough set approach to knowledge-based decision support. Eur J Oper Res 99:48–57, Elsevier
Pawlak Z (1998a) Rough set theory and its applications to data analysis. Cybern Syst 29(7):661–688
Pawlak Z (1998b) Reasoning about data: a Rough set perspective. In: Polkowski L, Skowron A (eds) Rough sets and current trends in computing. Lecture Notes in Computer Science, Vol.1424:23–34. Lecture Notes in Artificial Intelligence, Springer-Verlag, Berlin Heidelberg
Pawlak Z (1999) Rough classification. Int J Human-Comp Studies 51:369–383, Academic Press
Pawlak Z (2002) Rough sets and intelligent data analysis. Inform Sci 147:1–12, Elsevier
ProSoft (1999) Rose (Rough Set Data Explorer) Version 2.0– User's Guide. http://www-idss.cs.put.poznan.pl/software/rose/main.html. Accessed 24 Sep 2010
Sanchez-Sesma FJ (1985) Diffraction of elastic SH-waves in wedges. Bull Seism Soc Am 75(5):1435–1446
Soetanto DP, Van Geenhuizen M (2007) Technology incubators and knowledge networks: a rough set approach in comparative project analysis. Environ Plann B: Plann Design 34:1011–1029
Tertulliani A (2000) Qualitative effects of local geology on damage pattern. Bull Seism Soc Am 90(6):1543–1548
Tertulliani A, Riguzzi F (1995) Earthquakes in Rome during the past one hundred years, In: Boschi E et al (eds) Earthquakes in the past: multidisciplinary approaches. Annali di Geofisica XXXVIII(5–6):581–590
Walczak B, Massart DL (1999) Tutorial—Rough sets theory. Chemometr Intell Lab 47:1–16

Chapter 7
Spatial Multi-Criteria Risk Assessment of Earthquakes from Bucharest, Romania

Iuliana Armaş and Mircea Rădulian

Abstract The current paper is a piece of exploratory research, having as its purpose the evaluation of seismic risk spatial patterns for Bucharest, the capital of Romania. In this paper, risk (expressed in terms of risk indices) is calculated as a combination of hazard and vulnerability information, based on the semi-quantitative technique of multi-criteria spatial analysis. The scale of analysis was considered at the level of the census units (2002). The hazard component consists of the probability of occurrence of two scenarios: (1) an occurred earthquake that is a typical hazard case for the Vrancea region (event of 30 August 1986, $M_W = 7.1$) and (2) a computed seismic hazard for the largest magnitude earthquake expected to be generated in Vrancea ($M_W = 7.7$). In both cases, the seismic hazard is expressed in terms of peak ground acceleration (PGA) values. The total vulnerability index is obtained using four groups of indicators: natural environment, physical, social and economical indicators. The overall vulnerability to seismic hazard is assessed by comparing total vulnerability with the compound capacity factor. The resulting risk indices are relative, expressed by numerical values, ranging between 0 and 1, that do not have a direct meaning of expected losses. The spatial pattern of risk shows a significant increase in the case of the maximum expected event as compared with the 1986 earthquake case, particularly in the central part of the city. As a general conclusion, only the strongest earthquakes are really dangerous for Bucharest.

Keywords Earthquake risk · Risk indices · Multi-criteria spatial analysis · Spatial patterns

---

I. Armaş (✉)
Faculty of Geography, University of Bucharest, 1 Balcescu Bd.,
Sector 1 010041, Bucharest, Romania
e-mail: iulia_armas@geo.unibuc.ro

M. Rădulian
National Institute for Earth Physics, Măgurele, 12 Călugăreni str.,
077125, Ilfov, Romania
e-mail: mircea@infp.ro

M. Bostenaru Dan et al. (eds.), *Earthquake Hazard Impact and Urban Planning*, Environmental Hazards, DOI 10.1007/978-94-007-7981-5_7,
© Springer Science+Business Media Dordrecht 2014

7.1 Introduction

Social and economic costs caused by natural disasters are increasingly burdening international communities. Between 1998 and 2010, natural hazards provoked in Europe alone nearly 100,000 fatalities and an overall loss of about EUR 200 billion (Munich Re Group 2001; EM-DAT 2010; EMSA 2010). Earthquake events for the same period rank second in terms of fatalities and third in terms of overall losses. Globally, in only the last decade, over half a million people died due to earthquakes. Dramatic examples are the recent events in Haiti, Japan and New Zealand, that claimed thousands of lives and caused widespread damage. Half of the world's largest cities lie in the vicinity of faults known to produce catastrophic earthquakes (Tucker et al. 1994).

The gravity of earthquake impacts in urban environments is demonstrated in the vulnerability of buildings, infrastructure, people and the natural habitat.

As a consequence, the issue of natural risk in the perspective of increasing urban vulnerability remains as a priority for the 3rd millennium (e.g., Parker and Mitchell 1995; Uitto 1998; Alexander 2000; Wisner et al. 2004; Pelling 2007; Wenzel et al. 2007). There are many ways in which risk, and the individual components of risk, have been defined in literature and the definitions of risk continue to multiply due to the multifaceted characteristic of this concept (Risk Definition Committee of the Society for Risk Analysis—SRA 1987; Blaikie et al. 1994; Davidson 1997; Bolin and Stanford 1998; HAZUS 2000; Bollin et al. 2003; UNDP 2004; Kasperson et al. 2005; Thywissen 2006; Bollin and Hidajat 2006; Westen 2010).

Wahlström et al. define seismic risk in 2004, as "the components seismic hazard, seismic vulnerability, and value of elements at risk (both, in human and economic terms)." Starting in January 2009, the GEM (Global Earthquake Model) Foundation aims to develop global standards for earthquake risk assessment, connecting isolated initiatives, based on worldwide active fault and seismic source database projects, as well as developing earthquake inventory, exposure and consequences databases, as a basis for vulnerability estimation and for the development of a software called OpenQuake (www.globalquakemodel.org).

On the European scale, although the theoretical part of risk analysis has been well developed, there is currently a lack of a common approach to earthquake risk assessment (EEA Technical Report 2010). Several European research entities have developed a variety of risk assessment tools. The scientific risk literature offers a large range of methods and analytical techniques to estimate risk in a qualitative or quantitative manner, such as decision analysis, systems analysis, statistical techniques, experiments, cost-benefit analysis etc. Earthquake risk assessments have been developed in the EC-funded project, RISK-UE (Mouroux and Le Brun 2007), where Bucharest was also a participant city. In 2006, Tyagunov et al. mapped the seismic risk for Germany on a quantitative basis, using the expected losses of damage to residential buildings. An overview of earthquake risk assessments in urban environments can be found in Erdik 1991; Tucker et al. 1994. The availability of data is different depending on the country, thus uniform assessment techniques such as

GEM are an ambitious goal. In practice for Romania it was not possible to complete the data for the PAGER (Prompt Assessment of Global Earthquakes for Response) project of the USGS, while countries such as Italy or Germany could extensively comment on their entry data for, used for example in the mentioned project of the seismic risk map for Germany.

The paper presented is a piece of exploratory research, having as its purpose the evaluation of seismic risk spatial patterns for Bucharest, the capital of Romania. To achieve this goal, a semi-quantitative method was adopted and used in a GIS environment, due to its flexibility and possibility of dealing with a great amount of data derived from several sources, some of which are quantitative, while others are qualitative.

The most important aspect in selecting a semi-quantitative approach was the above mentioned data availability and limitations, because risk analysis requires a large amount of multi-disciplinary information, in many cases difficult to obtain for Bucharest. The resulting map is expressed in terms of risk indices as a combination of hazard and vulnerability information.

Bucharest was chosen for this assessment (Fig. 7.1), because it is controlled by the seismic hazard (about 140 km epicentre distance) of the Vrancea region (Radulian et al. 2006). Bucharest is located in southern Romania and is situated about halfway between the Danube and the Subcarpathians. The city has a high vulnerability to earthquakes, due to a heterogeneous built stock from functional and architectural points of view and the decay of infrastructures that increase the susceptibility to the impact of seismic hazard.

Combining these attributes with the seismic hazard induced by the Vrancea source, Bucharest has been ranked as the 10th capital city worldwide in terms of seismic risk (Bonjer et al. 2003; Arion et al. 2004). The dwellings are part of 32 residential areas divided up into six sectors.

The rest of this chapter is structured in three sections: the first section briefly describes the conceptual and methodological framework; the second section outlines the outcomes of the hazard, vulnerability and risk assessment. The conclusion is the final section. The study area section was especially omitted from this study, as well as an overview with important earthquakes recorded in Bucharest, because an extensive description of Bucharest environmental characteristics and an integrated seismologic analysis is offered by another chapter in this book (see Chap. 3).

7.2 Methodology

7.2.1 Background

This approach adheres to the conceptual structure of risk given by Davidson (1997) and adopted later by Bollin et al. (2003).

In a quantitative manner, risk is defined as the expected losses and damages due to a particular hazard event for a given area and reference period (Crichton

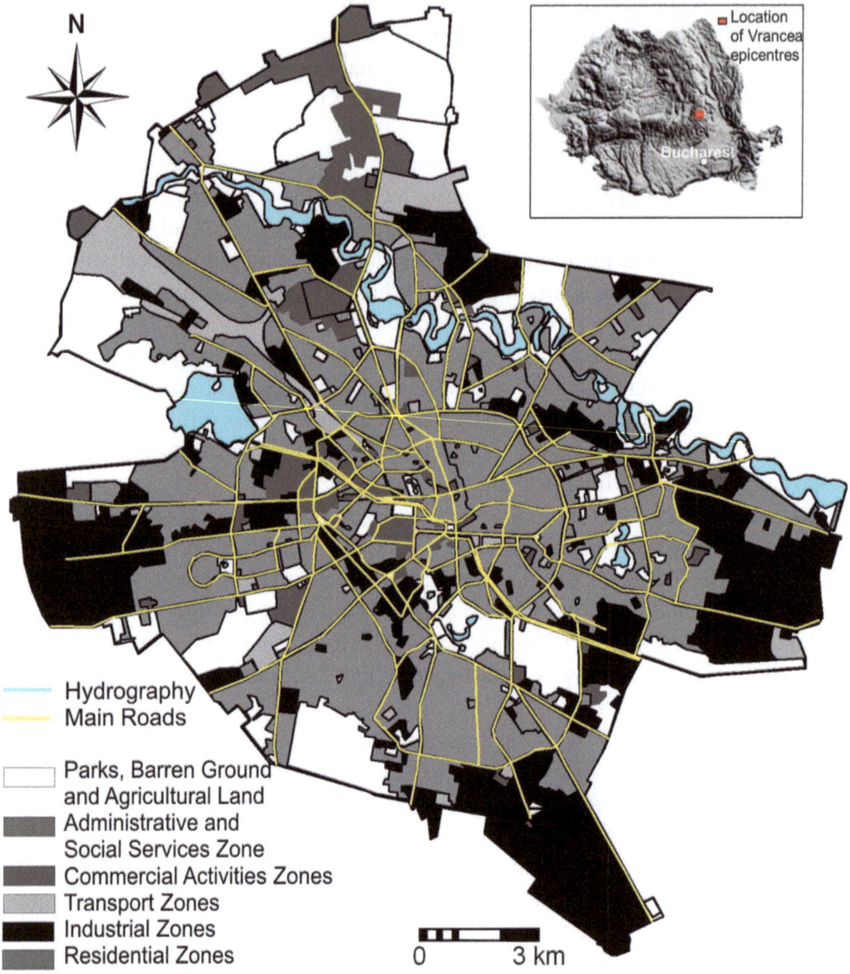

Fig. 7.1 Functional zoning of Bucharest—after the AGC Busman Publishing Plan, 2008

1999; Alexander 2000; Rashed and Weeks 2002; Cardona 2003; Schneiderbauer and Ehrlich 2004).

In a qualitative manner, risk is the expression of the relation between hazard, vulnerability and capacity, in certain conditions of exposure (Segudovic 2006; Westen 2010).

It can be expressed by the equation:

$$\text{Risk}=\text{Hazard}*\text{Vulnerability}/\text{Capacity} \tag{7.1}$$

The hazard component was defined by the probability of occurrence of an earthquake with a given magnitude within a specified interval of time. For vulnerability, we adopted the definition of Pelling (2007) according to whom vulnerability can be broken into three components: exposure, capacity, and resilience. This study considers only the first two components of vulnerability. Exposure is a consequence of geographical location and the characteristics of the built environment. Capacity reflects the strengths of the society to withstand and cope with the impact of a disaster (Thywissen 2006).

During the last several decades significant efforts have been made to develop qualitative and semi-quantitative GIS-based methods for risk assessment (e.g., Yusuff et al. 2001; Xiong et al. 2007). The weighting mechanism is the main difference between qualitative and semi-quantitative approaches and weights serve to establish the strength of evidence (Abella and Westen 2007; Seager 2004).

Given the limitations and diversity of the data basis required, in this study we applied a multi-criteria analysis. To spatially perform multi-criteria analysis, we used the SMCE (Spatial Multi-Criteria Evaluation) module of the Integrated Land and Water Information System (ILWIS)-GIS, which is implemented as an extension of the raster operations (ITC 2001; Westen 2010). The method is based on Saaty's widely adopted analytic hierarchy process (Saaty 1977, 1980; Saaty and Vargas 2001; Vargas 1990), assessing the performance of possible alternatives related to a set of criteria considered important. The alternative results in the one with the greatest value being selected. In a spatial context, alternatives represent the different spatial extent of the criterion (i.e. pixels in the maps). In this research, we performed a problem analysis, designing alternatives to support decision-making (i.e. considering a single column of maps in a 'criteria tree'). This means that the only alternatives evaluated in this study are the pixels of a single series of maps representing the various spatial and non-spatial criteria selected to compose the earthquake risk index.

The decision process implemented in the SMCE application is in detail explained in Abella and Westen (2007). The problem definition is hierarchically structured in a 'criteria tree' that includes how the selected criteria are grouped, normalized and weighed. The process is performed in a matrix that contains the criteria in one axis, and the possible alternative(s), on the other one. The value of each cell (pixel) in the matrix is computed by multiplication of the normalized value (between 0 and 1) of the criterion for the particular alternative, multiplied by the weight related to the criterion. The spatial representation of criteria consists of indicator raster maps introduced in the matrix. The outputs are more 'composite index maps,' which contain the accumulated suitability for all criteria, defined and processed as specified in the 'criteria tree'. The composite index maps can be further introduced in the analysis and combined (in another 'criteria tree') using different methods, such as the pair-wise comparison and outranking techniques (see next section).

More information about the multi-criteria method, its strengths and weaknesses can be found in Herwijnen (1999), Belton and Stewart (2002), Yoe (2002), Barzilai (2001), Figueira et al. (2005), Abella and Westen (2007), Westen (2010); Carver (2011).

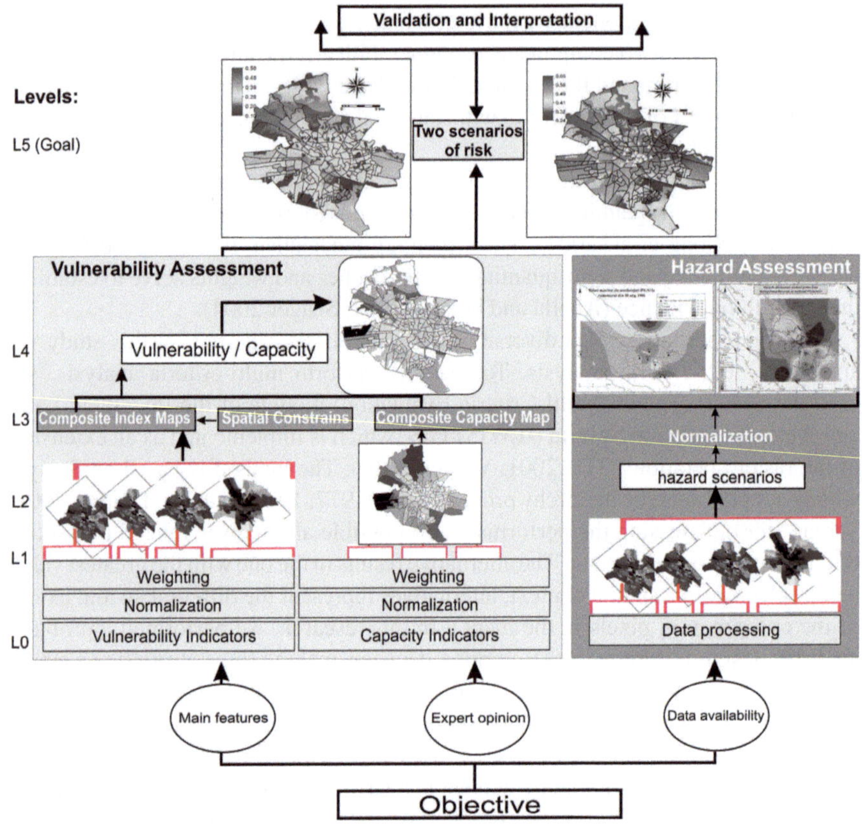

Fig. 7.2 Methodology outline

7.2.2 *Design of Criteria Trees and Processing Steps*

Bearing in mind the research goal, the risk index was composed of the intermediate levels 'hazard index' and 'vulnerability index' (level 4 in Fig. 7.2).

A selection of indicator raster-maps which measure the performance of the sub-criteria hazard and vulnerability were identified, grouped, normalized and weighted (from L_1-level upwards, see Fig. 7.2) in different 'criteria trees'.

7.2.2.1 The Hazard Index

The hazard analysis was based on two scenarios: (1) the observed seismic hazard for a typical case (the Vrancea event of 30 August 1986, $M_w = 7.1$) and (2) the computed seismic hazard for the maximum expected earthquake ($M_w = 7.7$). In both cases, the seismic hazard was expressed in terms of peak ground acceleration (PGA) values.

The earthquake of 30 August 1986 was chosen because it is the best instrumentally recorded major (magnitude greater than 7) event of Vrancea. Ten instruments that were operating in 1986 in the Bucharest urban area provided valuable seismic ground motion recordings.

The hazard distribution for the maximum expected earthquake was computed using the procedure proposed by Mărmureanu et al. (2010).

All the instrumental data recorded for Vrancea's greater force events available for the Bucharest area were processed statistically in order to evaluate the pattern for the site amplification factors and associated fundamental periods of oscillation. Then, the peak ground motion parameters induced by the maximum expected earthquake in different possible scenarios were computed using the neo-deterministic approach (Panza et al. 2011 and references herein). This approach consists in a complex hybrid waveform modeling (e.g., Panza et al. 2001) that combines modal summation (e.g., Panza 1985; Romanelli et al. 1996) with finite-difference techniques. The peak ground motion parameters computed deterministically in a reference site were afterwards extrapolated to other sites using the site amplification factors obtained statistically. Finally, corrections that took into account the non-linear effects of the superficial soils were applied.

The resulting hazard maps were introduced in the ILWIS software, rasterized, normalized and used in a pair-wise comparison with the composite vulnerability index map.

7.2.2.2 The Vulnerability Index

Although major progress has been made in defining vulnerability, methodologically there is no well-excepted or standard method available to assess it within its multifaceted nature. Basic assumptions are usually made on the influence of physical infrastructure. Virtually no attempt has been made to assess the potential integration of existing social vulnerability assessment methods with engineering methods of building and infrastructure vulnerability assessment. One of the main obstacles to this has been the lack of a common language between engineers and social sciences, and the lack of understanding as to how qualitative methods (predominant in social sciences) and quantitative methods (predominant in engineering) can be integrated.

In this research, the vulnerability profile for the city used relevant spatially based environment, social, and economic statistical data (e.g., Turner and Kasperson 2003; Cutter et al. 2003; Dwyer et al. 2004). As Chang stated in 2010, statistical data remain an under-utilised source of information, being essential for calibrating computer models, even if aggregate trends may mask important disparities (Chang 2010). The scale of analysis relevant for this purpose was at the level of the census units (2002).

A series of statistical reduction procedures were used to compute statistical variables into complex quantitative indicators.

The selection of indicators going into an index was based on the Delphi method, according to their relevance for the aim of the study, availability of data and statistical results (Armaş 2012).

Selected indicators were introduced as raster layers in the SMCE-module of ILWIS software. To obtain the raster dataset from these multiple heterogeneous sources, various pre-processing steps were used, e.g. converting the vector layers into raster maps, resampling at the same resolution and projection and normalization procedures in order to aggregate the indicators at indices level. A further requirement was to include weights, because different indicators and sub-indices show differences in the levels of importance. Rank order methods and pairwise comparison were used to weight the relative importance of alternatives in the groups of sub-criteria, based on expert knowledge (L_2 and L_3 in Fig. 7.2). More on the analytical steps and about normalization and weighting methods in measuring vulnerability at city level can be found in Armaş, 2012.

Comparing the total vulnerability to the compound capacity factor assessed the overall vulnerability to seismic hazard. Local coping capacity was reflected in the sites for medical/emergency facilities, level of preparedness (as expressed by mean distance to fire stations and police stations); distance from green and barren areas, and level of awareness (as expressed by degree of education). Capacity values were inversely calculated to the overall vulnerability (i.e. capacity indicators are unfavourable in relation to vulnerability). We also consider that influences of this features gradually decreases with the increase of the distance from these locations, according to the first law of geography (Tobler 1970).

7.2.2.3 The Risk Index

The earthquake risk was computed in the case of two scenarios of seismic hazard typical for Vrancea earthquakes, by combining the hazard index and the overall vulnerability index. In the case of a maximum earthquake scenario, we used the direct ranking method to favour hazard criteria more heavily than vulnerability ones, based on expert opinion. A weight of 0.6 was ascribed to hazard and a weight of 0.4 to vulnerability. A semi-structured reflective interview was used to personally interview specialists from the physical geography and seismology scientific fields. We tried to reduce the individual bias thus collecting numerous opinions. In the case of 1986 earthquake type scenario, an equal weighting was used for vulnerability and hazard. The resulting risk index map is relative, expressed by numerical values, ranging between 0 and 1, that do not have a direct meaning of expected losses. The final risk maps were not classified into qualitative classes, because normalized values allowed a better representation of the spatial variability of earthquake risk.

Figure 7.2 presents an overview of the methodological steps of the analysis.

7.3 Results and Discussions

7.3.1 Seismic Hazard Assessment

The seismic activity in Romania is dominated by the earthquakes generated at intermediate depths (60–180 km) in the Vrancea region, located at the South-Eastern

Carpathians arc bend. This area represents the junction of three tectonic units: East-European plate, Intra-Alpine subplate and Moesian subplate. A confined seismically active volume of high-velocity lithosphere is descended in the upper mantle where five shocks with magnitude $M_w > 6.5$ are produced on average per century, as shown by a catalogue recorded for an interval of 6 centuries (Oncescu et al. 1999). The maximum-recorded magnitude is attributed to the event of 1802 ($M_w = 7.9$). In the last century, four shocks with magnitude above 7 were recorded, characterized by large damage over extended and densely populated areas.

Note the time invariance of the seismicity rate for small-to-moderate events with around 15 events/month with magnitude greater than three and the predominance of the focal mechanism of reverse faulting type. More than 90 % of earthquakes are characterized by such mechanism and all the major shocks (M_w greater than 7) are generating rupture along a NE–SW almost vertical fault plane. The hypocenters distribution is elongated along the same plane (parallel to the Carpathians Arc) showing an atypical geometrical configuration (close to a planar distribution despite the three-dimensional structure of the high-velocity lithosphere going down in the mantle). The clear-cut features of the geometry configuration and failure process for this seismic region raise a lot of questions and debates still unsolved.

A key feature of Vrancea earthquakes in terms of Bucharest urban risk studies, is given by the predominance of long period radiation (1–2 s) for strong shocks, which is responsible for damaging effects on tall buildings, of 8–12 floors. By comparison, low buildings are much less affected by strong earthquakes (e.g., Mândrescu 1978; Mândrescu and Radulian 1999, 2004, 2007, 2008; Zaharia et al. 2008).

Seismic waves recorded in Bucharest for earthquakes with origin in Vrancea display many times higher amplitude for the radial component (approximately on the N–S direction) in comparison with the transversal component (almost on the E-W direction). This effect can be explained by the dominance of the reverse fault type focal mechanism with a rupture plan oriented nearly vertically NE–SW (Grecu et al. 2003; Radulian et al. 2006).

In the present paper we considered two scenarios for the Vrancea earthquakes, one reproducing the observed acceleration distribution in the Bucharest area as recorded during the 1986 Vrancea event ($M_w = 7.1$), the other as computed for the maximum expected event ($M_w = 7.7$). The distribution of PGA expected for the largest Vrancea earthquake is a result of considering all the available PGA values as recorded in the Bucharest area in 1977, 1986 and 1990 and the local site effects. It takes into consideration the local structure effects, including non-linear behaviour of the soil. Both probabilistic and deterministic approaches are combined to assess the seismic hazard values.

The distributions of the recording sites and of the recorded PGA values (maximum acceleration on horizontal components in cm/s^2) for the earthquake of 30 August 1986 are represented in Fig. 7.3. The larger acceleration value, as recorded on N–S or E–W component, is taken into account. Figure 7.4 represents the seismic micro-layering map of the Bucharest metropolitan area in terms of maximum acceleration (PGA) values in cm/s^2, as computed for the maximum expected Vrancea earthquake ($M_W = 7.7$).

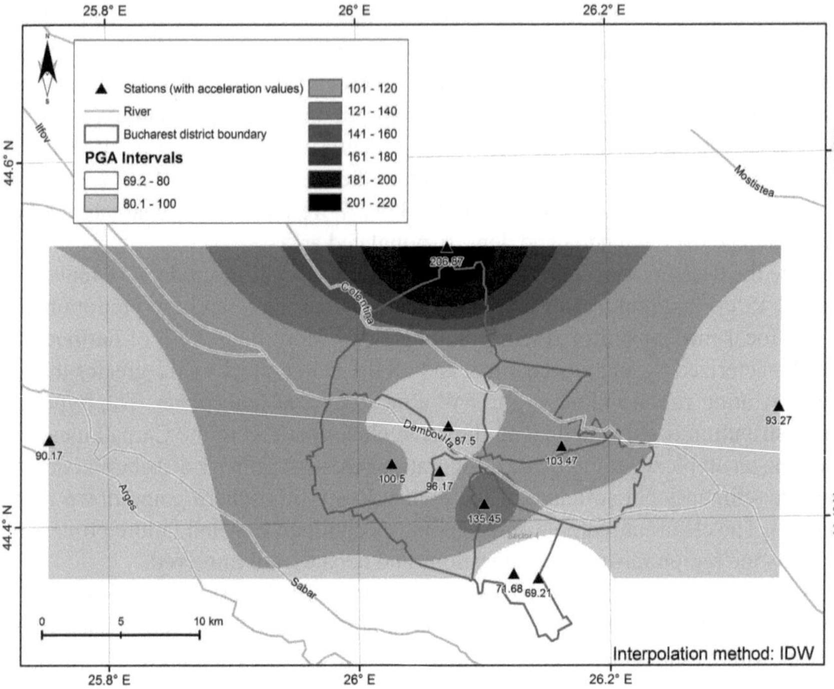

Fig. 7.3 Distribution of the PGA values recorded on the Bucharest metropolitan area during the Vrancea earthquake of 30th August 1986 (cm/s2) ($M_W = 7.1$)

High PGA values characterize a strip crossing the city area on a NE–SW direction. The largest values (close to 300 cm/s^2 for the maximum earthquake) are expected at the NE and SW edges. The lowest values characterize the SE corner of the city (Sector 4). An area with relative lower values is noticeable in both figures, in the central part of the city. Despite the sharp contrast in Fig. 7.4 between the SE and NE zones, the variation in PGA values is less than a factor of two. The strong maximum to the north for the 1986 event is due to a single acceleration recording (EREN station) whose operation proved to be highly questionable (for example, Mândrescu and Radulian 2008). As a rule, the areas of local increase/decrease relative to the general trend in Fig. 7.3 related to single station PGA value should be carefully considered, taken into account the inherent statistical fluctuations of the acceleration values.

7.3.2 Vulnerability Assessment

Complex vulnerability assessment of Bucharest was presented in Armaş (2012). In this chapter, we offer an overview of those main results corroborated with new findings, focusing on physical, social and economic aspects.

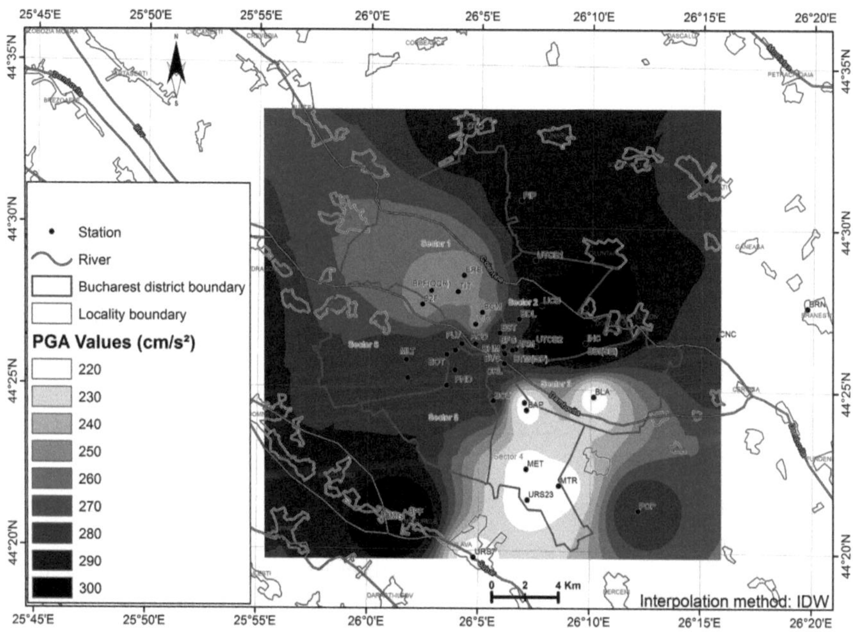

Fig. 7.4 Seismic micro-layering map of Bucharest for the maximum PGA values in cm/s^2 (M_W = 7.7)

7.3.2.1 Analysis of Building Stock Vulnerability Sub-index

Studies and observations undertaken after the strong earthquakes in 1940 (Mw = 7.7) and of 1977 (Mw = 7.4) have shown that damages are much higher for tall buildings than for small ones and that destroying effects are very low when the earthquake magnitude is smaller than 7.

The buildings' response to a specific seismic motion (given by the seismic source, propagation and site) is different depending on the structural characteristics. Thus, the same building (with an identical project) may respond differently on different sites and can present different levels of damage. Also, two buildings with different frame structures located on the same site can have different levels of damage.

Structures respond differently to the intermediate depth Vrancea earthquake induced vibration: the more rigid ones and those moderately flexible (with vibration periods up to about 1.4 s) have a stronger response in terms of acceleration, sometimes more flexible structures (with periods between about 1.4 s and 1.8 s) have a stronger response in terms of speed while very flexible structures (with periods over 1.8 s) have a stronger response in terms of displacements (e.g., Sandi 1986; Lungu et al. 1997).

Recommendations relating to compliance of the load bearing structure to lateral forces are given in normative documents for seismic design used in Romania since

Table 7.1 Structural typologies and height of buildings used in the study. (Based on the RISK-UE classification)

Structural type	No. of floors	height
M1.1	1–2	H ≤ 6
	3–5	6 < H ≤ 15
M1.2, M1.3, *M3.1*	1–2	H ≤ 6
M3.2, *M3.3*, M3.4	3–5	6 < H ≤ 15
M4, M5	6+	H > 15
M2	1–2	H ≤ 6
RC1, *RC2*, *RC3.1*	1–3	H ≤ 9
RC3.2, RC4, *RC5*	4–7	9 < H ≤21
RC6	8+	H > 21
S1, S2, S3, S4, S5	1–3	H ≤ 10
	4–7	10 < H ≤ 25
	8+	H > 25
W	1–2	H ≤ 5.5
	3+	6 > 5.5

1960. Similar recommendations are found in all seismic design codes of structures at international level.

The March 4, 1977 earthquake was an opportunity for reviewing the effectiveness of seismic design codes as well as for their radical improvement based on new knowledge gained. The most important change was the calibration of seismic design forces based on the first seismic records of the earthquake from the Vrancea source obtained at INCERC Bucharest.

The structural typology adopted in this study was defined on criteria based on the system used in the projects RISK-UE (2001–2004, 5th Framework Programme), World Housing Encyclopedia (http://www.world-housing.Net,EERI) and the German project SFB 461 'Strong Earthquakes' which developed the tool EQSIM—focusing on earthquakes in Vrancea (Armaş 2012; Bostenaru 2006). For the city of Bucharest, we designated the following dominant structural classes: M3.1—Masonry structure with timber floors; M3.3—Masonry structure with concrete-steel composite floors; RC1—Reinforced concrete frame structure; RC2—Reinforced concrete structure with reinforced concrete structural walls; RC5—Reinforced concrete structure with large prefabricated reinforced concrete wall panels (Table 7.1 and Fig. 7.5).

M3.1—Masonry structures with timber floors—have a different seismic vulnerability, depending on the number, size and position of openings. This building type is found in large number in Bucharest, being built in the nineteenth century and early twentieth century. Most of these buildings are in good condition, showing a good execution and maintenance.

The same characteristics in terms of seismic vulnerability are presented by type M3.3—Masonry structures with concrete-steel composite floors. Also many buildings in Bucharest correspond to this type, mostly representing luxury homes built in the second half of the nineteenth century and the beginning of the twentieth century.

The experience of observed damage to historical buildings, made without consideration of seismic action, show that their damage or collapse depended most heavily

Fig. 7.5 Examples of dominant structural typologies in Bucharest. (Photos, Aldea et al. 2010; HERA project www.hera.ase.ro)

on their structural compliance, on the quality of construction materials and quality of execution. Thus, buildings with a regular shape in their plan and with a balanced and rational order of the load bearing system, both in the horizontal and vertical plan, have suffered generally less damage (Lungu et al. 1999, 2004b).

RC1—Reinforced concrete frame structure—has a large diversity of frames. Frames with reduced lateral rigidity lead to high drifting in earthquakes, causing great damage to non-structural elements.

RC2—Reinforced concrete structure with reinforced concrete structural walls—had good response in case of earthquakes, suffering light structural damage. The only major damage occurred in structural walls not well enough reinforced to take over shear force.

RC5—Reinforced concrete structure with large prefabricated reinforced concrete wall panels—was used for the first time in Romania during 1950–1960, for buildings of five floors. During 1961–1963, their usage was extended to buildings of eight floors, and from 1973 also to buildings of nine floors. These typologies of buildings performed well during the earthquake of March 1977, when only in a few cases were some minor damages seen.

The overall building stock vulnerability composite map reflected the density of buildings per census unit, the density of inhabitants in buildings, average height of buildings per census unit, average age of buildings per census unit and building type indexes (structure and construction materials). The vulnerability of the RC building stock depends also on the buildings' age (whether or not following seismic codes/provisions), for example the newer ones are more resistant. Unfortunately, there was no information about the age of each building at the city level.

The spatial distribution designates the reduced vulnerability category for the western (sector 6) and southeastern areas (sector 3) of the city. Low building densities define these locations with most of the reinforced concrete structure buildings constructed in the 1940–1990 period (Fig. 7.6). In opposition, high vulnerability levels are shown for the northern (sector 1 and sector 2) and central parts of the city. The building stock in these areas is defined by high density values and high average height levels, most of the buildings being constructed in the interwar period and before 1977, with dominant M3.1 and M3.3 structural types and large commercial spaces at the ground floor.

The historic city centre (having as main location Lipscani street and, respectively, New St. George Church/the current point zero of the capital) is highly exposed to seismic risk considering the antiquity of the edifices together with the population densities (Fig. 7.6)

7.3.2.2 Analysis of Social-Economic Vulnerability Sub-index

In the assessment of social vulnerability, computed statistical variables and indexes were selected according to domain knowledge and data availability. We followed the inductive, statistical approach (relying on statistical relationships to infer leading factors) to identify social and economical vulnerability patterns and to obtain independent quantitative indicators. In this research, we used the varimax rotation strategy to make distinct the principle components (PC), and selected predictive indicators based on their PC's eigenvalue greater than 1.0. Next, we selected those quantitative indicators whose loading was maximum, making them useful for the vulnerability prediction (Armaş 2012).

The following step was to aggregate the scores of the variables in which the selected factors were saturated and to include them as significant criteria into the multi-criteria vulnerability analysis.

The overall social-economical vulnerability sub-index is shown in Fig. 7.7

From the overall spatial analysis based on social and economic criteria, we conclude that the downtown areas are the most vulnerable. These areas are occupied by a less favoured population (low incomes of under 150 € per month, high unemployment rate, a high density of people living in collective residences). This aspect represents a direct cause of the communist economic policy conducted since the early 1950s and consisting in the construction of apartment buildings for the working-class that generated large collective ensembles that would host the industrial labour force. The localization of large industrial units in the peripheral area of the city, situated along

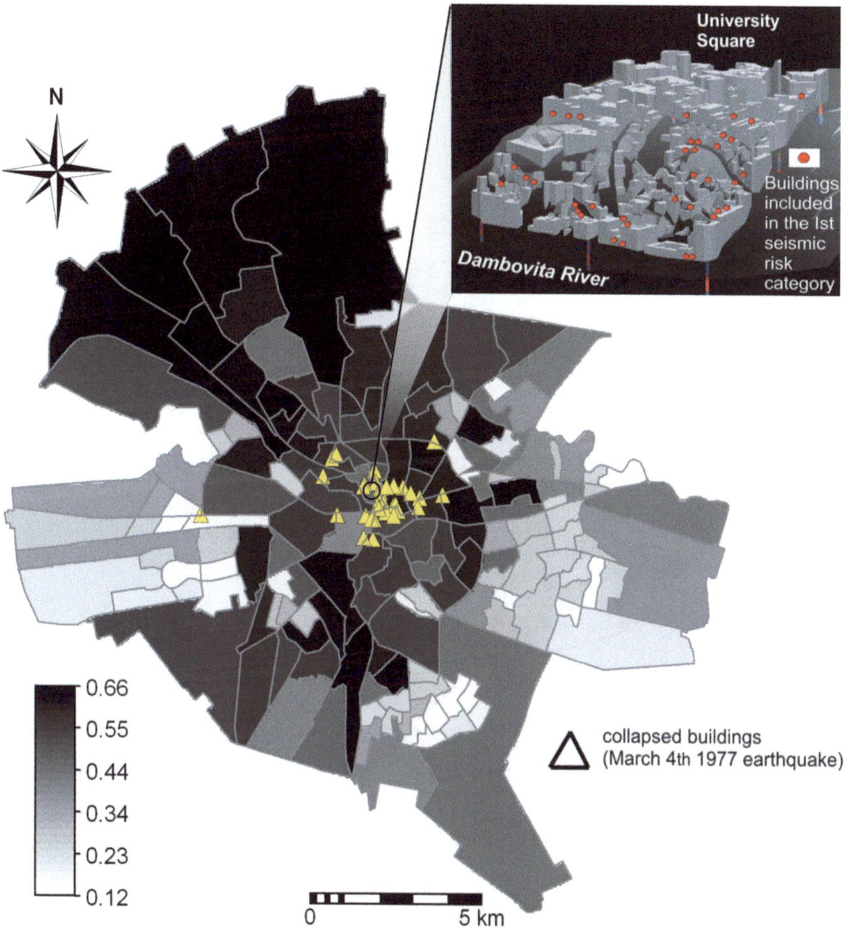

Fig. 7.6 The overall building stock vulnerability composite map (the 3D image in the *right* corner represents the historic centre, lying on the Dâmboviţa meadow, on the scarp and on the top of the 80 m terrace)

the axis of penetration in Bucharest preceded by collective residential areas, and the industrial decline in the 1990s are important factors responsible for the current situation.

A recurrent phenomenon in the urban area in the last 10 years refers to the location of high quality residential areas (housing type villa) in the marginal areas of the city, being part of the suburbanization process of the capital. This phenomenon consists of establishing high-income population located in areas situated at the limit of the city. In some cases, the preference for suburban areas in terms of comfort is not justified, the urban housing quality decreasing by the lack of urban equipment, utilities and accessibility to urban services.

Fig. 7.7 Composite social-economical vulnerability sub-index with spatial constrains (in *white*)

7.3.2.3 Capacity and Overall Vulnerability

The overall vulnerability to seismic hazard was assessed by comparing the total vulnerability to the compound coping factor (Fig. 7.8).

The resulting distribution map shows a radial spatial pattern, with values increasing from the central areas of the city towards the peripheral areas. Although the downtown areas are characterized by high values of vulnerability due to the ageing of the built structures, the preparedness and awareness levels record high values as well; consequently, these areas are rated as showing average values of overall vulnerability. Nevertheless, the areas lying along the Dâmboviţa River and its terraces (also including the historical city center area) appear as a distinctive narrow band with high vulnerability, especially due to the highest recorded values of the natural environmental vulnerability to seismic risk. Moderate values of the overall vulnerability were registered for the areas in the 6th sector, the Berceni area in the 4th sector, as well as the Primăverii-Domenii area in the 1st sector, with low ratings of their social and structural vulnerability, as well as average scores for the coping capacity.

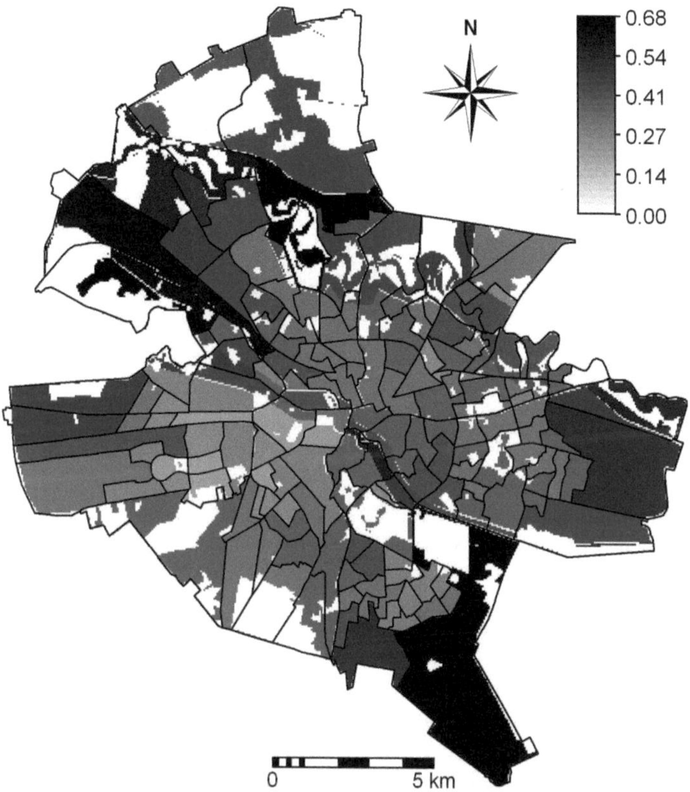

Fig. 7.8 Overall vulnerability

At the other end of the spectrum, high and very high values of the vulnerability are by-products of a combination between low scores of coping capacity and high values of economic and social vulnerability.

7.3.3 Risk Assessment

The analysis of the risk maps based on the two seismic hazard scenarios shows the major impact played by the magnitude size of the Vrancea earthquakes on the Bucharest metropolitan area. The risk index significantly increases in the case of the maximum expected event (Fig. 7.9b), as compared with the 1986 earthquake case (Fig. 7.9a), particularly in the central part of the city. In this way, we can explain why no major damage was discerned in Bucharest in 1986, in contrast with the 1977 destructive event ($M_w = 7.4$).

Note that, while in the central part of the city the risk index varies considerably from one scenario to the other, the difference is less important at the border areas. In

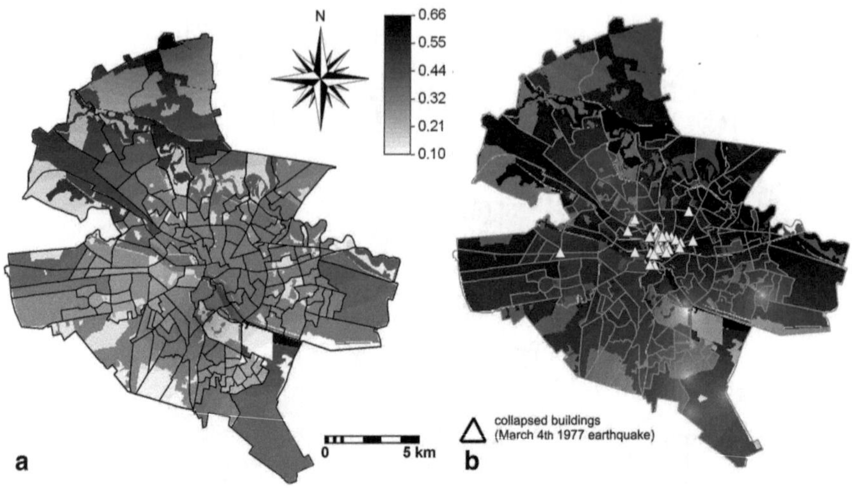

Fig. 7.9 Spatial distribution of seismic risk for two Vrancea earthquake scenarios. **a** is a typical scenario for an observed major event and **b** is a scenario for a maximum expected event

fact, the second scenario induces clearly a change in the risk distribution, emphasizing a higher rate of increase in the central part of the city as compared with the increasing rate at the border areas. This pattern of the spatial distribution of the seismic risk is in correspondence with the morphological configuration of the city, characterized by a fluvial topography in a flat landscape, with a general NW-SE orientation.

It is interesting to note that variation of PGA values which are less than a factor of 2 (from 220 to 300 cm/s^2) over the study area lead to variations in the risk index by a factor of 3, or even greater. The maximum risk is centred along the valleys of Dambovița and Colentina rivers and the terraced interfluves, especially in the southern part, near Colentina field.

The minimum risk is observed in the southern part of the city, in the 4th sector, mainly in Berceni area, situated on Berceni field, with a reduced vulnerability of the building environment, with good access to the barren areas, parks and hospitals, but with a higher social and economical vulnerability, as a consequence of restructuring the economical sector after 1990.

The possibility to obtain independent data to test the validity of risk prediction is very low. In this approach, we compared the spatial distribution of risk values with losses registered during the major event that occurred in 1977, when 32 tall buildings collapsed in the central part of the city. All the collapsed buildings fell on pixels corresponding to the highest risk index values of the study area (21 % of the collapsed buildings fell on pixels with the highest risk index values, 75 % on pixels with very high risk index values and 4 % on pixels with high risk index values).

The results represent the situation at the level of data provided by the 2002 census. We think that the increase in socio-economical vulnerability and the changes that occurred at the level of building environment during the last decade imply a significant increase of the influence of these factors in the seismic risk equation.

7.4 Conclusions

As many papers published over the last 10 years have concluded, the application of multi-criterial methods provides a significant improvement in the decision process and understanding of environmental risk alternatives. In this study, the semi-quantitative technique of the multi-criteria spatial analysis defines a seismic risk pattern of Bucharest in agreement with field observations. Bucharest comprises a heterogeneous surface from the functional and architectural points of view. Relevant examples are: residential houses and houses with mixed functions (of small shop type) from the nineteenth century, edifices of cultural and administrative-banking functions from the beginning of the twentieth century, as well as the large residential ensembles built between WWI and WWII or post-war before 1977. The imprint of hazards gives the particularity of the territory since the seventeenth century, which contributed to the changes of the city's urban image. The variety of the building environment (taking into consideration variables, such as structure, height, age and surface) and the specific socio-economical dynamics of the urban environment in connection with sensitivity to the Vrancea earthquakes emphasizes, from the entire range of urban risks, the highest vulnerability to seismic risk.

The hazard in the Bucharest area is controlled by the relatively distant earthquakes (about 140 km epicentral distance) occurring in the Vrancea region. The largest damage is noticed for tall and flexible buildings; likewise, there is another singular case, Mexico City, where distant earthquakes affect tremendously the tall buildings. However, the explanation is essentially different for the two cases: for Bucharest there is a coincidence of predominant period of the source with the fundamental period of the geological deposit beneath the city, while in Mexico City the destructions are firstly determined by the effects of the local site conditions with amplifications of more than 5 times for 1–2 s period (Whitman 1986).

The convergence of results validates the adopted method. And this, having in mind that a semi-quantitative method is partly subjective, the results may vary, depending on the experts' knowledge and subjective judgments and the accuracy of available data. Problematic issues are how one constructs the decision hierarchy, facts that influence outcomes of the multi-criteria method, and the roughness of the scale used in pair-wise comparisons.

The risk index computed in Bucharest for two Vrancea earthquake scenarios (one typical for an observed major event, the other for a maximum expected event) shows the critical impact of event magnitude (only the largest earthquakes are really dangerous) and the higher risk in the central part of the city.

We are aware that the census-designated areas influence the results of the statistical analysis and are not suitable for a requirement of high-accuracy location for disaster planning. This study is only a first step in a successful disaster risk reduction strategy based on safety land-use planning and risk-mitigation measures that will significantly reduce the costs in case of earthquake disasters occurrence in Bucharest.

Acknowledgements This study was carried out as a component of the HERA and REVEAL research projects financed by Romania National Authority of Scientific Research, having Prof. Dr. I. Armaş as PI. We are grateful to Dr. Al. Gavriş for useful suggestions and support.

References

Abella CEA, Westen van CJ (2007) Generation of a landslide risk index map for Cuba using spatial multi-criteria evaluation. Landslides 4:311–325

Alexander D (2000) Confronting Catastrophe—New Perspectives on Natural Disasters. Oxford University Press, Oxford, p 282

Arion C, Vacareanu R, Lungu D (2004) WP10—Application to Bucharest, RISK-UE. An advanced approach to earthquake risk scenarios with applications to different European towns. At ftp.brgm.fr/pub/Risk-UE

Armaş I (2008) Social vulnerability and seismic risk perception. Case study: The Historic Center of The Bucharest Municipality/Romania. Nat Hazards 47(3):397–410

Armaş I (2012) Multi-criteria vulnerability analysis to earthquake hazard of Bucharest, Romania. Nat Hazards. doi:10.1007/s11069-012-0209-2

Barzilai J (2001) Notes on the analytic hierarchy process. In Proceedings of the NSF Design and Manufacturing Research Conference, Tampa, Florida (pp 1–6)

Belton V, Stewart T (2002) Multiple criteria decision analysis: an integrated approach. Kulwer Academic Publishers, Boston

Blaikie P, Cannon T et al (1994) At risk: natural hazards, people's vulnerability and disasters. London etc., Routledge

Bolin R, Stanford L (1998) The northridge earthquake: vulnerability and disaster. Routledge, New York

Bollin C, Cárdenas C, Hahn H, Vatsa KS (2003) Natural Disaster network; Disaster Risk management by Communities and Local Governments, Washington, d.C.: Inter-American Development Bank

Bollin C, Hidajat R (2006) Community-based disaster risk index: pilot implementation in Indonesia. In: Birkmann J (ed) Measuring vulnerability to natural hazards-towards disaster resilient societies. UNU-Press, Tokyo

Bonjer KP, Grecu B, Rizescu M, Radulian M, Sokolov V, Mandrescu M, Lungu D, Moldoveanu T (2003) Assessment of site effects in downtown Bucharest by recording of ambient noise, moderate and large intermediate depth earthquakes from vrancea focal zone, Proceedings Int. Conf. Earthquake Loss Estimation and Risk Reduction, 24–26 Oct., 2002, Bucharest, Romania

Bostenaru Dan M (2006) Wirtschaftlichkeit und Umsetzbarkeit von Gebäudeverstärkungsmaßnahmen zur Erdbebenertüchtigung. Grundlagen und Lösungsansatz unter besonderer Berücksichtigung der Situation in Bukarest, Rumänien [Applicability and economic efficiency of building strengthening measures for seismic retrofitBasics and solution approach under special consideration for the situation in Bucharest, Romania] http://www.shaker.de/de/content/catalogue/index.asp?lang=de&ID=8&ISBN=978-3-8322-5727-9Shaker, Aachen

Cardona OD (2003) Indicators for Disaster Risk Management. First Expert Meeting on Disaster Risk Conceptualization and Indicator Modelling. Manizales, March 2003. http://idea.manizales.unal.edu.co/ProyectosEspeciales/adminIDEA/CentroDocumentacion/DocDtales/documentos/01%20Conceptual%20Framework%20IADBIDEA%20Phase%20I.pdf. Accessed 24 Jan 2006

Carver SJ (2011) Integrating multi-criteria evaluation with geographical information systems. Int J Geogr Inf Sci 5(3):321–339. doi:10.1080/02693799108927858

Chang SE (2010) Urban disaster recovery: a measurement framework and its application to the 1995 Kobe earthquake. Disasters 34(2):303–327

Crichton D (1999) The risk triangle. In: Ingleton J (ed) Natural disaster management. Tudor Rose, London, pp 102–103

Cutter SL, Boruff BJ, Shirley WL (2003) Social vulnerability to environmental hazards. Soc Sci Q 82:242–260. doi:10.1111/1540-6237.8402002
Davidson R (1997) An Urban earthquake Risk Index, The John A Blume Earthquake Engineering Center, department of Civil engineering, report No 121, Standford: Stanford University
Dwyer A, Zoppou C, Nielsen O, Day S, Roberts S (2004) Quantifying Social Vulnerability: A methodology for identifying those at risk to natural hazards, Geoscience Australia Racord 14
EEA (2010) European Environment State and Outlook Report 2010. www.eea.europa.eu/soer. Accessed 1 Dec 2010
EM-DAT (2010) The International Disaster Database, Centre for research on Epidemiology of Disasters—CRED. www.emdat.be. Accessed 10 Sep 2010
EMSA (2010) European Maritime Safety Agency EMSA. www.emsa.europa.eu. Accessed 10 Sep 2010
Erdik M (1991) Urban Earthquake Hazard, Risk and Mitigation, Invited Paper, Proc., 5th International Conference on Soil Dynamics and Earthquake Engineering, Sept. 23–26, University of Karlsruhe, Germany
Figueira J, Greco S, Ehrgott M (2005) Multiple criteria decision analysis: state of the art surveys. Springer, New York
Grecu B, Popa M, Radulian M (2003) Seismic ground motion characteristics in the Bucharest area: Sedimentary cover versus seismic source control. Rom Rep Phys 55:511–520
HAZUS 99 (2000) Natural Loss Estimation Methodology, available online at www.fema.gov/hazus/hazus99.htm Accessed 1 Dec 2010
Herwijnen M van (1999) Spatial Decision Support for Environmental Management. Vrije Universiteit, Amsterdam, p 274
ITC (2001) ILWIS 3.0 Academic—User's Guide. Enschede, Netherlands, ITC
Kasperson JX, Kasperson RE, Turner BL II, Schiller AMD, Hsieh W (2005) Vulnerability to global environmental change. In: Kasperson JX, Turner BLII, Schiller AMD, Hsieh W (eds) Social contours of risk. II: risk analysis corporations and the globalization of risk. Earthscan, London, pp 245–285
Lungu D, Cornea T, Aldea A, Zaicenco A (1997) Basic representation of seismic action. In: Lungu D, Mazzolani F, Savidis S (eds) Design of structures in seismic zones: Eurocode 8- Worked examples. TEMPUS PHARE CM Project 01198: Implementing of structural Eurocodes in Romanian civil engineering standards. Bridgeman Ltd., Timisoara, pp 1–60
Lungu D, Aldea A, Moldoveanu T, Ciugudean V, Stefanica M (1999) Near-surface geology and dynamic properties of soil layers in Bucharest. In: Wenzel F, Lungu D (eds) Vrancea earthquakes: tectonics, hazard, and risk mitigation. Kluwer Academic Publishers, pp 137–148
Lungu D, Aldea A, Arion C, Cornea T, Petrescu F, Vacareanu R (2004b) RISK-UE—Bucharest Synthesis Report
Mândrescu N (1978) The Vrancea earthquake of March 4, 1977 and the seismic microzonation of Bucharest, Proc. 2nd Inter. Conf. Microzonation, San Francisco, 1, 399–411
Mândrescu N, Radulian M (1999) Seismic microzoning of Bucharest (Romania): a critical review. In: Wenzel F, Lungu D, Novak O (eds) Vrancea earthquakes: tectonics, hazard, and risk mitigation. Kluwer Academic Publishers, pp 109–122
Mândrescu N, Radulian M, Mărmureanu G (2004) Site conditions and predominant period of seismic motion in the Bucharest urban area. Rev Roum Geophysique 48:37–48
Mândrescu N, Radulian M, Mărmureanu G (2007) Geological, geophysical and seismological criteria for local response evaluation in Bucharest urban area. Soil Dyn Earthq Eng 27:367–393
Mândrescu N, Radulian M, Mărmureanu G, Ionescu V (2008) Integrated research of the geological, geophysical and seismological data for local response evaluation in Bucharest urban area, Publishing House of the Romanian Academy, Bucharest, pp. 136
Mărmureanu G, Cioflan CO, Mărmureanu A (2010) Studies on local seismic hazard (microzonation) of the Bucharest metropolitan area. Seismic microzonation map. Ed. Tehnopress, Bucharest (in Romanian)
Mouroux P, Le Brun B (2007) Presentation of RISK-UE project. Bull Earthq Eng 4(4):323–339

Munich Re Group (2001) Annual report: how much risk can the world take? Munich, Germany
Oncescu MC, Marza VI, Rizescu M, Popa M (1999) The Romanian earthquake catalogue between 984-1997. In: Wenzel F, Lungu D, Novak O (eds) Vrancea earthquakes: tectonics, hazard and risk mitigation. Kluwer Academic Publishers, Dordrecht, pp 43–47
Panza GF (1985) Synthetic seismograms: the Rayleigh waves modal summation. J Geophys 58:125–145
Panza GF, Romanelli F, Vaccari F (2001) Seismic wave propagation in laterally heterogeneous anelastic media: theory and applications to the seismic zonation. Adv Geophys (Academic press) 43:1–95
Panza GF, Irikura K, Kouteva M, Peresan A, Wang Z, Saragoni R (eds) (2011) Advanced seismic hazard assessment. Pure Appl Geophys 168. doi:10.1007/s00024-010-0179-9
Parker D, Mitchell JK (1995) Disaster vulnerability of megacities. GeoJournal 37(3):295–301
Pelling M (2007) The vulnerability of cities. Natural disasters and social resilience. Earthscasn Publications Ltd, London
Radulian M, Mândrescu N, Grecu B (2006) Seismic ground motion variability over the Bucharest area. Acta Geodaetica et Geophysica Hungarica 41(3–4):361–368
Rashed T, Weeks J (2002) Assessing vulnerability to earthquake hazards through spatial multicriteria analysis of urban areas. Int J Geogr Inf Sci 2003. 17(6):547–576
Risk Newsletter (1987) "Defining 'Risk'", editorial note, Risk Newsletter, 7(3):5
RISK-UE (2001–2004) An advanced approach to earthquake risk scenarios with applications to different European towns, Fifth Framework Programme of the European Commission, 2000–2004
Romanelli F, Bing Z, Vaccari F, Panza GF (1996) Analytical computations of reflection and transmission coupling coefficients for Love waves. Geophys J Int 125:132–138
Saaty TL (1977) A scaling method for priorities in hierarchical structures. J Math Psychol 15: 234–281
Saaty TL (1980) The analytic hierarchy process. McGraw-Hill International Book Company, New York
Saaty TL, Vargas LG (2001) Models, methods, concepts & applications of the analytic hierarchy process. Kluwer Academic Publishers
Sandi SH (convener) (1986) Vulnerability and Risk Analysis for Individual Structures and for Systems. Report of EAEE WG 5/10 to the 8-th European Conference on Earthquake Engineering, Lisbon
Schneiderbauer S, Ehrlich D (2004) Risk, hazard and people's vulnerability to natural hazards. A Review of Definitions, Concepts and Data. European Commission Joint Research Centre. EUR 21410 EN
Seager TP (2004) Understanding industrial ecology and the multiple dimensions of sustainability. In: O'Brien and Gere Engineers (eds) Strategic environmental management. Wiley, New York
Segudovic H (2006) Qualitative risk analysis method comparison, INFIGO Information Security, available online at http://www.infigo.hr/files/INFIGO-MD-2006-06-01-RiskAsses_ENG.pdf. Accessed 20 Dec 2011
Thywissen K (2006) Core terminology of disaster reduction. Measuring vulnerability to Natural Hazards Towards disaster resilient societies. J. Birkmann, UN University Press
Tobler WR (1970) A computer model simulation of urban growth in the Detroit region. Econ Geogr 46(2):234–240
Tucker BE, Erdik M, Hwang CN (eds) (1994) Issues in Urban earthquake risk. Kluwer Academic Publishers, Dodrecht
Turner BL, Kasperson RE (2003) A Framework for Vulnerability Analysis in Sustainability Science. Proceedings of the National Academy of Sciences of the United States of America (PNAS). 100(14), 8074–8079
Tyagunov S, Grünthal G, Wahlström R, Stempniewski L, Zschau J (2006) Seismic risk mapping for Germany. Nat Hazards Earth Syst Sci 6:573–586
Uitto JI (1998) The geography of disaster vulnerability in megacities. Appl Geography 18(1):7–16

UNDP—Bureau for Crisis Prevention and Recovery (2004) Reducing disaster risk: a challenge for development. In: Pelling M, Maskrey A, Ruiz P, Hall L (eds) A global report. John S. Swift Co
Vargas LG (1990) An overview of the analytic hierarchy process and its applications. Eur J Oper Res 48:2–8
Wahlström R, Tyagunov S, Grünthal G, Stempniewski L, Zschau J, Müller M (2004) Seismic risk analysis for Germany: methodology and preliminary results. In: Malzahn D, Plapp T (eds) Disasters and society—from hazard assessment to risk reduction. Logos Verlag Berlin, pp 83–90
Wenzel F, Bendimerad F, Sinha R (2007) Megacities—megarisks. Nat Hazards 42:481–491
Westen C von (ed) (2010) Multi-hazard risk Assessment, Distance education course, Guide book. Enschede, Netherlands, United Nations University—ITC School on Disaster Geoinformation Management (UNU-ITC DGIM)
Whitman R (1986) Are the soil depositions in Mexico City unique? The Mexico earthquakes -1985. Factors involved and lessons learned. In: Cassaro MA, Romero EM, editors. Proceedings of the international conference. Mexico City: ASCE; 1986. p 163–77
Wisner B, Blaikie P, Cannon T, Davis I (2004) At risk: natural hazards, people's vulnerability and disaster. Routledge, London
Xiong Y, Zeng GM, Chen GQ, Tang L, Wang KL, Huang DY (2007) Combining AHP with GIS in synthetic evaluation of eco-environment quality—a case study of Hunan Province, China. Ecolog Model 2009:97–109
Yoe C (2002) Trade-off analysis planning and procedures guidebook. Prepared for Institute for Water Resources. U.S. Army Corps and Engineers
Yusuff RM, Yee KP, Hashmi MSJ (2001) A preliminary study on the potential use of the analytical hierarchical process (AHP) to predict advanced manufacturing technology (AMT) implementation. Robot Comput-integr Manuf 17(5):421–427
Zaharia B, Radulian M, Popa M, Grecu B, Bălă A, Tătaru D (2008) Estimation of the local response using Nakamura method for Bucharest area. Rom Rep Phys 60(1):131–144

Chapter 8
The Urban System of Crotone, Italy, Facing the Earthquake Impact

Agostino Goretti, Sabino Vetta, Francesco Palmieri, Fabio Adamo, Massimo Berlingeri and Luigi Palmieri

Abstract A methodology is proposed that can predict, with a systemic approach, the status of the economic, social, cultural and administrative components of an urban centre in normal growth and in seismic crisis. This is essential to evaluate the ability of a city to fulfill its functions after a disaster. In the model, the following sub-systems are considered as part of the urban system: Social, Residential, Cultural, Health, Education, Energy, Economic, Administrative and Emergency. Each one of the above sub-systems includes many elements at risk and is influenced by the other sub-systems. As an application example we look at Crotone municipality, South Italy, where data on active faults, soil, buildings, infrastructures, lifelines, and population have been extensively collected in the past, within the PIC-Urban II project. Damage scenarios are proposed for earthquakes with different return periods. The functionality of the considered sub-systems is also evaluated.

Keywords Seismic scenario · Loss · Urban system · GIS

8.1 Introduction

In a conventional analysis of an earthquake impact, 'physical' damage, such as structural damage to buildings, is usually the subject of evaluation. Consequences of the 'physical' damage, such as economic damage, casualties, and unusable buildings are often estimated. The expected damage to lifelines is sometimes included in the analysis. However, very rare are the analyses that provide an estimate of the cultural loss due to damage to the historical buildings and almost non-existent is the evaluation of the social impact of the seismic event due to social discomfort, loss of

A. Goretti (✉)
Seismic and Volcanic Risk Office, Civil Protection Department,
Via Vitorchiano 4, 00189 Rome, Italy
e-mail: agostino.goretti@protezionecivile.it

S. Vetta · F. Palmieri · F. Adamo · M. Berlingeri · L. Palmieri
Planning and Land Management, Crotone Municipality,
Piazza Resistenza 1, 88900 Crotone, Italy

M. Bostenaru Dan et al. (eds.), *Earthquake Hazard Impact and Urban Planning*,
Environmental Hazards, DOI 10.1007/978-94-007-7981-5_8,
© Springer Science+Business Media Dordrecht 2014

identity, job suspension, disruption of essential services, etc. Similarly, many other aspects, such as institutional disruption, alteration of external relationships, loss of efficiency in rescue activities, more or less capability of the system to sustain the impact with internal or external resources, lifelines interactions, etc., may have a significant influence on the seismic impact and recovery.

In other words the seismic performance of an urban system, considered as an ensemble of interacting sub-systems that make the citizen's normal life course possible, is seldom evaluated.

In the paper a methodology is proposed that is able to predict, with the help of a systemic approach, the status of the economic, social, cultural and political components of an urban centre in normal growth and in a crisis period. This is essential in order to evaluate the ability of the city to fulfill its functions. In the model the following sub-systems are considered: Residential, Cultural, Health, Education, Energy, Economic, Administrative and Emergency. Each one of the sub-systems includes many elements at risk and is influenced by the remaining sub-systems. As an example, the Residential system (buildings) affects the Road system because the collapsed buildings may block roads, and, in turn, the Road system affects the Emergency systems, since the Hospital may not be within easily reached distance of the damaged buildings. The consequences to the Social system (casualties) are also evaluated, but not included in the Urban System due to the difficulties of considering both casualties and system functionality.

An application of the model is made with reference to Crotone municipality, South Italy, where data on active faults, soil, buildings, infrastructures, lifelines, roads, and population have been extensively collected in the past, within the PIC-Urban II project. Damage scenarios are performed for earthquakes with different return periods. Results are shown at different territorial scale: individual buildings, Istat census tracks and larger macroareas. The functionality of several sub-systems is also evaluated. We believe that the proposed approach can be very useful both to local civil protection authorities and to urban planners.

8.2 The 'Crotone Project'

The Axis I 'Multifunctional and environmentally sustainable reurbanization' of the PIC URBAN II program, implemented by Crotone Municipality, provided a series of measures directed to the improvement of life quality in the urban context. Measure 1.1 'Safety of urban settings exposed to seismic risk' is directed to the seismic risk assessment and mitigation of the Crotone urban area. In order to reach this target, Crotone Municipality signed an agreement with the Seismic Risk Office of the Italian Civil Protection Department and the University of Calabria. The cooperation between the three bodies has led to the so-called 'Crotone Project' (Goretti et al. 2007). The activities of the project have been subdivided into the tasks shown in Fig. 8.1, which also contains the costs of each task and the total cost of the project.

Fig. 8.1 Tasks and costs of the 'Crotone Project'

TASK	ACTIVITY	COST (KEuro)
1	GIS	162
2	URBAN SYSTEM	80
3	HAZARD	20
4	SITE EFFECTS	40
5	VULNERABILITY	150
6	SCENARIO	20
7	VALORISATION PLAN	145
	Total	617

Fig. 8.2 Target area of the 'Crotone Project' and Istat 2001 census tracks

The target area of the 'Crotone Project' is represented by the urban area of Crotone municipality. In order to consider the actual built-up area and the ISTAT2001 census zones, the target area has been slightly enlarged and is shown in Fig. 8.2.

Besides the Crotone municipal administration, several local technicians have been involved in the survey of geological, building and infrastructure data, in the evaluation of soil response, in the assessment of strategic buildings, and in the implementation of the GIS (Palmieri et al. 2014). In this way, the 'Crotone Project' increased seismic awareness of local authorities and technicians, and their know-how in a 'building capacity' approach.

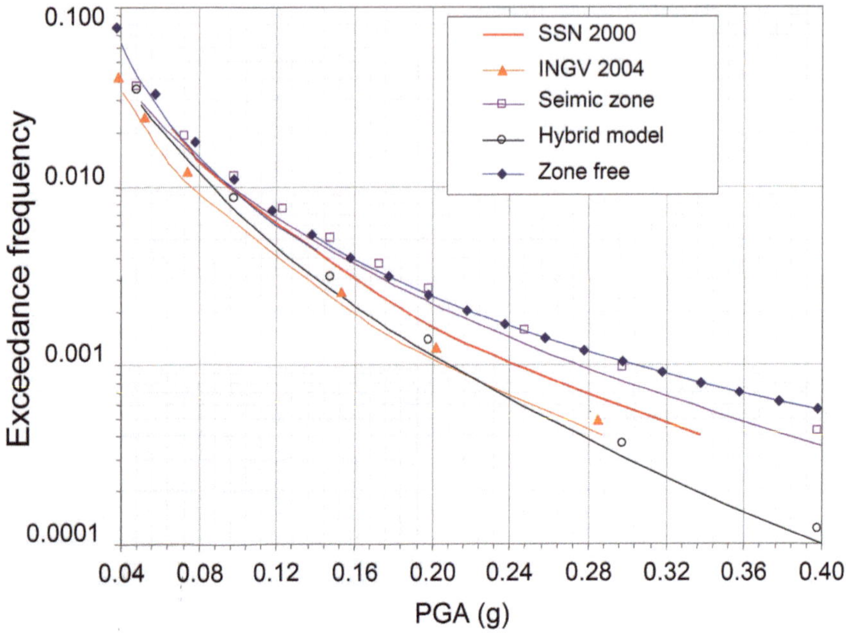

Fig. 8.3 Comparison of the hazard in terms of PGA

8.3 Seismic Hazard and Site Effects

The seismic hazard has been evaluated according to three different probabilistic methodologies (Galli and Lucantoni 2014). The first one (Cornell 1968) makes use of the seismogenetic zones around Crotone, the second one is based on a zone free model (Woo 1996), and the third one is a hybrid model (Galli and Lucantoni 2003) where the catalogue of seismicity has been integrated with data on active faults obtained from paleoseismic studies. Figure 8.3 shows the obtained hazard curves in terms of Peak Ground Acceleration (PGA). It can be seen that the hybrid model is less conservative because (i) the active faults are quite far from Crotone and (ii) the return periods of destructive earthquakes in the hybrid model are longer because they are based on paleoseismic studies and not on the completion period of the catalogue.

The seismic hazard has been disaggregated (Bazzurro and Cornell 1999) by magnitude and distance. Disaggregation is based on computation of the relative contributions to seismic hazard of different seismogenic zones characterized by their geometries, recurrence relationships and maximum magnitude. Those contributions are typically expressed in terms of probability density functions of M, R and ε, conditional to the level that PGA, for which the hazard is being disaggregated, is exceeded. Epsilon is defined as the number of logarithmic standard deviations by which the logarithmic ground motion deviates from the median predicted by an appropriate attenuation relationship. The analysis of the M, R and ε distributions allows us to identify some values of the variables giving the largest contribution to disaggregated

Table 8.1 Hazard disaggregation results

PGA (g)	Return period (yr)	Magnitude	Distance (Km)	Epsilon	Contrib. (%)
Model based on seismogenetic zones					
0.082	72	5.1	13.6	– 0.033	79.8
0.182	475	5.4	9.1	0.574	93.6
0.337	2,475	5.6	6.4	1.045	98.4
Hybrid model					
0.071	72	4.6	10.8	0.08	72.0
0.161	475	6.5	15.9	– 0.22	18.6
0.274	2,475	4.8	4.9	1.474	64.4

hazard or considered representative in some other statistical sense (e.g., mean or modal values of M, R and ε).

The magnitude-distance pairs selected for different return periods (72, 475 and 2,475 years) are shown in Table 8.1. From these pairs, artificial accelerograms have been generated with a non-stationary procedure.

Geological, geotechnical and geophysical data have been collected through a survey campaign within the Crotone urban area (Scerra 2014). All the data have been implemented in the GIS (Palmieri et al. 2014). The survey included:

- Geognostic campaign (15 surveys);
- 3 samples per survey, and analysis of lythologic static and dynamic characteristics;
- Down-hole analysis (7 survey);
- Microtremors measurement according to SASW (Park et al. 1999) and ESAC (Ohori et al. 2002) methodologies;
- Microtremors measurement according to Nakamura (1989) methodology (Mucciarelli and Albarello 2011).

From the previous data, shear waves velocity (Vs) profiles and damping profiles have been evaluated in several sites. The target area has been then subdivided into six homogeneous zones, in each of which the soil properties and the soil response have been considered uniform. The zones are shown in Fig. 8.4. Red zones have the highest hazard level, while grey zones have the lowest one.

Numerical simulations on 1D (PROSHAKE 2000) and 2D (QUAKE 2004) models have been carried out based on the soil model and on the seismic inputs previously described. The analyses (Giordano and Naso 2014) provided, for each zone identified in Fig. 8.4, both the amplification factors and the 5 % damped elastic response spectra for seismic events with different return periods (72, 475 and 2,475 years). Response spectra, Sa(T), are reported in Fig. 8.5 for two different areas of the municipality.

8.4 Residential Buildings

The typological and vulnerability classification of the residential buildings followed a multilevel approach. The first level is based on the ISTAT 91 data, which counts 5,045 buildings within the whole target area, distributed in vulnerability classes as follows (Bramerini et al. 1995):

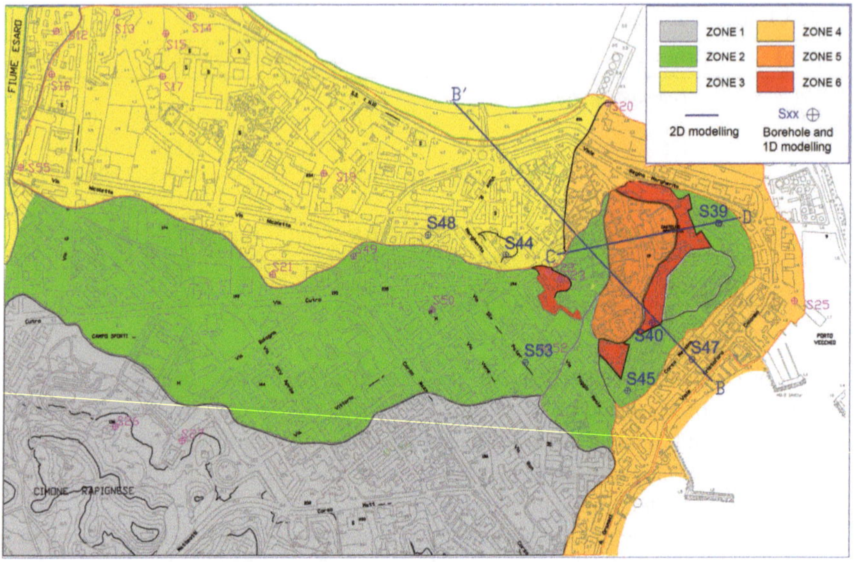

Fig. 8.4 Homogenous zones for site specific response spectra

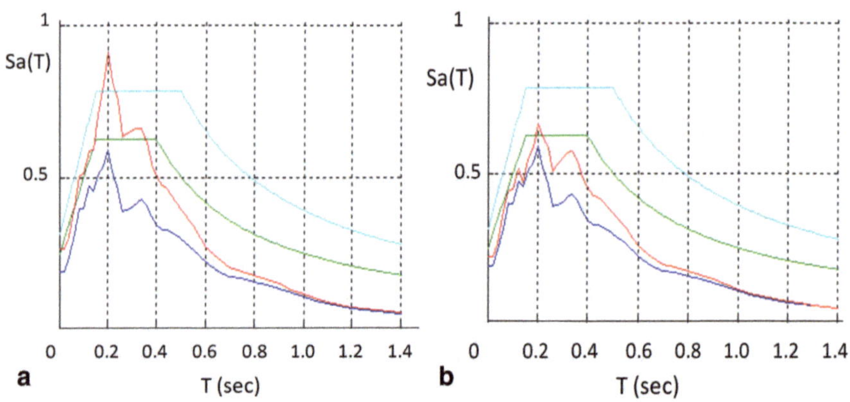

Fig. 8.5 Site specific response spectra for *Tr* 475 years. *Blue line* bedrock, *red line* site, *green line* 2003 seismic code on bedrock, *cyan* 2003 seismic code on soil type C. Left side refers to the *green* zone in Fig. 8.4, right side refers to the *red* area in Fig. 8.4

Masonry—Poor quality—A	7.9 %
Masonry—Medium quality—B	5.9 %
Masonry—Good quality—C	3.8 %
Reinforced concrete	82.4 %

At the second level, the whole Provincial territory has been subdivided into 22 homogeneous zones or macroareas, of which only 13 belong to the target area of the Crotone Project. In every macroarea, the distribution of the typological features, such as the construction period, the number of floors, the average floor surface, the main seismic vulnerability factors, was collected.

Fig. 8.6 Masonry building in the historical core

At the third level, every building has been surveyed and data have been collected with an inspection form, specifically drafted for this project, containing various information on the resistant structure, floor surface and vulnerability factors. In the survey, 19 local technicians trained on purpose (engineers, architects and surveyors) have been involved. Each survey team has been provided with a map containing the polygons of building aggregates and their numbering in the GIS (Palmieri et al. 2014). During the survey, technicians subdivided the aggregates into buildings according to the actual building configuration. This was particularly complicated in the historical core of Crotone, where masonry buildings were built in the past as the result of a process of building aggregation (Fig. 8.6). Outside the historical core, 6–7 storeys RC buildings prevail (Fig. 8.7).

The building data have been inserted in the same GIS where soil data were recorded. Buildings in the GIS were represented by polygons, obtained from aerophotogrammetric data. After raster-vectorial conversion, 4665 polygons were present in the target area, but only 4,234 were buildings. Technicians collected information on 2966 buildings. Collected data have been computerized in an ACCESS database and then linked to the GIS building polygons. By filling the fields 'Aggregate-code' and 'Building-code' in the inspection form, and by their concatenation, a unique association between the inspected buildings and the building polygons in the GIS was obtained.

Vulnerability has been attributed to each building, according to two different approaches: (i) damage probability matrix (DPM) (Braga et al. 1982), and (ii) capacity curves and response spectrum (Vulcano and Mazza 2014).

In the DPM approach, two main classes of buildings have been considered: masonry and reinforced concrete buildings. Within each class, 5 vulnerability classes have been defined: (1) low, (2) middle-low, (3) middle, (4) middle-high, and (5) high vulnerability. Basic vulnerability for masonry buildings has been based on vertical

Fig. 8.7 RC buildings in the surroundings

and horizontal bearing structures, while vulnerability modifiers have been based on masonry mortar, dimension and location of openings, chains, type of roofing, seismic upgrading or retrofitting. For reinforced concrete buildings, basic vulnerability has been based on construction age, while vulnerability modifiers have been based on planimetric and altimetric irregularity, presence of soft storeys, short columns, corrosion, seismic upgrading or retrofitting. When vulnerability data (level III data) were not available within the inspection form, the macroarea's characteristics (level II data) have been assumed. Figure 8.8 shows the spatial distribution of reinforced concrete and masonry buildings in the historical core and its surroundings. Figure 8.9 shows the distribution of reinforced concrete (2,990) and masonry (1,239) buildings into the 5 vulnerability classes in the whole investigated area.

The damage distribution for each vulnerability class and macroseismic intensity has been obtained through the so called Petrini model (Petrini and Guagenti 1989) assigning to each of the previously defined vulnerability classes a distribution of the vulnerability index. It is worth saying that the Petrini model provides an estimate of the economical damage (Damage Factor) when the building is affected by ground acceleration. In order to obtain the DPM, that is the physical damage when the building is affected by macroseismic intensity, the damage factor has been converted into physical damage, according to Di Pasquale and Goretti (2001). Similarly the ground acceleration has been converted into MCS macroseismic intensity, according to Margottini et al. (1987).

The physical damage is expressed in a scale that ranges between 0 and 5, according to the EMS-98 macroseismic scale. $D=0$ is associated to the intact building, while $D=5$ is associated to the totally collapsed building. However, in our approach, it is the only damage classification that follows the EMS98 scale, since the seismic

Fig. 8.8 Building type in the historical core and its surroundings

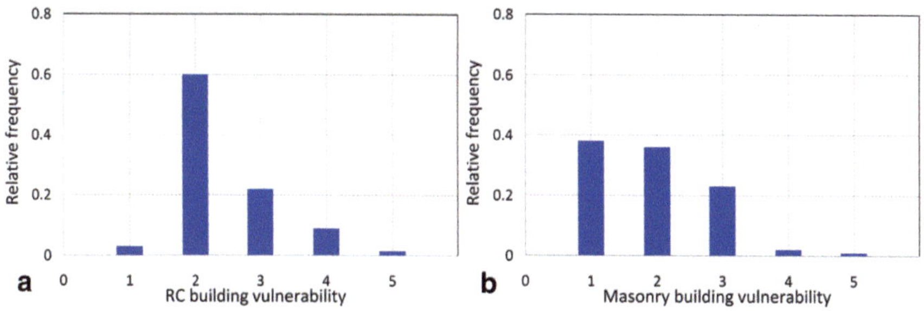

Fig. 8.9 Distribution of RC (2,990) and masonry (1,239) buildings into the 5 vulnerability classes

intensity follows the MCS scale. In other words, the seismic vulnerability is expressed in terms of DPMs that link the MCS macroseismic intensity to the physically observed damage, defined as in the EMS98 scale. This is due mainly for historical reasons, being the Italian catalogue expressed in terms of MCS intensities, and it is not an inconsistency since damage grades are not defined in the MCS scale.

Even if most of the reinforced concrete buildings were built in the 1960's and 1970's, they proved to be less vulnerable than the masonry buildings. This is due to the poor quality of the masonry and to the scarce maintenance of the masonry buildings. It has also to be said that, on average, the Petrini model provides, for reinforced concrete buildings, a lower vulnerability index and hence a lower vulnerability.

8.5 Population, Strategic Buildings, Infrastructures and Lifelines

Residential population has been obtained from the national census (ISTAT 2001) that provided for each census section the number of resident men and women. The total residential population in the target area is about 50,000 people. Within each census section, the population has been associated to the residential buildings on the basis of the ratio of the building surface to the total building surface within that section.

All public buildings have been georeferenced in the GIS. An initial analysis was based on the LSU1 database (MLPS et al. 1999), which counted 290 buildings. This initial database has been later updated because of the many differences that were found, above all due to delocalized functions. A specific vulnerability assessment has been carried out for a restricted number of relevant buildings: local authorities have considered schools as the key category for this type of analysis. In particular, the assessment involved 7 local structural engineers, specifically trained on the assessment of reinforced concrete buildings. Each engineer analyzed 2 school buildings. The assessment has been carried out through non-linear static analysis and SeismoStruct (2004) software. The vulnerability of the remaining strategic and relevant buildings has been assessed in the same way as for the residential buildings.

The road network has been defined through a graph containing 985 nodes and 1,220 oriented arcs. Nodes and arcs have been surveyed with a first level form containing information on street and sidewalk width, and on infrastructures such as bridges or retaining walls. Every single element has been implemented in the GIS. Road graph analysis will allow evaluating the safe routes to recovery areas.

Similarly the water supply network has been implemented in the GIS through a graph containing 1,027 arcs. Information, such as pipe material, pipe diameter and depth, has been associated to each arc.

8.6 Seismic Scenario

Seismic scenarios have been obtained through a specific GIS application. The analysis has been carried out for residential buildings, strategic buildings and population. While waiting for mobility flow data, the seismic event hour has been assumed to be at night, considering all the population present in the residential buildings. Seismic scenarios have been elaborated in terms of macroseismic intensity in MCS scale. In each homogeneous zone, with respect to the expected response spectrum, the effective peak acceleration (EPA) has been initially determined:

$$EPA = \frac{1}{25}\frac{1}{(T_2 - T_1)}\int_{T_1}^{T_2} S_a(\xi = 0.05, T)dT. \tag{8.1}$$

In the above equation S_a represents the 5 % damped acceleration spectrum, while T_1 and T_2 have been respectively assumed as 0 and 0.5 s. The EPA values have been

Table 8.2 Losses for different scenarios

	Tr = 72 yrs	Tr = 475 yrs	Tr = 2,475 yrs
Unusable buildings	236	1,595	2,700
Not repairable buildings	3	214	765
Collapsed buildings	0	93	502
Buildings with D = 0	3,505	583	0
Buildings with D = 1	340	852	219
Buildings with D = 2	327	1,757	1,558
Buildings with D = 3	55	734	1,477
Buildings with D = 4	1	210	473
Buildings with D = 5	0	93	502
Average damage (0–5)	0.4	2.4	3.8
Injured	2	882	4,103
Victims	0	202	930
Homeless	2,250	17,132	30,788
Relative cost of repair	0.01	0.15	0.34
Repair cost (million Euro)	37	485	1,109

then converted into macroseismic intensities according to the relation (Decanini et al. 2002):

$$\log_{10} EPA = 0.594 + 0.197 I_{MCS}. \tag{8.2}$$

For the three seismic events with return period $T_r = 72$, 475 and 2,475 years, the maximum macroseismic intensity in the target area was found respectively equal to I_{MCS} = VII, IX and X.

Primary building vulnerability in terms of physical damage has been obtained through DPM as described in the vulnerability section. The possible consequences of the physical damage (secondary vulnerability) have been defined in terms of functional and economic damage (Di Pasquale and Goretti 2001). The number of unusable, non-repairable and collapsed buildings and the number of homeless, injured and other victims have been evaluated.

Table 8.2 reports the scenario results in terms of losses for the whole target area. The seismic event with return period $T_r = 72$ years shows consequences that can be faced with local resources (Galanti et al. 2005). The seismic event with return period $T_r = 475$ years cannot be faced with local resources and needs additional regional, or most probably national, resources. The seismic event with return period $T_r = 2,475$ years is very catastrophic, and it is probably not suitable to carry out an urban planning in the face of such an event. From Table 8.2 the great number of homeless can be highlighted. In the seismic events with return period $T_r = 72$ and $T_r = 475$ years, the homeless respectively amount to 5 % and 35 % of the total population in the target area. This result is consistent with the observed homeless in recent Italian seismic events. The results of the scenario can be shown per single building or they may be aggregated per census section or homogeneous zones. In the first case, the model provides the probability of observing a well defined consequence, such as building unusability or collapse. In the second one, the model provides the expected

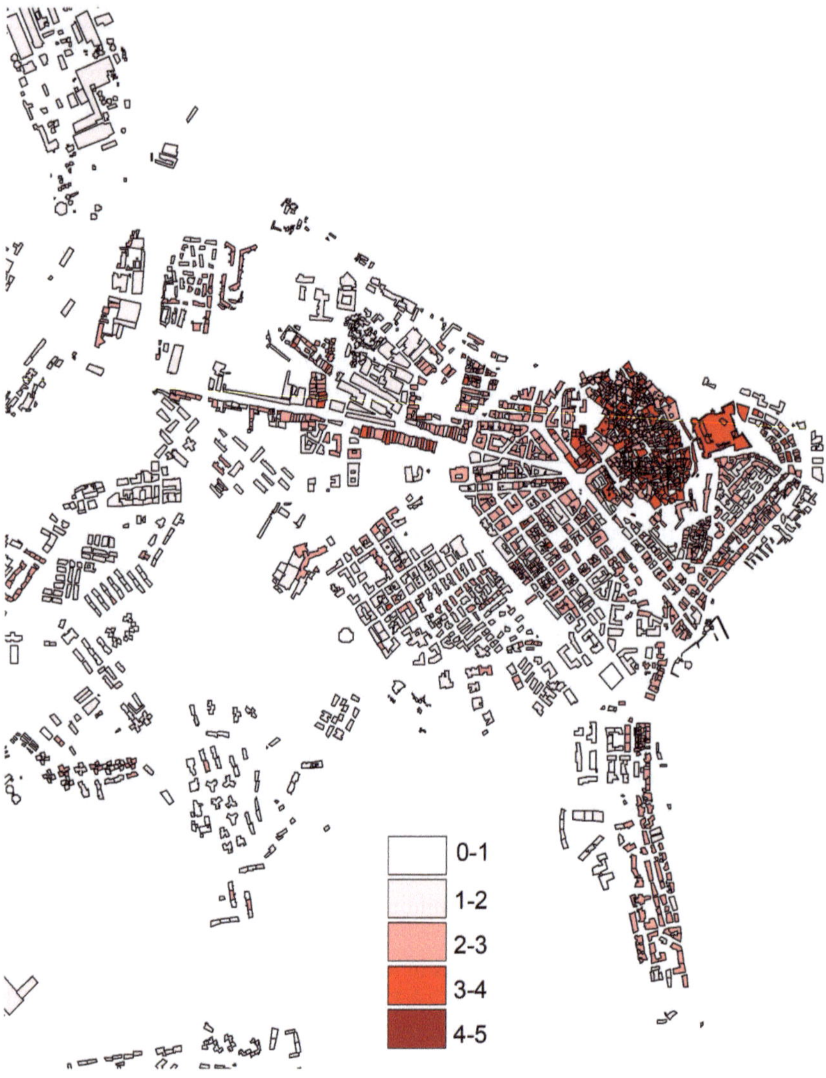

Fig. 8.10 Mean damage distribution in case of a 475 years return period event (0 = intact building, 5 = totally collapsed building)

number of buildings, in the aggregated area, with a well-defined consequence, such as unusability or collapse.

Figure 8.10 shows the mean expected damage in each building, expressed in a scale that ranges between 0 and 5, in case of a seismic event with return period of $T_r = 475$ years. It can be seen that the most damaged buildings are located within the historical core, where masonry buildings, often with bad mortar, prevail. Results also

Table 8.3 Building use within the considered sub-systems

Sub-system	Building use
Residential	Residential buildings
Administration	Public Offices, Court
Emergency	COM & COC, fire brigade station, Carabinieri and police stations, army
Health	Hospitals (public and private), doctors on call
Education	Schools, university
Energy	Stations, sub-stations
Economic	Commercial or craftsmanship activities on more than 25 % of the floor area, industries
Cultural	Churches, museums, libraries

show that the zones where most of the homeless are expected are those characterized by high population density and not by high building vulnerability.

Road network analysis is still under way and will take into account the interaction between buildings and road network as a consequence of partial or total building collapses and of building shoring and propping. The analysis will be performed according to the model reported in Goretti (2005) that considers both the short term interaction in the road network connectivity and the long term interaction in the road network serviceability.

8.7 The Systemic Approach

In the systemic approach the following sub-systems have been considered: Residential, Administration, Emergency, Health, Education, Energy, Economy, and Cultural. The consequences to the Social system (victims, injured and homeless), evaluated in the seismic scenario as described in the previous paragraph, have not been included in the Urban System analysis due to the difficulties of considering both casualties and system functionality. Each building in the target area has been associated to at least one sub-system. In case of multi-function building it may happen that one building belongs to more than one sub-system. Examples would be the case of a commercial store on the first storey and residential dwellings on upper stories, or the case of a school that may become a Center for Emergency management (COM or COC) in case of disaster. The building functions considered in the different sub-systems are reported in Table 8.3.

In order to analyze the target area with a systemic approach, 19 homogenous areas have been identified. They have been defined differently from the macroareas. In fact, the latter ones are based on the only building types distribution in the area, while the former ones are based on the main functions performed in the area. The homogenous areas considered in the systemic approach are shown in Fig. 8.11.

Within each sub-system, each building has been ranked with an associated value ranging from 1 to 3, that reflects the importance of the functions that are accomplished in those buildings with respect to the sub-system activities (Masure and Lutoff 2003).

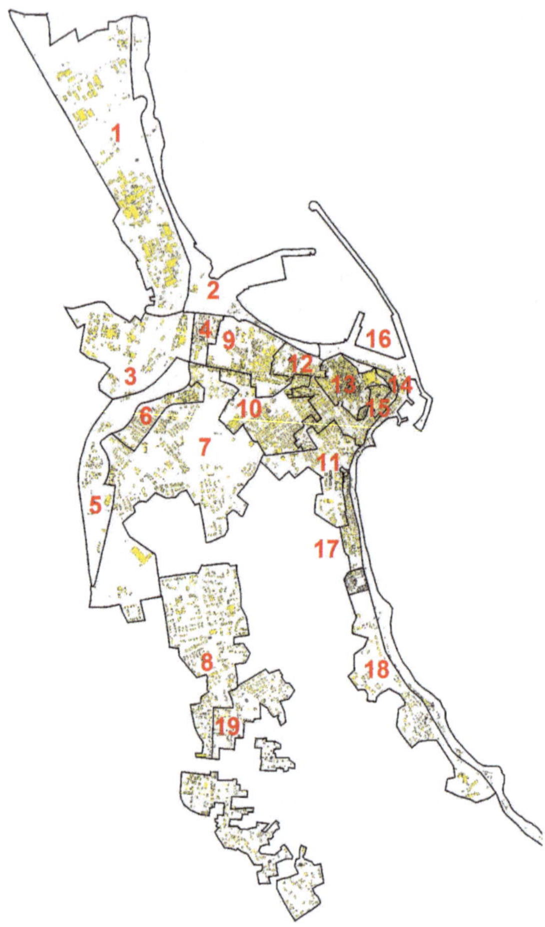

Fig. 8.11 Homogeneous areas (19) used in the systemic approach

For each sub-system, the relative importance of each homogeneous area with respect to the whole target area is initially evaluated in the pre-event phase. It is given by the following equation:

$$RI_{KS} = \frac{\sum_{j=1}^{Nb} I_{jks} w_{js}}{\sum_{j=1}^{Nb} I_{js} w_{js}}. \tag{8.3}$$

where RI_{ks} = relative importance of area k in sub-system s, N_b = number of buildings in the whole target area, w_{js} = importance of building j within sub-system s, $I_{jks} = 0$, but 1 if building j belongs to area k and sub-system s, and $I_{js} = 0$, but 1 if building j belongs to sub-system s.

The RI_{ks} values range continuously from 0 to 1. When an area k has a null RI_{ks} value in a specific sub-system s, it means that, in that area, the building functions are not associated to sub-system s, that is sub-system activities are not performed

Table 8.4 RI_{ks} values in the pre-event situation. *Yellow* cells indicate the prevailing vocation of each area

Area	Residential	Economic	Cultural	Health	Educational	Energy	Administrative	Emergency
0	*0.11*	0.00	0.00	0.00	0.00	0.00	0.00	0.00
1	0.00	*0.09*	0.00	0.00	0.03	0.00	0.00	0.00
2	0.01	0.06	0.03	0.00	0.05	0.00	0.00	*0.31*
3	0.00	0.00	0.00	0.00	0.00	0.00	0.00	0.00
4	0.02	0.01	0.00	0.14	0.01	0.00	0.00	0.00
5	0.00	0.02	0.00	0.00	0.00	0.00	0.00	0.00
6	0.00	*0.33*	0.00	0.00	0.00	0.00	0.00	0.00
7	0.10	0.03	0.02	0.03	0.17	0.00	0.18	0.00
8	0.12	0.02	0.02	0.02	*0.25*	*1.00*	0.00	0.00
9	0.01	0.00	0.02	0.00	0.08	0.00	0.00	0.00
10	0.08	0.10	0.02	*0.64*	0.09	0.00	*0.24*	0.00
11	0.05	0.02	0.13	0.03	0.04	0.00	0.18	0.00
12	0.14	0.14	0.00	0.07	0.04	0.00	0.21	0.00
13	0.04	0.04	0.07	0.03	0.10	0.00	0.00	0.00
14	*0.19*	0.04	*0.58*	0.03	0.10	0.00	0.09	0.22
15	0.03	0.01	0.00	0.00	0.00	0.00	0.00	0.00
16	0.04	0.00	0.08	0.00	0.04	0.00	0.00	0.00
17	0.04	0.01	0.03	0.00	0.00	0.00	0.09	0.19
18	0.00	0.08	0.00	0.00	0.00	0.00	0.00	0.28

in that area. On the contrary, when $RI_{ks} = 1$ it means that all the functions of that sub-system are located in that area. As a measure of the concentration of a specific sub-system function, the following indicator can be considered:

$$C_s = N_a Var_k[RI_{ks}]. \tag{8.4}$$

where N_a is the number of assumed areas and Var_k is the variance with respect to the areas. C_s is again an indicator that ranges from 0 (subsystem s equally spread on all areas) to 1 (sub-system s concentrated in one area).

The RI_{ks} values in the pre-event situation for Crotone municipality are reported in Table 8.4.

In the post-event evaluation, each building has a probability of being unusable due to the seismic event. The residual functionality of each building is then the probability of being usable after the event, that is one minus the probability of being unusable. The relative importance indicators for each area, in the post-event, $RI^p{}_{ks}$, can then be evaluated as:

$$RI^p_{ks} = \frac{\sum_{j=1}^{Nb} I_{jks} w_{js} p_j(usability/event)}{\sum_{j=1}^{Nb} I_{js} w_{js}}. \tag{8.5}$$

where P_j(usability/event) is the probability of building j to be usable after the seismic event.

While the sums over all the areas of the RI_{ks} in the pre-event phase, provide 1, the sums of the $RI^p{}_{ks}$, over all the areas, provide the residual functionalities of the

Table 8.5 RI_{ks} values in the post-event situation (Tr = 475 years). *Yellow* cells indicate the prevailing vocation of each area. The lower line indicates the residual functionality of each sub-system

Area	Residential	Economic	Cultural	Health	Educational	Energy	Administrative	Emergency
0	*0.09*	0.00	0.00	0.00	0.00	0.00	0.00	0.00
1	0.00	0.06	0.00	0.00	0.02	0.00	0.00	0.00
2	0.01	0.03	0.02	0.00	0.03	0.00	0.00	*0.21*
3	0.00	0.00	0.00	0.00	0.00	0.00	0.00	0.00
4	0.01	0.01	0.00	0.09	0.01	0.00	0.00	0.00
5	0.00	0.01	0.00	0.00	0.00	0.00	0.00	0.00
6	0.00	*0.24*	0.00	0.00	0.00	0.00	0.00	0.00
7	0.08	0.02	0.01	0.03	0.14	0.00	0.15	0.00
8	*0.09*	0.01	0.02	0.01	*0.19*	*0.79*	0.00	0.00
9	0.01	0.00	0.01	0.00	0.06	0.00	0.00	0.00
10	0.05	0.06	0.01	*0.44*	0.06	0.00	*0.16*	0.00
11	0.03	0.02	0.09	0.02	0.03	0.00	0.13	0.00
12	0.08	0.08	0.00	0.04	0.02	0.00	0.13	0.00
13	0.03	0.03	0.05	0.02	0.08	0.00	0.00	0.00
14	0.06	0.02	*0.12*	0.02	0.04	0.00	0.05	0.04
15	0.01	0.00	0.00	0.00	0.00	0.00	0.00	0.00
16	0.02	0.00	0.03	0.00	0.02	0.00	0.00	0.00
17	0.02	0.00	0.01	0.00	0.00	0.00	0.06	0.11
18	0.00	0.05	0.00	0.00	0.00	0.00	0.00	0.18
	0.59	0.65	0.38	0.68	0.71	0.79	0.67	0.54

sub-systems. The actual model considers the same w_{js} in the pre- and in the post-event phase, although in principle they can differ due to the different importance of the functions performed in normal time and in emergency in the same building. Moreover, at the moment, no interaction among sub-systems is considered.

The RI_{ks} values in the post-event situation (T_r = 475 years) for Crotone municipality are reported in Table 8.5, together with the sums of the $RI^p{}_{ks}$ (last row). Comparing Table 8.4 and Table 8.5, it can be deduced that, while in the pre-event phase the most important area for the residential sub-system was the area number 12, the earthquake causes areas 1 and 8 to become more important. In the other sub-systems the areas that were prevailing in the pre-event are still prevailing in the post-event, but with a residual functionality that ranges between 0.38 and, showing that certain sub-systems are more vulnerable than others. The results of this application reflect the higher vulnerability of the cultural sub-system, as shown also by recent Italian earthquakes.

8.8 Conclusions

The Measure 1.1 'Safety of urban settings exposed to seismic risk' of the PIC URBAN II program is directed to the seismic risk assessment and mitigation of the Crotone urban area. The activities focused on different subjects related to the seismic assessment of the urban system, including: hazard analysis and selection of scenario seismic events, identification of homogeneous areas for the determination of the site-specific ground motion (site effects), inventory of residential and strategic buildings, inventory of infrastructure and lifelines, assessment of physical vulnerability of residential and strategic buildings, evaluation of the functional and economic consequences of physical damage on individual objects and network systems. All the collected data and all the results of the analysis have been implemented in a dynamic GIS system.

All activities were performed in close collaboration between the Municipality of Crotone, the Seismic Risk Office of the Italian Civil Protection Department, and the University of Calabria. Many local technicians such as geologists, surveyors, architects, engineers and computer scientists have also been involved in the project. The greatest difficulty was found, as it often happens in such projects, in the collection, standardization and computerization of data.

The results of the analysis indicate that the frequent earthquake ($T_r = 72$ years) has no consequences, except in terms of homelessness, and can be addressed with local resources. The rare event (Tr = 475 years) has a greater impact, both in terms of casualties and economic losses, and demands for additional, regional or national resources to cope with the disaster.

A holistic model able to handle the urban center seismic behavior has been then presented. The methodology is able to predict, with the help of a systemic approach, the status of the economic, social, cultural and administrative components of the urban centre in normal growth and in crisis period. This is essential in order to evaluate the ability of the city to fulfill its functions. To this end, the following main sub-systems were considered in the analysis: Residential, Cultural, Health, Education, Energy, Economic, Administrative and Emergency. The consequences on the Social system (casualties) have also been evaluated, but not included in the Urban System due to the difficulties of considering both casualties and system functionality. The application shows that very detailed data are required for this type of analysis.

A positive result of this project was the significant raising of awareness in the local administration and in the local technicians on seismic assessment and mitigation programs. The knowledge transfer to local technicians proved to be very effective. This was made possible through the continuous involvement of local actors in the project.

References

Bazzurro P, Cornell A (1999) Disaggregation of Seismic Hazard. Bull Seism Soc Am 89:501–520

Bramerini F, Di Pasquale G, Orsini G, Pugliese A, Romeo R, Sabetta F (1995) Rischio sismico del territorio italiano. Proc. VII Italian Conference on Earthquake Engineering, Siena, Italy

Braga F, Dolce M, Liberatore D (1982) A Statistical Study on Damaged Buildings and an Ensuing Review of the MSK-76 scale, 7 ECEE, Athens, 65–84

Cornell CA (1968) Engineering seismic risk analysis. Bull Seism Soc A 1–86, m., 58, 1968: 1583–1606

Decanini LD, Mollaioli F, Oliveto G (2002) Structural and seismological implications of the 1997 seismic sequence in Umbria and Marche, Italy. In: Oliveto G (ed) Innovative approaches to earthquake engineering. WIT press, Italy, pp 229–323

Di Pasquale G, Goretti A (2001) Vulnerabilità Funzionale ed Economica negli Edifici Residenziali Colpiti da Recenti Eventi Sismici Nazionali, Proc. X Italian Conference on Earthquake Engineering, Potenza-Matera, Italy (in Italian)

Galanti E, Goretti A, Foster B, Di Pasquale G (2005) Managing civil protection. In: Oliveira CS, Roca A, Goula X (eds) Assessing and managing earthquake risk. Kluwer ed, The Netherlands

Galli P, Lucantoni A (2003) Analisi di pericolosità sismica con metodi ibridi dell'area crotonese, SSN Internal Report (in Italian)

Galli P, Lucantoni A (2014) Pericolosità sismica di Crotone. In: Rubettino (ed) Il progetto Crotone (in press) (in Italian)

Giordano F, Naso G (2014) Risposta sismica locale del centro storico di Crotone. In: Rubettino (ed) Il progetto Crotone (in press) (in Italian)

Goretti A (2005) A probabilistic model able to handle interaction among road network, buildings and emergency services in urban areas, IX International Conference on Structural Safety and Reliability, 19–23 June, paper 366

Goretti A, Naso G, Giordano F, Vulcano A, Mazza F (2007) Scenari sismici a scala urbana. I primi risultati del "Progetto Crotone", Proc. XII Italian Conference on Earthquake Engineering, Pisa, Italy (in Italian)

ISTAT (2001) 14 Censimento Generale della Popolazione e delle Abitazioni. http://dawinci.istat.it/MD/

Margottini C, Molin D, Narcisi B, Serva L (1987) Intensity vs acceleration: Italian data, Proc. Workshop on Historical Seismicity of Central-Eastern Mediterranean Region, Rome, 27–29 October 1987

Masure P, Lutoff C (2003) Report title: "Methodology on Urban System Exposure (USE) Assessment to Natural Disasters", Report number, 82 pages (30 Figs., 14 Tables, 1 Appendix)

MLPS DPC, GNDT (1999). Censimento di Vulnerabilità degli Edifici Pubblici, Strategici e Speciali nelle Regioni Abruzzo, Basilicata, Calabria, Campania, Molise, Puglia e Sicilia, Progetto Lavori Socialmente Utili (D.L. n 232 del 14.06.95) (in Italian)

Mucciarelli M, Albarello D (2011). Profili di Vs da misure di microtremori. In: Rubettino (ed) Il progetto Crotone (in press) (in Italian)

Nakamura Y (1989) A method for dynamic characteristics estimation of subsurface using microtremor on the ground surface. QR Railway Tech Res Inst 30:25–33

Ohori M, Nobata A, Wakamatsu K (2002) A comparison of ESAC and FK methods of estimating phase velocity using arbitrarily shaped microtremor analysis. Bull Seism Soc Am 92:2323–2332

Palmieri F, Adamo F, Berlingeri M, Palmieri L (2014) Il Sistema Informativo Territoriale del Comune di Crotone. In: Rubettino (ed) Il progetto Crotone (in press) (in Italian)

Park CB, Miller RD, Xia J (1999) Multichannel analysis of surface waves. Geophysics 64:800–808

Petrini V, Guagenti E (1989) Il Caso delle Vecchie Costruzioni: Verso una Nuova Legge Danni Intensità, Proc. IV Italian Conference on Earthquake Engineering, Milan, Italy (in Italian)

PROSHAKE (2000) EduPro Civil Systems, Inc. 5141 189th Avenue N.E. Sammamish, WA 98074 USA

QUAKE (2004) GEO-SLOPE International Ltd.1400, 633–6th Avenue S.W. Calgary, Alberta T2P 2Y5, Canada

Scerra A (2014) Il modello geologico-tecnico del sottosuolo di Crotone. In: Rubettino (ed) Il progetto Crotone (in press) (in Italian)

SEISMOSTRUCT (2004) Release 3.0, Seismosoft. www.seismosoft.com

Vulcano A, Mazza F (2014) La vulnerabilitàdegli edifici residenziali. In: Rubettino (ed) Il progetto Crotone (in press) (in Italian)

Woo G (1996) Kernel estimation methods for seismic hazard area source modelling. Bull Seism Soc Am 86(2):353–362

Part III
Risk Management

Chapter 9
Principles for the Evaluation of Risk Governance and the Measurement of Performance

Urbano Fra Paleo

Evaluation flourishes in eras of accountability...

Noblit and Eaker (1989)

Abstract Risk governance is an evolutive step from risk management propelled by the increasing complexity of social and natural processes, and interactions between the two systems to deal with uncertainty. We should not look to governance evaluation as a straightforward process of performance measurement and neutral and objective exercise, but as a complex social practice where multiple and diverse interests meet to reach a consensus based on formally established criteria. The evaluation of risk governance should be a process of dialogue and consensus among different stakeholders that will serve to identify the weaknesses and strengths of a shared political system and of each policy, programme and organization that contributes to risk reduction or escalation in a certain region. In order to serve these purposes, evaluation of risk governance has to adopt a participatory approach where fundamental dimensions, components and criteria of risk governance are examined; in order to assist social and organizational learning and increase risk awareness of all stakeholders.

Keywords Public participation · Evaluation · Policy · Agents · Actors · Lessons learned

9.1 Introduction

Governance is not just about governing; it concerns the dynamic patterns, relationships, and processes resulting from the decentralization of decision-making with multiple actors intervening in a changing context of increasing complexity and uncertainty (Fra Paleo 2010, 2011). The reflexive society is involved in political processes which explicitly set goals and avenues to achieve them. This viewpoint evokes the

U. Fra Paleo (✉)
Land Laboratory, School of Engineering, University of Santiago de Compostela,
Campus universitario, 27002 Lugo, Spain
e-mail: urbano.fra@usc.es

M. Bostenaru Dan et al. (eds.), *Earthquake Hazard Impact and Urban Planning*,
Environmental Hazards, DOI 10.1007/978-94-007-7981-5_9,

© Springer Science+Business Media Dordrecht 2014

concept of governance as a continuous political process formulated by Jachtenfuchs and Kohler-Koch (2003).

Despite concerns about the misuse or abuse of the term -Jachtenfuchs and Kohler-Koch (2003) sustain that "Usually, a rapidly increasing usage of a concept goes hand in hand with a loss of precision of its meaning"-, it has evolved to become a paradigm for political analysis and practice, and has an identifiable interdisciplinary character. These authors argue for narrowing the concept down and consolidating some basic principles to differentiate between a general use of the concept of governance in discourse and the narrower academic use: neither should all kind of goal-oriented actions be included in the notion nor should the concept of governance determine its substantive outcomes (Jachtenfuchs and Kohler-Koch 2003); a perspective whereby, setting differences between good and bad governance, might inspire dispensable moral values (Fra Paleo 2010). Ironically, the first viewpoint -adopted in the corporate sphere- has revitalized the concept while the latter has had an impact on advancing of accountability and financial governance in developing countries. The third principle put forth by Jachtenfuchs and Kohler-Koch, regarding a restriction of the domain of the nation-state, seems to disregard those global processes taking place in the international arena which are shaping global governance, especially in the economic and the environmental sectors, along with the devolution of power and responsibilities to regional governments. Governance is thus no longer national but multi-level (Marks 1992).

Governance is the multifaceted, systemic approach adopted to deal with complexity and rapid change based on the principles of multi-level, multi-agent, and multi-sectoral decision-making. Although Bache and Flinders (2004) contend that the concept of multi-level governance already holds both vertical and horizontal dimensions, with the second they actually refer to multi-agent relationships –between institutional and civil society organizations- but not to intra-governmental interactions. This implies that the notion of multi-level governance is not comprehensive by any means and should just refer to the vertical interface.

Risk governance is "...the translation of the substance and core principles of governance..." (Renn 2008) to the mitigation of risk and the reduction of disasters. The heart of risk governance is, without doubt, disaster response as effective response to disasters happens to be a priority for communities after a catastrophic event. Relentlessly, increased complexity in disaster response developed into more organized disaster management, which unfolded to emergency preparedness and management after plans and appropriate resources were arranged to respond more effectively (Fra Paleo 2005). When pro-active measures began to be incorporated, all policies focusing on risk were framed within the context of risk management, then conceptually integrated -together with other political processes- in the framework of risk governance (Fig. 9.1). The weaknesses shown by former, more restrictive and reactive frameworks, along with the observed increasing human and economic costs of disasters have forced a gradual broadening of the scope to go beyond emergency preparedness. Still, current modes of risk governance are characterized by skewed power relationships with a strong emphasis on emergency preparedness and response as virtually unique dimensions.

Fig. 9.1 Unfolding of risk governance

Governance does not implicitly contain evaluation, for each of the multiple modes of governance emphasizes particular combinations of constituents. Although reciprocal feedbacks and influences between decision-making and evaluation substantiate a mutualistic interaction, benefiting organizational and societal learning, but just evaluation could not exist independently from governance. Their close association and unbalanced dependency speaks in favor of a symbiotic mutualism. Treib et al. (2007) propose two policy dimensions to make things clearer: the type of instruments applied and the approach to their implementation, and leave aside the politics and polity dimensions. And, although not explicitly as well, because the politics dimension of actors participating is not taken into consideration for purposes of classification, every mode of governance (coercion, targeting, framework regulation, and voluntarism) reveals a certain capacity for incorporating evaluation as a measure of the level of implementation, or as a built-in regulatory tool.

9.2 Evaluation

Often misunderstood, the concept of reflexive modernization refers -according to Beck et al. (2003)- to an increased radicalization of modern society; not only to a transformation of the inherited institutions but also of the principles of modern society, in transition towards a reflexive second modernity. This concept of reflexivity was firstly proposed by Giddens (1986) who understood it "... as grounded in the continuous monitoring of action which human beings display and expect others to display." It represents a rationalization of action in the author's framework and, hence, monitoring and evaluation become intelligible and operational ground for human action. However, rationality is construed by social groups –particularly elites- and may well be biased towards the needs of decision-makers or, alternatively, to the demands of varied stakeholders and constructed in a multi-rational mode (Hanberger 2006), if provisions are made to perform democratic evaluation.

With a different point of view, Campbell (1988) looks –in an essay formerly published in 1971- at innovative policies, designated as experimental, as political processes which should be evaluated in order to get feedback and understanding of what works and what does not, what is accepted or rejected, and what should be revised. Evaluations can be thus understood in the context of Campbell's experimenting society. Hanberger (2006) sums up a different perspective to the reflexive and the experimenting society and thinks that, in the haste of change, those emerging policies give rise to uncertainties regarding the roles of organizations and actors taken over and, accordingly, accountability is increasingly demanded by societies to provide a means of checking. Decision-making requires, in the context of governance, feedback processes and bonds between system constituents which act as a structuring element of the control system. Evaluations return information of and to the political processes, helping to modulate their operation. The more multiple, solid, dynamic and democratic they are, the more robust governance will become. Because governance has an intrinsic systemic nature, evaluation can measure both structure and function, components and processes, when assessing the multiple dimensions of risk governance.

Evaluation is about measuring performance. This is its primary aim when it is being designed or implemented, although likewise, it may well serve other basic functions. The evaluative output assists in checking whether policy goals have been achieved or whether programs operate successfully. Thus, it contributes to improving policy design and program implementation, prioritization -particularly in drawing up budgets (Mackay 1998)-, enhancing the institutional setting and the organizational capability, and identifying the changes the country or region being evaluated should put into practice to reorganize governance. It should be noted that whether standalone, or non-periodic, evaluation does not lead to significant improvement and only multiple, accumulative evaluations help to track progress and lead to evolutionary governance. In this framework performance is understood as goal-oriented rationality, which allows advancing the efficiency and effectiveness of public policies, i.e. achieving the planned objectives at the minimum cost possible. In his paper on performance management, Davies (1999) rejects a narrow interpretation, and calls for expanding the notion of performance from program to organizational so as to include the environment when policies and programs are created, approved, scheduled, funded and implemented.

In order to respond to address the aims of evaluation, additional questions regarding the evaluative process shortly arise, particularly referring to the method, functions of evaluation and implementation of results. The CDCP (1999) has formulated these as three questions:

- What is the best way to evaluate?
- What is being learned from the evaluation?
- How will lessons learned from evaluations be used to make public efforts more effective and accountable?

There is a wealth of literature on program evaluation and approaches supporting varied methods, but much less can be found referring to governance evaluation. Besides, a profusion of categories is used to measure performance (Mackay 1998), such as the measurement of projects, programs or sectoral policies, which all fall under the broad concept of monitoring and evaluation (M&E). To respond to the first challenge CDCP (1999) formulates further expedient questions that help to set out the path of governance evaluation:

- What will be evaluated? i.e. the object of evaluation
- What aspects of the program will be considered when judging program performance? i.e. the dimensions to be evaluated
- What standards must be reached for the program to be considered successful? i.e. criteria
- What evidence will be used to indicate how the program has performed? i.e., level of performance
- What conclusions regarding program performance are justified by comparing the available evidence to the selected standards? i.e. evidence details.

However, if we wish to completely clarify the evaluation method we still need to identify and describe the actors under evaluation: who the evaluators are, what roles they play, how they interact; and, as pointed out by Treib et al. (2007), this brings politics in.

9.2.1 The Functions of Evaluation

Engel and Carlsson (2002) contend that governance evaluation is helpful because it focuses not only on decision making processes but also on institutional development to improve societal learning, which are two of the key functions of evaluation. Mackay (1998) maintains –by adopting a narrower definition of the concept of governance- both international organizations and the development community recognize the importance of institutional development and capacity building to increase the efficiency of development policies.

However, a distinction should be made between the subsidiary functions of the evaluative process and those served by the evaluation output, because they fulfill different purposes at different times, short and mid-term, multiplying the number of functions. Hanberger (2006) exhaustively enumerates a range of functions: rational feedback, political legitimation of government, justification of policies, empowerment of social groups, enhancement of political debate, and societal and institutional learning. However, rational feedback should not be considered a side or parallel effect but the core objective of evaluation, as it has been described. Subsequently Hanberger summarizes the functions of evaluation as three: rational feedback, enhancement of learning and accountability.

So, to clarify, a clear distinction must be made between democratic progress as an aim of evaluation and improved policy efficiency and effectiveness, the accountability of policy-makers and managers regarding operations performance, and transparency as goals of evaluation and finally, performance measurement as the objective of evaluation.

Evaluation is also part of the strategy in response to the declining trust in state in general terms and government in particular, conceived by the ruling political power to increase legitimacy and restore trust (Hanberger 2006). Conversely, other social actors or political parties may use output as an argument against adopted policies, and see evaluation as an opportunity to delegitimize government or, in some instances, the state itself –arguing against the administration's inefficiency in resolving problems and providing services.

Similarly, mixed and divergent goals are recognizable in the context of political justification, as we will see when referring to the functions of output. They range from simple justification of previously made decisions to more complex validation of the appropriateness of measures, and support for binding actions particularly when adopted in a multi-actor context (Hanberger 2006). In the context of public participation, or of participatory democracy in broader terms, evaluation turns out to be a platform for incorporating disadvantaged groups in decision-making or even promoting their self-determination (Hanberger 2006).

9.2.2 *The Functions of Evaluation Findings*

Engel and Carlsson (2002) believe that –apart from feedback enhancement- the goals of evaluation go much further, the principal outcome being learning at the organizational and societal levels -rather than at the individual level. But in Morabito's opinion (Morabito 2002) few evaluation methodologies seek to learn from the evaluative inquiry process or benefit from integrating the knowledge and beliefs of stakeholders. The application of evaluation outputs does not have a history of success, and as such questions the applicability of results. Very skeptically, Stame (2006) points out that the track record is dismal, and puts forth factors such as the adoption of resistance strategies by public administrators under evaluation, the goal displacement of the process and delayed cognitive advancement, as the root causes for past failures. However, Rose (2008) believes that progress can be recognized both on tangible advances in performance and in "... sufficient evidence of success in efforts to do so ...", substantiated by the effects of diffusion of innovation.

More comprehensively Engel and Carlsson (2002) identify four principal applications of the evaluation results, discounting disuse. The first two, the ritual and legitimizing uses are of particular concern to scholars and practitioners, because they possibly reflect a misuse of the output. The former describes a symbolic condition of evaluation, which serves as an act of propaganda and hides a lack of any political will. Concerning legitimation, Engel and Carlsson refer to the use of evaluation as a façade, concealing approaches and decisions already made in a disconnected political arena, or as a validation of the appropriateness of the measures taken. O'Sullivan and D'Agostino (2002) question the collaborative role of program personnel in using the output of evaluation. Since staff is not commonly involved in evaluative processes, administration personnel usually do not understand or do not have access to evaluation output and, therefore, are not aware of the lack of genuine impulse. Engel and

Carlsson (2002) also categorize two more –supportive- uses of evaluation findings. Output from the process may be conceptually used, supplying thoughts and ideas that help to adapt structures and processes but without completing a radical transformation. The highest level of application consists of an instrumental utilization of results, testing their conclusions and mainstreaming practices to the conclusions. In terms of non-utilization causes are manifold, but the lack of understanding of its applicability in particular, a shortness of political relevance for policy-makers or simply unawareness of the process stand out.

Success factors and barriers play an important role in speeding up or slowing down the implementation of results (Mackay 1998), or even the realization of repeated evaluations. Among the success factors are the leading role of innovators –coming from finance or planning departments- in encouraging or supporting sustained evaluation, for instance by way of incentives, having a multi-sectoral approach and making evaluation a part of the decision-making course of action (Mackay 1998). While progress depends on individual organizational action, the political environment may disincentivate evaluation, particularly within a culture which lacks auditing, accountability, and performance measurement (Mackay 1998), and ultimately lead to a rejection by decision-makers, arguing against its poor usability and high cost (Davies 1999). This stance reflects an elitist attitude that mistrusts sharing power and accountability and, consequently, favors disuse or weak support, providing fresh arguments for abandoning the evaluative practice.

9.3 The Political Nature of Evaluation. Evaluation as a Political Act

Noblit and Eaker (1989) reject the supposedly appealing idea of evaluation as intrinsically neutral and objective, of being just a technical issue, and strongly support its political dimension, arguing that "... not only the outcomes of evaluation, but the evaluation process itself is political and the decision to subject a program and its participants to evaluation is itself a policy decision." Inspired by the logic of positivism -Davies (1999) calls for an "independent input"-, some social scientists think evaluation can be performed on the basis of the scientific method, and distance themselves from the political dimension (Greene 2005). But Noblit (1984) contends that it is "... inherently political because it wishes to establish the bases of judgment for others and moreover to replace those that might otherwise be employed ..." In this sense it is evolutionary, for it alters the foundations for decision-making.

Despite the political potential of evaluation in strengthening policy design, enhancing organizational capability and societal empowerment, Hanberger (2006) questions whether it is always used as a democratic instrument. Leaving aside the kind of evaluation approach implemented, Hanberger believes that the political context plays a major role in the instrumentalization of evaluation for deflecting public perception away from hot issues and serving hidden political agendas. Evaluators are constrained by the concerns and interests of evaluatees or members from the evaluated organizations, and are asked to negotiate the questions to be addressed and the interests to be served (Greene 2005). This certainly makes their task more difficult.

Far from some expert views and decision-maker presumptions, evaluation is not simple nor straightforward but a complex social process where results will not be easily and directly usable (Noblit and Eaker 1989). Since the functions of evaluation are not limited to only measuring the merit or worth of programs (Greene 2005), describing their significance for various stakeholders but also valuing and judging, again here –according to Greene- evaluators enter the political arena because they adopt standards and criteria to deliver judgments. Apart from their knowledge, skills and abilities, participants bring their personal beliefs and values, their own interests -what Patton (1997) terms adaptative character- and hidden agendas to the process, and take into consideration and appraise the consequences evaluation might have for themselves and other parties alike. Uphoff and Combs (2001) warn against what they call paradigm traps, since paradigms that are instrumental in facilitating our understanding of the environment also may become a barrier for innovation and the acceptance of progress. It would probably be helpful to let stakeholders unlearn what they know, but this would not be simple and evaluators might not be willing to do so.

Finally, results from social research and evaluations alike are used selectively by groups as political ammunition (Weiss 1977) in support of, or as a critique of, current practices to support particular stances. The implications are particularly intricate when policy-makers receive results which point to limited goal achievement and, therefore, do not gain extra support (Hanberger 2006) from participating stakeholders nor find viable directions to take.

That is why Sanders (2002) believes evaluation is a marginal political activity, not integrated within the daily operations of standard policy design and political processes, because it challenges the contemporary *status quo* and institutional settings. Until governments comprehend the merits and the strength of evaluation, it "is unlikely to be sustainable" (Mackay 1998). While the need for evaluation grows, two opposing pressures are anticipated (Stame 2006). One that pushes for advancement in evaluation and the other which seeks to undermine the scope of future evaluations. Davies (1999) contends that decision-makers are in this last sector, arguing evaluations do not meet the expectations regarding the application of results and are too costly. Disuse of evaluations being the ultimate result.

Therefore, mainstreaming evaluation in decision making should be a priority for the advancement of governance, and this implies firstly demonstrating its value to governments and society and secondly, its contribution to politics, although efforts made to encourage awareness have hitherto failed (Sanders 2002). The opportunity for institutionalizing evaluation by assigning additional human or material resources to build capacity, instead of effectively embedding it in the culture of governments and organizations, has been warned against by Sanders (2002). This author puts forth three levels of integration: capacity building, institutionalization and mainstreaming, linked to organizational culture and development. The highest level of integration made by a system entails understanding its value contribution to the improvement of the organization or enhancing its effectiveness through repeated processes as an accepted behaviour, and using its results in policy formulation and implementation. In order to move forward in having evaluation as a core value, Sanders (2001) proposes five stages, ranging from baseline awareness of its importance to the organization, to evolved self-sustained design, organization and structure; and including compliance, obligation and commitment.

From this political standpoint three dimensions of evaluation are identifiable. The first notion is that evaluation is embedded in politics and intrinsic to manifold political processes, secondly, that it revolves around decision-making from start to end, and finally, that evaluation is the basis for decision making, for building options. This implies that, far from being simple or straightforward, evaluation is sinuous and subjective, for multiple stakeholders intervene in a single or in multiple phases endorsing the conflicting interests at stake. Once gathered, participants have to negotiate and bargain in order to reach a consensus judgment. The major output from evaluation will be a report which may include evidences, conclusions or recommendations. But Noblit and Eaker (1989) contend that a more salient outcome is a political rearrangement of the parties that will be the primary basis for future action. Differences among evaluators can be distinguished when analyzing the roles played during or at each stage of participation. As Smith (1989) states "... no facts without values, and different values can actually lead to different facts". Thus, output should not be a simple aggregation of the opinions of the participants but a consensus among different points of view, identifying agreements and disagreements. Every decision becomes intrinsically political.

Although measuring democratic performance and assessing governance may not be coincident processes, both actions are closely associated and some of the criteria used will be common or else inform about the other concept. The European Union is distinctive in this sense since it evaluates potential candidates for EU accession both in terms of governance and democratic character (Rose 2008). Thus, having democratic elections is a necessary but not a sufficient condition, as the Copenhagen criteria for admission have to be met too. These criteria are extensive (Rose 2008) and this means that they cover a wide range of governance dimensions. Instead of synthetic indices, Rose (2008) speaks in terms of profile, because "Unlike a democracy index, which reduces a country to a single score, a profile presents a rounded picture of a country's system of governance" based on the indicators used. Hanberger (2006) casts doubts over the contribution of any kind of evaluation to democracy, and furthermore, we might say governance, for it is very dependent on the approach adopted to measure.

9.3.1 The Societal Learning Approach. Evaluation as a Learning Process

Mackay (1998) believes that the main precondition for developing an evaluation is country demand. This is an indicator of a certain institutional framework and of actual incentives not easily found in developing countries. However evaluation processes are, either in favorable or unsuitable conditions, supportive statements for improving governance and opportunities for institutional (Engel and Carlsson 2002) and non-expert evaluator learning (O'Sullivan and D'Agostino 2002) and, accessed to learn about the evaluation process and gain new skills.

Engel and Carlsson (2002) judge that the growing emphasis on learning in the course of evaluation illustrates a shift in the goals of evaluation, since the role of the evaluator has been altered. "From a distant, research-oriented person trying to systematize the known and unearth the hidden, he or she becomes a process facilitator whose greatest skill is to design and organize others' learning effectively." (Engel and Carlsson 2002). Unquestionably, knowledge is not uniquely in the realm of academy or expertise –"we don't know how to do it all" and "... should be treated as a shared organizational resource rather than as an instrument of hierarchical power ..." (Davies 1999)— as, otherwise, local knowledge would be ignored. If the purpose is to facilitate societal learning the evaluation process must be inclusive and integrate all stakeholders, "recognizing their complementary roles in reaching development targets" (Engel and Carlsson 2002). The outcome is a strengthening of social capital and empowerment via the acquisition of new skills, upgrading the knowledgebase, enhancing problem-solving capacities, promoting cooperation and supporting group identity. However it looks like not all the evaluators can play that role of facilitator, and specific strategies have to be adopted in order to promote the learning of all members by separating roles and assigning an actor the function of facilitator in the transfer of knowledge.

9.3.2 Actors, Stakeholders and Public Participation

The key constituents of evaluation, according to Hanberger (2006), are evaluation focus, the inclusion of stakeholders, the process of exchanging ideas and the role of the evaluator. It is observable how major emphasis is put on the participating agents, as well as what to assess, for this is a critical decision in dealing with how to assess. Said approaches display multiple configurations of constituents with variable responsibilities assigned.

Noblit and Eaker (1989) have examined the strategic nature of evaluation credibility, highly dependent on how knowledge is produced and who generates knowledge. Thus, suitable evaluation design and the inclusiveness of key stakeholders are necessary — although not sufficient- factors for success. Little attention has been paid to "... develop commitment of participants to the evaluation ..." (Noblit and Eaker 1989), however, because evaluation is an innovative idea in terms of triggering new social processes and it is not only a question of the number of participants but mostly of the quality of participation. Some theorists also argue that trust among the participants is an essential precondition in order to perform successful evaluation and vigorous governance, whilst others contend that trust emerges throughout the evaluation process as the result of continued social interaction (Noblit and Eaker 1989; de Marchi and Ravetz 1999).

It seems critical to pay close attention to how the process is conducted and, above all, to draw attention to stakeholders' input, bringing together legitimate scientific learning and practitioner expertise, along with traditional and local knowledge. The interaction and interplay among groups of participants throughout the evaluation process will produce crossover and agreement but also bias. This could be the case when the sponsor and evaluatee are from the same organization, a circumstance which requires design mechanisms to neutralize the influence over evaluators. According to Davies (1999) administration personnel should not be neglected in performance management, as usually occurs, and accordingly in evaluation, since they are key drivers of successful organizational change, which is built on trust, leadership and ownership. Cousins and Earl (1995) support the idea of using evaluation results well by getting buy-in and, in turn, by having program personnel participating in the evaluation (Alkin and Christie 2004). Ultimately, participation is a necessity for increasing the possibility of utilization. Elsewhere Davies (1999) argues that performance management and implementation of evaluation results cannot be forced on people, for it may lead to goal displacement or, worse, diminish the relevance of the program rather than expand its development.

The rejection of participation in evaluation does not seem to be exclusively an anti-democratic type of behaviour but also intrinsic to elitist democratic evaluation (Hanberger 2006). This is seen as an exercise of excessive inclusion and dialogue for -"Inclusion, dialogue and deliberation are not considered to be vital democratic values"- and expressions of structural weakness, and top-down processes using prescribed policy implementation are preferred. Beyond doubt, participation may remain an empty word unless actual practice and procedures "... are set in place to make it real and effective ... " (de Marchi 2003). For how can participatory attributes of governance be measured without adopting a participatory approach?

9.3.2.1 Collaborative Evaluation

Various terms are in use to designate evaluation processes with multiple, same-group or distinct-group, participants where dialogue and negotiation takes place, and diverse approaches can be adopted to interact and reach consensus among evaluators. The notion of collaborative evaluation points to a reinforcement of bridges within the assemblage of evaluators and stakeholders (Cousins and Earl 1995), while the concept of participatory evaluation highlights how other sectors are moving away from expert-based evaluation. Participation and collaboration may reach various degrees, depending on the instruments and resources available, the commitment of promoters to make it participatory and the commitment of stakeholders to cooperate or collaborate. But, according to O'Sullivan and D'Agostino (2002), collaborative or participatory evaluations respond to the same principle–well rooted in empowerment-, and are applied interchangeably in literature when referring to a method for interpreting program outcomes and the development of new insights into program operations or effects, "... or especially organizational processes and consequences." (Cousins and Earl 1995).

With a common pragmatic perspective both Greene (2005) and JCSEE (1994) propose separate strategies to make evaluation valuable. However, Greene diminishes the impact evaluation methods have as a distinguishing factor because "... what importantly distinguishes one evaluation from another is not methods, but rather whose questions are addressed and which values are promoted ..." It is hardly arguable that each approach to evaluation does not bring its own worldview and promote particular values. In contrast, the Joint Committee on Standards for Educational Evaluation (JCSEE 1994) elaborated a set of 30 goal-oriented standards which seek to answer the question "Will this evaluation be effective?" and serve participants as guiding, although not stringent, principles to follow throughout the evaluation course.

Without question participatory evaluation is essential for increasing risk awareness among citizenry and administration staff if we wish to modify the perception of human relationships with their environment and behaviour and attitudes towards risk. Thus, a basic principle of risk governance evaluation should be inclusiveness.

9.4 The Architecture of Risk Evaluation

The framework for the evaluation edifice has as its key constituents the unit of analysis, the scope of evaluation and the review mode.

9.4.1 The Unit of Analysis

The first step in an evaluation is to identify what will be analyzed and, whose performance needs to be measured. This unit of analysis is geographical because administration levels, from the nation state to the local municipality or the city, have a set of responsibilities and spatial dimension. Although it should not be forgotten that all risk policies converge at the community level, with multiple top-down and bottom-up interactions. Policies and regulations are commonly translated into mandates for lower level administrations, and these, in turn, are suppliers of information about implementation for the higher levels. Thus, decisions span multiple organizations horizontally at the same level and, manifold levels of government vertically. Organizations and processes at one level closely or loosely interact with each other at other levels and these cross-level interactions are at the crux of multilevel analysis (Heinrich and Lynn 1999).

From a global perspective, Leonard (2005) judges that "... the primary unit of analysis is the spheres of authority that constitute the global governance pattern ...", putting the focus on the power relationships, going away from the individual to the organizational agents. Whilst at the lesser level the unit of analysis in corporate governance is the firm, the same as the individual agent in agency theory (Solomon 2007). Risk governance evaluation requires a different angle in order to fulfil the aim

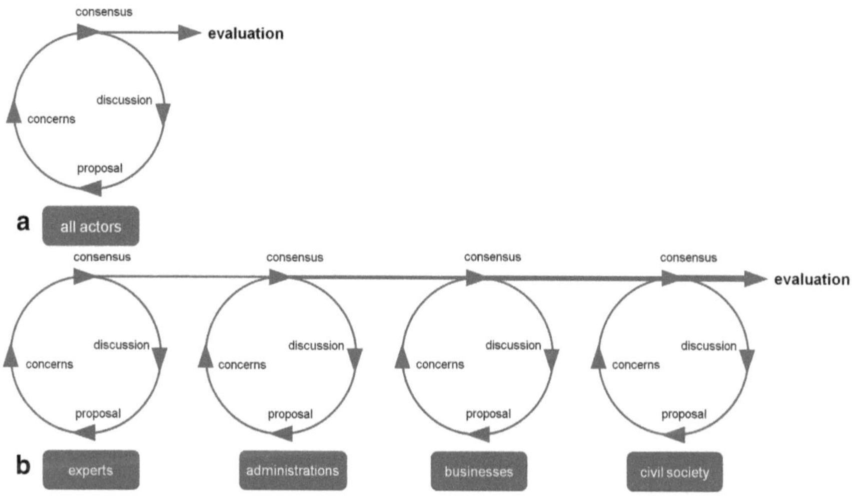

Fig. 9.2 Two alternative approaches to multiagent evaluation. **a** melting pot, and **b** serial evaluation

of measuring performance, because the unit of analysis needs to be not only geographical but also organizational. When analyzing governance in higher education Kennedy (2007) argues that "... the unit of analysis is the university's governance structures and processes ...", and in welfare reform it is the program (Sandfort 1999) "... and by extension, the whole organization". CDCP (1999) purposely identifies programs as the broad object of evaluation in the health system, understanding them as organized public action, because it allows the application to almost any public activity, including projects, services, investigations, campaigns, interventions, and administrative systems.

Evaluation of risk governance has to focus on a certain geographical area where a particular set of programs, institutions and organizations interact among them and with social and natural processes to develop a landscape of risk.

9.4.2 *What Should be Evaluated?*

This question raised by Kaufmann and Kraay (2007) provides some answers which helps when distinguishing governance rules from governance outcomes, and the perception of outcomes. This does not seem to be a trivial feature for it brings to the forefront the gap between theory and practice, and the perception of performance. It looks like the three levels require specific approaches and methods, for the object of analysis and the stakeholders differ, although indicators are ordered in a continuum where bright lines do not simply show up (Kaufmann and Kraay 2007). They find virtues in measuring rules, for their clarity and straightforward assessment, but also drawbacks. Among these, particularly notable are the subjective nature of judgments

Fig. 9.3 Classical conceptualization of risk management

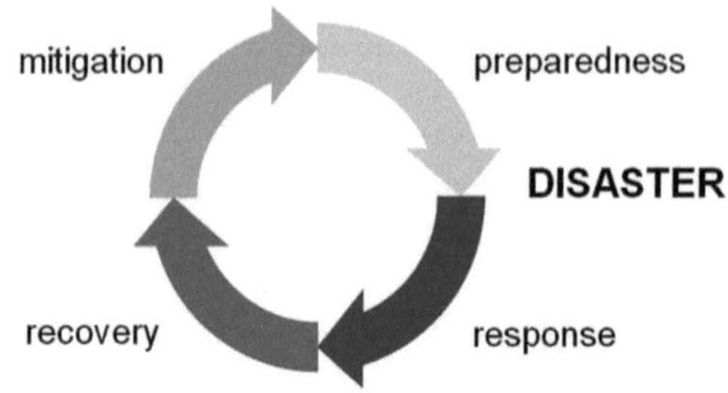

–not unique to this approach-, and the normative character of rule-based indicators of governance. Individual subjectivity may be neutralized to some extent by bringing both intraspecific and interspecific diversity to an evaluation panel, within the same group and with a range of evaluators from experts to lay people.

The prescriptive methodology also creates hindrances for reform, since it is difficult to identify which measure amongst all should be implemented first or which of them would be more effective in prompting changes. Moreover, Kaufmann and Kraay (2007) warn about the risk of the reform illusion, which consists in making isolated changes of rules or procedures to show progress, but with little real impact on action. Conversely, establishing links between outputs and measures turns out to be a quite problematical evaluation task. However, it is not the purpose of this chapter to examine the drawbacks of outcome-based evaluation, as well as the gap between *de iure* rules and *de facto* implementation. The last shortcoming indicates that an isolated evaluation does not suffice and that multiple evaluations would help to construct a more complete picture of the design-impact-perception continuum.

Among the attributes of programs, organizations and processes to be examined and evaluated, commencing with their focus and goals and following with their outcomes, are the degree or stage of development, the level of completeness, the scale of integration, and the capacity to perform. Performance measurement may go even further by seeking to identify the relationship between inputs and outputs, and outputs and outcomes, i.e. delineating the links between practices and decisions, on one hand, and with outcomes, on the other.

9.4.3 The Evaluation Methodology

Following the line of reasoning proposed by Greene (2005), who contends that the evaluation method is not the most differentiating factor but criteria and values, necessary steps to establish basic operating principles without overlooking the evaluative process. In terms of criteria, these have to be comprehensive enough to maximize their inclusiveness, which might sometimes make them unavoidably redundant, since the intricate structure of risk governance does not simply allow for separating dimensions in the risk management continuum (Fig. 9.3) from disaster recovery to risk reduction, as from emergency preparedness to disaster response.

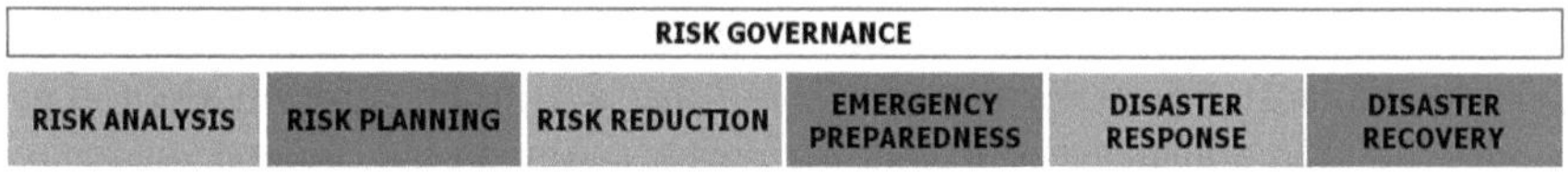

Fig. 9.4 Key dimensions of risk governance to be adopted in evaluation

The organization of the broad spectrum of criteria into different levels helps to differentiate between multiple public policies, and facilitates the evaluation process to experts, public officials or lay people on the basis of how they understand policies and their insertion in the general context of risk governance. The highest level of evaluative organization is defined by risk dimensions (Fig. 9.4), big blocks of constituents that depart from the classical phases of the disaster cycle (Fig. 9.3), from risk analysis through to risk planning, risk reduction, emergency preparedness, disaster response, and disaster recovery; and, again, risk analysis.

The risk analysis dimension deals with measuring advancement in understanding of the extant hazards, both natural and technological, and the environmental changes that are taking place, conceivably through monitoring and the application of models to foresee future scenarios, in the identification of vulnerabilities of population and physical capital, and -ultimately- in the perception and comprehension of the risk by communities and society at large. Finally, not less important is to measure whether a certain community has reached –or has taken reasonable steps to reach- any agreement upon the importance of the multiple risks and set preferences for action.

The evaluation of risk planning concerns determining the level of development and integration of all policies fostered to deal with risk, comprehensively understood, within the framework of integrated natural resource management. This includes planning for emergency and mitigation, as separate strategies or integrated in risk planning; along with the cohesion of sectoral policies, including economic development, urban and regional planning, and the management of resources, such as agricultural land, forestry and fisheries, the management of air, water and soil quality, water resources, biodiversity, ecosystems and landscapes.

The risk reduction dimension is, together with the former, a major component of the crux of proactive approaches and here the goal is to identify which programs have been implemented to downsize the various components from the risk equation, -hazard, exposure and vulnerability-, and to measure their degree of maturity. The two dimensions –planning and reduction- are closely associated, although there may be plans, or even very well elaborated plans, which have not been made effective. The programs run are multiple, including those which seek to control some dimensions of the hazard, increase resistance of public infrastructures and facilities, buildings or industrial structures, design and enforce zoning in order to avoid or make exposure compatible with land uses, relocate population or infrastructures, reduce vulnerability in areas of risk, as well as promote public awareness, adoption of self protection measures or insurance purchasing.

Emergency preparedness is a unique dimension of risk governance because it is perhaps the most universally developed, as it happens to be a priority even in regions with scarce financial or technical resources. Among the components within this dimension to be evaluated is monitoring, which focuses on the deployment of networks of instruments and personnel to carry out data capturing and the interpretation of measurements of environmental variables, forecast and estimation of event probabilities, early warning systems to alert communities, the structure and operation of civil protection systems and other public services needed in emergency situations -such as health systems, security and the repair of lifelines-, and evacuation infrastructures and systems.

A principal output of the World Conference on Disaster Reduction in 2005 was the Hyogo Framework for Action (ISDR 2005) which articulated strategic goals and priorities for the period 2005–2015. This compilation of oriented policies, with a high degree of scope and inclusiveness, shapes an *acquis hyogoien*, or body of accumulated learning translated into principles and commitments that are necessary to reach the ultimate goal of disaster reduction in a certain region. Policies and programs oriented to risk mitigation and emergency management do not operate on a blank slate nor cannot work in an adverse political context, but they emerge within, and are very dependent on, the governance system of the region or country in question; what Renn (2008) terms institutional arrangements and political culture. This supports the inclusion of the fifth dimension that will help to measure wide-ranging attributes of governance. This includes the institutional and organizational framework which helps form the decisions made by policy-makers (Lion et al. 2003), citizens and markets; the environment required to implement those decisions, and the formal and informal arrangements adopted to codify and implement a policy (Lion et al. 2003).

Whilst these dimensions strongly focus on policy elaboration or programme application, there are two phases of the disaster cycle that deal with post-disaster operations, i.e. disaster response and recovery. This holds particularly true for disaster response, which enters into effect just immediately after the disaster event, while recovery programs work on the basis of a mid-term horizon. In the former case criteria to be appraised include operations of rescue, relief and assistance, and -in the case of developing countries- the management of external aid. Layers of knowledge about progression in the dimension of recovery to be evaluated comprise operations of rehabilitation, reconstruction, housing and relocation, financial support to victims, risk reassessment and, ultimately, revision of risk governance failures and strengths.

This approach to risk governance evaluation is more extensive than intensive, more qualitative than quantitative, more inclusive and comprehensive than narrow and in-depth, covering a wide range of dimensions, components/constituents and criteria/indicators to gain a better understanding of the system. Dimensions are families of constituents of risk governance, related by their lineage and commonalities, with a high degree of dependency, that represent the bricks of the whole risk governance structure. The last layer of the structure and elements of the evaluation process are the criteria, the benchmark by which development is measured. The approach of using dimensions, encompassing a policy perspective, together with components broken up into criteria, with a structural point of view, confers a multidimensional character onto evaluation. According to Rose (2008) it facilitates the identification of both strengths and weaknesses in risk governance, a concern of governments over political scientists, and brings into consideration the dynamic nature of governance, and its uneven development, as some components or processes are established earlier than others. The purpose is to orient actions addressing weaknesses and emphasize the need to introduce political reforms. The fact that new objects and subjects of evaluation occasionally emerge adds uncertainty and complexity to the process of evaluation itself (Lion et al. 2003), paired with the intrinsic complexity of governance.

Criteria are descriptive reference points for benchmarking the actual state or condition of a specific policy or program component. They have a qualitative character to allow for the elaboration of judgments and arguments for deliberation and negotiation among evaluators and ultimately reach consensus. Given that evaluation seeks to measure level of development –among other objectives-, each criterion is evaluated against a scale with five levels:

- No advances have been identified as regards to the criterion,
- Isolated measures have been adopted without responding to a structured programme,
- Programmes have been elaborated but have not been implemented through projects,
- Programmes have been elaborated and projects translate their principles. Both plans and projects are occasionally evaluated,
- Programmes and projects respond to a designed policy. Both programmes and projects are regularly evaluated.

Some governance measurements finish up with the elaboration of synthetic indices. Nonetheless Coppedge (2002) casts doubts on the appropriateness of reducing the multidimensionality of democracy or governance to the single quantitative dimension of an index, by averaging and diluting complexity. The Bertelsmann Stiftung's Sustainable Governance Indicators (SGI) use a structure of four levels (dimensions, categories, criteria and indicators/items) (Brusis 2009) and supports evaluation on both quantitative and qualitative indicators that are aggregated and, in due course, produce indices of status and management to rank countries according to their level of sustainable governance. Brusis believes that this "... micro-level evaluation of executive governance pinpoints observable deficiencies and may allow efforts toward improving the quality of governance to be guided in a more targeted way." However, emphasis on quantification, or the aggregation of numerical scores, may not satisfactorily contribute to produce useful recommendations or localize weaknesses or deficits that must be addressed. Indices are mostly usable for comparison between countries or for monitoring diachronic progress towards –or retreat from- higher levels of development, but do not provide sufficiently rich clues to orient reform or sustain argumentation within the panel.

9.4.4 The Evaluation and its Agents

Hanberger (2006) distinguishes between three types of evaluators according to the evaluation approach –and democratic orientation- implemented. While an elitist democracy-oriented evaluation (EDE) evaluator is asked to inform and provide public access to data, the participatory democracy-oriented evaluation (PDE) evaluator speaks for the people as an advocate, and the discursive democracy-oriented evaluation (DDE) evaluator will act as a mediator, informing about the misuse of arguments and evaluation. This perspective seems to only refer to the same class of evaluator, the expert, playing different roles in different contexts. As an example Donahue et al. (2000) have analyzed government performance within the Government Performance Project (GPP), based on an expert group of practitioners and scholars in each area of government management. They assessed each component of government, based on their research, experience and on the results from other evaluations. Most of the systems rely on an expert-based judgment; see Kaufmann and Kraay (2007) for a list and comparison of approaches, because they find multiple advantages in it: cost-effectiveness, comparability between countries and knowledge availability. However, if we do not pursue comparability among states, there is not a sufficient justification for expert-based evaluation. And the presumption of knowledge availability fails when dealing with de facto implementation, where lay people and administration staff may have a better understanding of impact.

If we agree that "... science can provide crucial information, but cannot determine correct policies" (de Marchi and Ravetz 1999) we can conclude that scientific and technical knowledge is very valuable but does not successfully bring the most advantageous solutions.

A different perspective would be to include various classes of evaluators (Fig. 9.2) –expert and non-expert- who represent different interests and views, particularly within the latter type of evaluation identified by Hanberger, the discursive democracy-oriented evaluation. The interaction between evaluators may be synchronic, with multiple groups of evaluators acting simultaneously–the melting pot approach- on the basis of the same criteria and arguments and reaching a global consensus. Contrarily, the participation may well be along successive segmented phases of the evaluation process –the serial approach-, or both combined. Advantages and weaknesses may be found in the three perspectives. Among the latter, in the first approach it has been found that in group meetings those with authoritative knowledge, with a higher rank position or some kind of control, adopt a leading role in the debates and suppress discrepant opinions or prevent them from emerging. Using the second methodology, the first participating group can establish well-rooted conclusions, which will rarely be questioned or eroded in the following phases, and bring path dependence to conclusions; while the latter evaluators may introduce arguments that will not have the chance to be discussed further because conclusions at that late stage do not allow for a return to the starting point. This can be resolved by using concatenated loops, where cycles of argumentation, debate and negotiation succeed each other whilst advancing towards closing consensus. The evaluation panel is thus made up of an association of different workgroups, or groups assigned with the same evaluation task (Morabito 2002) but with endogenous and exogenous diverse points of view.

With this approach, different actors play different roles towards same goals. While experts and practitioners have a wider understanding of the governmental framework and operations, public sector staff have a better understanding of programmed operations and administrative processes, and civil society groups are able to identify gaps and bridge the divide between government and citizenship. Sandfort (1999) believes that the role of the second group is more critical than anticipated since administration staff plays a powerful role in shaping policy within their organization. Public officials act as filters for public policies. In this context, experts also play the role of initiators, providing judgments and developing strong arguments to persuade other evaluators of their value. They investigate what works and what does not, identifying weaknesses in policy development and program implementation but will not question stated goals (Hanberger 2006), a task that citizens may perform more satisfactorily. The program implementation perspective brought by the other two groups of evaluators, administration staff and components of civil society, organized into multiple constitutions such as focus groups, citizen juries, consensus conferences or stakeholder fora (de Marchi and Ravetz 1999) is much different. Citizens may help identify the gaps between design and implementation, unbalanced achievement, and social, economic and environmental inequities and concerns. Organization stakeholders bring the perspective of program operation, for they recognize the barriers and opportunities for facilitating the translation of policies into action.

Provisions to preserve the credibility of evaluators (Engel and Carlsson 2002) should be carefully made because both trustworthiness and integrity are always at risk (Patton 1997) in order for the assessment not to be questioned. The steps towards this achievement are to be active in the presentation of their individual judgment, reactive whilst listening to other's concerns and adaptive in incorporating diverse interests, according to Patton. The role of coordinator thus becomes particularly critical; this agent not only has to plan, manage, and monitor the evaluative process, but also ensure continuity throughout the evaluation course. Great effort should be put in the selection of evaluators, avoiding bias in the composition towards coordinator's preferences, by including evaluators with skills, expertise and experience that will aid a swift understanding of the structure and process of evaluation, the use of up-to-date information, and the identification of strong arguments to sustain the formulation of an assessment. Evaluators should excel in communication, process facilitation, conflict resolution and negotiation skills (Engel and Carlsson 2002). But this will not always be the case and eventually the evaluative process should facilitate these achievements. Assembling the evaluator panel is a critical political issue. If the coordinator is directly engaged by the evaluatee he/she may feel the pressure–exerted or not- to soften conclusions and select a team that can be easily influenced. At this point evaluation quality might be at stake and only process transparency may help to prevent potential deviation.

9.5 Conclusions

Increased disaster risk calls for enhanced political action and greater functional integration to prevent higher human and material costs. In this context the governance approach is an option to address both complexity and the policy implementation gap. Its requirement of feedback processes and emphasized actor interaction drives the need for continued evaluation. Thus, risk governance evaluation adopts a highly dimensional character, with a large number of components which focus on a variety of policies and programmes, and diverse participants, nonetheless subsidiary to any social process. In their entirety, risk governance dimensions assist both with the categorization and measurement of public policies directly or indirectly focused on the mitigation of risk and the reduction of disasters and, ultimately, on strengthening social capital in terms of social learning and empowerment. Risk governance evaluation processes provide a great opportunity to integrate multiple goals with this same aim. Risk communication, a key theme in the social study of risk and a basic long-term strategy utilized to increase risk awareness and modify the perception and attitude of citizens, policymakers and public officials alike is an instance of evaluation. The hierarchical thematic structure of risk governance starts with dimensions, and follows with components and criteria to help identify gaps and overlaps in the outcomes of multiple policies within the logic of the risk cycle.

The objective of evaluation is measuring both the efficiency and effectiveness of a mode of risk governance by understanding the interactions between civil society, the private sector and the state in the elaboration of policies and in the implementation of programs. If we understand governance as a systemic approximation to the process of decision making, and if we see evaluation as a key component that provides feedback about processes that helps to adjust and reform, to increase transparency and accountability, and as an avenue to incorporate the multiplicity of stakeholders involved, then it is necessary to make evaluation processes participatory. Multiple agents will certainly increase the complexity of evaluation, but they might also satisfy requirements through participatory or cooperative evaluation. Different actors may play different roles in the process by bringing identifiable and specific knowledge which is negotiated and aggregated in an evolutionary mode.

The outcome of evaluation should be a profile of the geographical unit being evaluated that may help to identify the strengths and flaws which should be addressed by governments to enhance and make risk governance more functional. Evaluation is normative, in so far as it defines what is considered to be most appropriate, in accordance with the risk paradigm and the basis for political action and reorganization, and sets the path to prioritize goals and actions. In the end, it promotes a particular course of action.

Since the aim of risk governance is to increase the cohesion of policies and the integration of stakeholders towards the avoidance of hazard exposure (Fra Paleo 2009) and the mitigation of risk and reduction of disasters, the evaluation of risk governance has two principal objectives: to identify the structure and measure the cohesion of policies directly or indirectly dealing with risk and serve institutional development, and to enhance societal and organizational learning by increasing risk awareness. In order to progress towards advanced modes of risk governance, evaluation should be based both on consecutive evaluations -that help to track progress and regularly feed the system with fresh information- and on enhanced participative processes.

Acknowledgment This paper is based upon work developed within the research project supported by the Spanish Ministry of Science and Innovation under award number CSO2008-00560/GEOG given to the University of Santiago de Compostela. The views and conclusions contained in the paper are those of the author and do not necessarily represent those of the Spanish Ministry of Science and Innovation.

References

Alkin MC, Christie CA (2004) An evaluation theory tree. In: Alkin MC (ed) Evaluation roots: tracing theorist's views and influences. Sage, Thousand Oaks

Bache I, Flinders M (2004) Themes and issues in multi-level governance. In: Bache I, Flinders M (ed) Multi-level governance. Oxford University Press, New York pp 1–11

Beck U, Bonss W, Lau C (2003) The theory of reflexive modernization. Problematic, Hypothesis and research programme. Theory Cult Soc 20(2):1–33

Brusis M (2009) Designing sustainable governance indicators: assessment criteria and methodology. In: Bertelsmann S (ed) Sustainable governance indicators. Bertelsmann Stiftung, Gütersloh, pp 71–100

Campbell DT (1988) The experimenting society. In: Overman ES (ed) Methodology and epistemology of social science: selected papers. The Chicago University Press, Chicago, pp 290–314

Catania AC (1997) Learning, 4th edn. Prentice-Hall, Englewood Cliffs

[CDCP] Centers for Disease Control and Prevention. (1999). Framework for program evaluation in public health. MMWR 1999, 48(No. RR-11)

Coppedge M (2002) Democracy and dimensions: comments on Munck and Verkuilen. Comp Polit Stud 35(1):35–39

Cousins J, Earl LM (eds) (1995) Participatory evaluation in education: studies in evaluation use and organizational learning. Routledge, London

Davies IC (1999) Evaluation and performance management in government. Evaluation 5(2):150–159. doi:10.1177/13563899922208896

Donahue AK, Selden SC, Ingraham PW (2000) Measuring government management capacity: a comparative analysis of city human resources management systems. J Public Adm Res Theory 10(2):381–411

Engel P, Carlsson C (2002) Enhancing learning through evaluation: approaches, dilemmas and some possible ways forward. Background papers. 2002 EES Conference, Seville; October 10–12, 2002

Fra Paleo U (2005) Information and spatial data processing in risk management. In: Shahbazian E, Rogova G, Valin P (eds) Data fusion for situation monitoring, incident detection, alert and response management. Kluwer, Dordrecht, pp 760–773

Fra Paleo U (2009) On exposure to natural hazards: revisiting a neglected primal action. In: Fra Paleo U (ed) Building safer communities. risk governance spatial planning and responses to natural hazards. IOS Press, Amsterdam, pp 61–78

Fra Paleo U (2010) El avance de la gestión hacia la gobernanza del riesgo en un contexto de complejidad e incertidumbre. In: Ferreira JA, Neves D, Ferreira CS (eds) Recursos naturais, sistemas de informação geográfica e processos sociais. Universidade Federal de Viçosa, Viçosa, pp 43–57

Fra Paleo U (2011) El avance de la gestión hacia la gobernanza del riesgo en un contexto de complejidad e incertidumbre. In Ferreira JA, Einloft CJ, Gonçalves RL (eds) Recursos naturais, sistemas de informação geográfica e procesos sociais. Universidade Federal de Viçosa, Viçosa, pp 43–57

Giddens A (1986) The constitution of society. Outline of the theory of structuration. University of California Press, Berkeley

Greene JC (2005) Qualitative program evaluation: practice and promise. In: Denzin NK, Lincoln YS (eds) Handbook of qualitative research. Sage, Thousand Oaks

Hanberger A (2006) Evaluation of and for democracy. Evaluation 12(1):17–37

Heinrich CJ, Lynn LE (1999) Means and ends: a comparative study of empirical methods for investigating governance and performance. Fifth National Public Management Research Conference, Texas A & M University, College Station, Texas, December 3–4, 1999

[ISDR] International Strategy for Disaster Reduction (2005) Hyogo Framework for Action 2005–2015: Building the resilience of nations and communities to disasters. ISDR. A/CONF.206.6

Jachtenfuchs M, Kohler-Koch B (2003) Governance and institutional development. In: Wiener A, Diez T (eds) European integration theory. Oxford University Press, Oxford, pp 97–115

[JCSEE] Joint Committee on Standards for Educational Evaluation. (1994) Program evaluation standards: how to assess evaluations of educational programs, 2nd edn. Sage, Thousand Oaks

Kaufmann D, Kraay A (2007) Governance indicators: Where are we, Where should we be going? Policy Research Working Paper 4370. The World Bank, Washington

Kennedy DM (2007) Knowledge management and governance in higher education. actKM Online J Knowl Manag 4(11)

Leonard EK (2005) The onset of global governance: international relations theory and the international civil court. Ashgate, Aldershot

Lion C, Martini P, Volpi S (2003) Evaluation and governance: the Italian experience. International Conference of the Regional Studies Association, Pisa, 12–15 April 2003

Mackay K (ed) (1998) Public sector performance. The critical role of evaluation. The World Bank, Washington

de Marchi B (2003) Risk governance. Public participation and risk governance. Sci Public Policy 30(3):171–176

de Marchi B, Ravetz JR (1999) Risk management and governance: a post-normal science approach. Futures 31(7):743–757

Marks G (1992) Structural policy in the European community. In: Sbragia A (ed) Europolitics: institutions and policymaking in the "New" European community. The Brookings Institute, Washington

Morabito SM (2002) Evaluator roles an strategies for expanding evaluation process influence. Am J Eval 23(3):321–330

Noblit GW, Eaker DJ (1989) Evaluation designs as political strategies. In: Hannaway J, Crowson R (eds) The politics of reforming school administration. Falmer Press, London, pp 127–138

Noblit GW (1984) The prospects of an applied ethnography for education: a sociology of knowledge interpretation. Educ Eval Policy Anal 6(1):95–101

O'Sullivan RG, D'Agostino A (2002) Promoting evaluation through collaboration findings from community-based programs for young children and their families. Evaluation 8(3):372–387

Patton MQ (1997) Utilization focused evaluation: the new century test. Sage, Thousand Oaks

Renn O (2008) Risk governance. Coping with uncertainty in a complex world. Earthscan, London

Rose R (2008) Evaluating democratic governance: a bottom up approach to European Union enlargement. Democratization 15(2):251–271

Sanders JR (2001) A vision for evaluation. Am J Eval 22:363–366

Sanders JR (2002) Presidential address: on mainstreaming evaluation. Am J Eval 23(3):253–259

Sandfort JR (1999) The structural impediments to front-line human service collaboration: examining welfare reform at the front-lines. Soc Serv Rev 73(3):314–339

Smith JK (1989) The nature of social and educational inquiry: empiricism versus interpretation. Ablex, Norwood

Solomon J (2007) Corporate governance and accountability. Wiley, New York
Stame N (2006) Governance, democracy and evaluation. Evaluation 12(1):7–16. doi:10.1177/1356389006064173
Treib O, Bähr H, Falkner G (2007) Modes of governance: towards a conceptual clarification. J Eur Public Policy 14(1):1–20
Uphoff N, Combs J (2001) Some things can't be true but are: rice, rickets and what else? Unlearning conventional wisdoms to remove paradigm blockages. Cornell International Institute for Food, Agriculture and Development, Cornell University, New York
Weiss CH (1977) Using social research in public policy making. Lexington Books, Lexington

Chapter 10
Continuity and Discontinuity of Urban Form—The Issue of Risk

Tiberiu Florescu

Abstract The aim of this paper is to make a contribution to the representation, manifestation and evolution of urban forms as we currently understand them. We further hope to succeed as a theoretical and practical support for, in particular, future research in the broad scope of urban planning. By this we mean the study and determination of the particularities of the urban organism in the real plan of its evolution, in other words of the character and identity which are attributes inextricably linked to the urban form. Beginning with urban topology and the topological transformations given by movement, projection and deformation, the work proceeds to urban morphogenesis, which is a result of the dynamics of urban forms and the generator of their complexity. Discontinuities in urban systems are caused by inherent risks and vulnerability. These risks create the context that will be dealt with in this chapter on "continuity and discontinuity". Discontinuities can be static or dynamic, the latter relating the urban form to the context. Risks are a relating element between form and context, relating continuity and discontinuity in the evolution of the urban organism. Risks are considered in detail, whether natural or anthropic, with a special view on those in the urban environment. One of the original contributions of the author in this context is the development of formulas regarding risk in the context of the static state of urban organisms and in their metabolic form, thus relating the risk of individual elements, depending on the discontinuities that affect vulnerability.

Keywords Discontinuity · Urban form · Static form · Metabolic form · Morphogenesis and context

Based on T. Florescu (2009) Formă şi transformare urbană, Editura Universitară "Ion Mincu", Bucharest (in Romanian) 170 pp. This chapter is edited by Maria Bostenaru Dan.

T. Florescu (✉)
Department of Urban and Landscape Design, "Ion Mincu" University of Architecture and Urbanism, Academiei 18–20, 010014 Bucharest, Romania
e-mail: tiberiuflorescu@gmail.com

M. Bostenaru Dan et al. (eds.), *Earthquake Hazard Impact and Urban Planning*, Environmental Hazards, DOI 10.1007/978-94-007-7981-5_10,

© Springer Science+Business Media Dordrecht 2014

10.1 Introduction

Goethe was the first to use the concept of morphology, which means the study of form and structure of organisms and their specific structural features; then it was promoted in bioscience, which connects it to the definition of urban organism.

We will deal with the issue of "risk" as a particular application of the reflections on continuity and discontinuity of the urban form. From the definition of the concept of "form", a model can be built for the urban form and transformation, continuity and discontinuity in the approach and evolution of the urban organism in order to redefine the urban form. Regarding the continuity and discontinuity of the urban form we start with differences of scale and proceed to particularities such as fractals in urban structure and architectural shape, as an introduction to the complexity of metabolic forms.

The city as a bearer of history, densely defined, marked by the congruence between spaces and social practices, has suffered some deep changes in the last centuries. Characterized by permanent changes in the elements that compose the urban structure, the evolution of the ratio continuity/ discontinuity was always marked. The two categories of urban forms defined in this paper are characterized by different developments with respect to time or context: a (basic urban form) is characterized mainly by continuity (topological transformations); another (urban form complex) class is characterized by discontinuity (dynamic changes).

Maintaining or changing the identity, whether it concerns progress or regress in the relationships established in the urban structure, with direct consequences on the configurative-space structure, it had always had a leading role in the realization and manifestation of the urban form.

The research is focused on the study and the nature of urban form, following two objectives:

- Identifying the general trends of evolution and dynamics that characterize the urban body in terms of specific urban form changes,
- Putting an accent on the moments of continuity/discontinuity in the evolution of urban form to enable them to track the growth of an urban body in relation to three crucial levels: internal complexity of urban form, development of urban form in relation to time and in relation to the context.

In this paper a theoretical study on the implications of the present and how it will impact knowledge in the field, as well as the applicability of research results, is an attempt to establish theoretical and practical issues for a modern approach to urban development and shape of the city organism, targeting a deeper understanding of their correlation, with major risks that influence the evolution of cities in the contemporary era.

10.2 The State of Crisis and Vulnerability of the Urban Form

Studying natural and man-made risks in urban areas cannot be addressed outside of the specific context of the development of the urban form, of the political and the sustainable development context. Recent developments of the context of risk associated with the concept of responsibility shows that it is superficial to address only the problem of "natural" risks, which means to be prepared to prevent just this type of risk. The integration in the general context of urban risk of important factors such as legal and institutional structures, as agents of urban risk reduction, acting through risk prevention policies, delimits in a precise manner the limits of acceptable risk. The question of acceptable risk thus proves to be an important stake of urban management and policies.

Man-made risks, which are based on natural hazards, when technological hazards come together with the natural setting, threaten many urban manifestations, especially in the cities that are located in developing countries, whose population is growing. Population growth causes the often uncontrolled and disordered growth of urban form and leads to random structure of irrational urban forms. It is even more necessary in this case to take measures to reduce and mitigate urban and urban planning risk.

10.2.1 Systemic View On The Urban Organism

The systemic understanding of things is a result of ecological understanding in the territory. Sustainability takes the systemic approach in view of comparing an optimal system for each criterion, so layers can be found. It must be analyzed according to global stability, versatility and adaptability of the dynamic systems. The cybernetic analysis-synthesis approach means the formation of a system according to its description, using more advanced knowledge. To exemplify how the cybernetic theory applies to complex urban situations we call such a system, employed by Alexander (1977), categorized by Bostenaru (2004) as a participative approach. An incursion into the steering structures of investigated urban systems is inevitable, because a specific functionality and some specific properties of a given system are taken in the cybernetic analysis-synthesis view. The cybernetic approach requires the building of models, which are simplified versions of systems, equivalent to these in some standpoints but easier to analyze. It can be talked about a cybernetic model, for example in the acceptation of Alexander (1977), defined in each case through a morphological (spatial) and an anthropological (functional) factors, when there are at least three systems: a system to be elucidated (an impressive space situation X, the goal, or condition), a known system (a typical architectural arrangement Z, the practice or answer) and a system to be discussed (the pattern Y, the behavior or pattern). In cybernetic, or computer language similar programming terms

INPUT variables IF goal THEN practice THUS behavior OUTPUT variables END IF the variables connect to other patterns, these are the combination rules.

The spatial and social organization of a complex planning artefact consists of elements, which are the cybernetic models called "patterns" of the real life systems and their combination rules. Generally speaking there are various models in relation to their validity type. A development model category, the result of which is a constant reflection of the environment, is a dynamic model (model behavior).

Any system is organic and displays interrelationships. Given that it is also vertical, horizontal and ordered, we can tell that this is possessing hierarchy. The goal, the stability and the differentiation determine its variability. The most evolved systems are adaptable through their entropy, feed-back and self regulation. The balance of an ecosystem is insured because any change drives a new feature of the system. The so-called s-functions drive in the direction of the current state being retained, while striving to avoid disfunctions in order to increase stable states. A flexible system, with its variety of stability conditions, allows positive action of the s-functions. Its entropy, a key feature of a system, is high.

To exemplify we can tell that the entropy of the urban system is essential for its dynamics. Given that this is in relationship with the number of fixed elements (the flexibility of the system), one of these being the street pattern, the problem of transport is more important (OECD report 1996). The sustainable city has to adapt its newly structured zones to this pattern. The economy isn't an one direction entropic system.

The urban organism as a complex system has two morphological elements which correspond to two inter-conditioned layers (Sandu 1988):

- At the phenomenological level we distinguish the content of the system, called "urban life". The "urban life" consists of all urban activities together with a precise location. These localized urban activities form the urban functional structure. Going to the essence of these urban activities as part of a system the urban functional structure becomes "the way of life". This way the spatial-functional layer is formed;
- At the physical level we distinguish the container of the system, called "urban frame". The "urban frame" consists of the totality of the spaces corresponding to the site where they take place. All spaces together, corresponding to the location where urban activities take place, form in functional spatial co-operation the urban spatial structure. Going to the essence the urban spatial structure this becomes "life frame". This way the spatial-perceptive layer is formed.

In a simplified comparison we can consider also the structure of a building, and the relationship between shape and function. The container consists of the load-bearing scaffold, constructively needed and the contents of the lived architectural space, related to the given program (such as housing for example). Such relationships have been subject of studies in tectonics in architecture, focusing on the sincerity of expression of the structure and the role of ornament.

Analytically, a system is defined through its elements and their relationships. Restructuring urban actions are reordering actions of the elements into urban textures. Urban textures are defined as sub-systems of the "life frame" elements which are able to respond to certain functions. The unit for analytic division into morphologic elements of the urban composition is a texture of such elements.

These measures, however, are often difficult to be timely structured in risk prevention policies, on the one hand because of the lack of prior studies on risk factors that threaten in these cases the elements of "urban life". Such risk factors express themselves mainly in the economic and the social field. Studies in this field are completed after a long time of preparation and which often do not provide immediately applicable results. On the other hand, it is necessary to take into account the two directions: identifying and analyzing risk factors, and then to analyze the conditions of reduction.

An example of such socio-economic studies is currently accomplished by the Global Earthquake Model (http://www.globalquakemodel.net/) corresponding module, and includes the analysis of literature on social vulnerability by the means of ontology (http://www.vuwiki.net/), in the sense computer science gave to this philosophic concept. Thus, ontology is a formal representation of a set of concepts within a domain and the relationships between those concepts. As such, it is related to the cybernetic understanding of a system by means of elements and relationships, building possibly one of its representations.

The peculiarities and specificities of each urban form requires careful study and precise delineation of risk factors which are leading to particular decisions from one situation to another, from one form to another, imposing measures and specific documentation, including planning the content, as shown in the regulatory documentation (Florescu and Ionita 1999).

10.2.2 Urban Morphogenesis

In the previous paragraph we introduced the systemic view on the urban organism in the sense of the theory of urban structures elaborated by Sandu (1988). This theory presents the urban organism as a combination in existence of the urban structure and the urban expression, which is in strong correlation with the concepts or urban form and urban phenomenon. The urban form is understood as an analytic stage, at a certain moment, of an urban organism, which builds the expression of the structure and its morphologic elements. It is at the same time the frame of a sequence of an urban phenomenon. The urban form can be understood as a socio-spatial system, the formal representations of the city being a new instrument for the study of the urban morphology.

We define in this work two categories of urban form which are characterized by different evolutions related to time or the context. The elementary urban form is characterized mainly by continuity (topological transformations). The complex urban form is characterized by discontinuity (dynamic transformations).

In terms of correlation of the occurrence of risk issues with the development of the urban form, through the understanding of the urban form as a formal discontinuity of the context, considering the overall plan of the space formally expressed in the context, the establishment (emergence) of an urban form follows three successive stages, regardless of its mode of occurrence (morphogenesis), by its nature or context features.

The support for the manifestation of urban form is a cut from the general framework of existence. The limit of this cutting is materialized by an apparent contour (limited) urban form, determining it spatially, being in close interdependence with it and being a fundamental component in the socio-spatial system represented by the urban form. If, formally, by definition, the urban form is a discontinuity in the spatial context, the relationship between it and the context is a bi-univocal relationship determination. Evolution and formal changes of the context directly influence urban form evolution, as evolution of these can influence and change local or general context. We define context as a support framework for the existence and manifestation of the urban form, a background on which it is materialized and expressed through a system of elements outside the limits established by the socio-spatial system of the urban form. Elements of this system are of different nature (natural, ecological, demographic, technological, human, etc.).

The context, a newly introduced element by us, has the following features:

- It is a coherent system of elements that have a particular spatial location, outside the urban form,
- This occupation of space in the territory is accompanied by a certain temporality, so the context has some limits, defined in space and time, regardless of its nature,
- Its nature (type) and its quality are important elements in defining the urban form (natural context, economical, political, administrative, historical, etc. context).

The general context to which the urban form relates is a formal space that includes the socio-spatial system of the urban form, with which it establishes permanent relationships. The context is therefore a formal reference space, against which we define urban form. The relationship between the complexity of the context and the urban form is one of direct proportionality in the sense that the discontinuities of a formal context with a certain complexity can give rise to complex urban forms and the context of a reduced scale and complexity of the basic urban form (urban architectural form).

Thus, in an initial phase, we distinguish a formal uniform space of the context, which has formal qualities of uniformity and continuity, a plan to which no other form (lump) is attached.

The quality of formal space gives it a dynamic character that determines that the evolution of some of its components gives rise to static discontinuities (expressed in physical, material forms), which materialize through the detachment of the supporting plan of the context, or of dynamic discontinuities, that can get in time the same formal manifestation, leading to the birth (appearance) of some components of the urban form.

The appearance and existence of these urban form elements before its establishment as a socio-spatial system, be it discussed in the historical perspective in the context of the first distant forms of housing or in that of adapting the environment to human needs in the context of the last century, through the appearance of cities on pre-established plans (planned cities as opposed to cities which grew organically in time), is a sine qua non condition for establishing the urban form, a vital link in urban morphogenesis.

The urban morphogenesis, process of trans-formation or creation of a new urban form, is approached in this work more than in the formal understanding of the phenomenon of the birth of the form considering the human component as being defining for the identity of the urban form. Through urban morphogenesis we will name the birth, occurrence of a new urban form, as a result of a need or due to the transformation of an existing urban form into a different urban form, with a different character. Urban morphogenesis means the appearance of a new urban morphological structure at a certain moment, or the different expression of the same morphological structure within the same urban form, through the relationship change between this and the served area or the number of users, which determines the initial urban form to express itself or to be perceived differently.

Drawing a comparison the studies of Thom (1984) regarding the morphogenesis in biology, and accepting the town is an urban organism, the appearance of a new urban form is characterized either through a development of the initial urban structure through evolution or through change (trans-formation) to one different of the initial urban form, either through the apparition of new urban morphologic elements in the territory which enrich the initial urban form, transforming it through an evolution process (positively or negatively).

It is important to consider the morphogenesis factors, especially in the possible context of the appearance of discontinuities in evolution, since these are the ones which, objectively or subjectively, influence in a vital way the evolution of the urban organism, making it to leave the path of a continuous, linear, perfectly controllable, easy to model or to foresee development, as usually the development of the city is wished to be. These perturbing factors of the linear evolution, which have to be considered for the description of the evolution of the urban organism, are the risk factors. In this chapter we approach the relationship between the urban morphogenesis and the notion of risk. Risk factors can appear in the interior of socio-spatial systems, either at the level of decision, of urban evolution or of urban design and planning, these being the internal morphogenesis factors for the urban form. Talking about the external factors, we will approach the notion of risk as one of the main factors to cause discontinuity moments of the context in the frame of the existence of the urban form, factors which lead to the appearance of trans-formation phenomena. It is important to define the way in which the risk factors, depending on the risk degree, characterize certain hazards and lead to the occurrence of particular types of disasters, and the way in which each of these disasters influence the urban form, determining its evolution in negative or sometimes positive way, through transformation in another urban form or through the creation of new opportunities for the evolution of the initial urban form.

As we did an incursion to the meaning of the term of morphology in the previous paragraph, we will do a short interpretation of the term of morphogenesis. Morphogenesis, similarly to the concept of container and content, can be applied at building scale and at urban scale. At building scale in the so-called "computational morphogenesis" Aou and Ohmori (2005) developed a computer based method based on the "evolutionary structural optimization" of Xie and Steven (1997) which meant the step-by-step removal of the inefficient parts of the initial structure, including also the extension of the necessary parts. This concept can be applied at urban scale for risk management considering vulnerable parts of the urban texture. Also, in the project "Semantic enrichment of 3D city models for sustainable urban development" (http://www.semcity.net/) morphology has been seen as an essential ingredient for sustainable development, including risk reduction, and the semantic enrichment, meaning overlapping ontology over the model, a mean for this. By these means we build on the findings of the previous paragraph on modeling of systems and of employing cybernetic means through ontology. However, computer based methods in morphogenesis are applied not only in the design of the load bearing structure, but also in parametric architecture (such as the one of Patrick Schumacher), but not only. Augustin Ioan made an exercise of morphogenesis of certain architectural language (Ioan 2012) of the architect Marcel Janco based on the extrusion of 2D shapes into 3D shapes. In a similar procedure Bostenaru developed a morphogenesis based model of the impact of the 1,755 earthquake on Lisbon in frame of the above mentioned project "Semantic enrichment of 3D city models for sustainable urban development". The 2D models depicted in pre-earthquake ceramics (azulejos) representation of Lisbon can be transformed into 3D representation of just the landmarks, while the common urban space can be built by agent based means developed by Patrick Schumacher following the principles of Rome's plans by Nolli (http:://nolli.uoregon.edu/).

An urban form may not appear immediately. Even in the case of building cities on a pre-established plan, there was an interim period of time between the conception, the development itself and when the newly created urban form began to function and to exercise its role in territory marking distinct phases in acquisition of urban identity.

To summarize, urban morphogenesis is the result of the dynamics of urban forms and generator of the complexity of these. Discontinuity is caused by risks and vulnerability at urban scale. Discontinuities can be static and dynamic, the latter relating the urban form to the context. Risks are a relating element between form and context, relating continuity and discontinuity in the evolution of the urban organism.

In the following we will develop formulas regarding the risk in the context, in the static form of the urban form and in the metabolic form of the urban form, thus relating the risk of the individual element to that of the urban form, depending on discontinuities, which affect the vulnerability.

10.2.3 Urban Form And Risk Mathematics

Each component of the urban form, along with the way it is obtained, characterized by a certain degree of vulnerability that marks its progress and defines its resistance (internal cohesion) and structural stability over time compared with the hazard existing on site. This is a systemic approach, a system being characterized by cohesion.

The product of these two sizes, one characteristic for the element (vulnerability) and one for the other external environment, which is the context (hazard), gives the size of the risk to which the component of the urban form is subjected in time, regarding the danger or negative or sometimes unwanted developments. Vulnerability is here understood as the predisposition of susceptibility of an element to being negatively affected by external causes.

$$R = H \times V$$

If we consider a certain size (whether it is constant or it changes over time, as increase or decrease) expressing the size of existing hazard on the area where elements of urban form take material shape that is included in the space of the formal system (context), the vulnerability associated with these elements characterizes each one individually.

The risk associated with that area is given by the sum of individual risks associated with the elements.

$$R_{a1} = H \times \left(V_1 + V_2 + \ldots + V_n\right)$$
$$R_{a1} = H \times \Sigma V_e,$$

where

V_e represents the vulnerability associated to each element

With the evolution and the emergence of cohesion relations established among these formal elements, initially disparate, a socio-spatial system is set up, which has a limit that incorporates these elements. This is the first stage of the formal identity, this system detaching formally from the context, becoming a formal system.

The first degree of complexity associated with this socio-spatial system is associated with the urban elementary form (urban architectural form), which in terms of information exchange with the outside is a static form, stable, insensitive to perturbations (Fig. 10.1)

If V_1, V_2,.... V_n represent the individual vulnerabilities of the elements, these enter in the computation of the risk for the investigated area proportionally with the number and the importance of these elements in the frame of the composition of the respective form.

$$R_{a2} = H \times V_1 \times a\% + H \times V_2 \times b\% + \ldots + H \times V_n \times z\%,$$

from which we get

$$R_{a2} = H \times (V_1 \times a\% + V_2 \times b\% + ... + V_n \times z\%).$$

Since a %, b %, ..., z % represent subunitary values,

$$V_1 \times a\% \leq V_1, V_2 \times b\% \leq V_2, \ldots , V_n \times z\% \leq V_n,$$

the sum of the terms in the parenthesis is smaller than the sum $V_1 + V_2 + \ldots + V_n$, so R_{a2} is smaller than R_{a1}.

In this case, the risk associated to the basic urban form is lower than the aggregate risks associated with all the constitutive elements, taken separately.

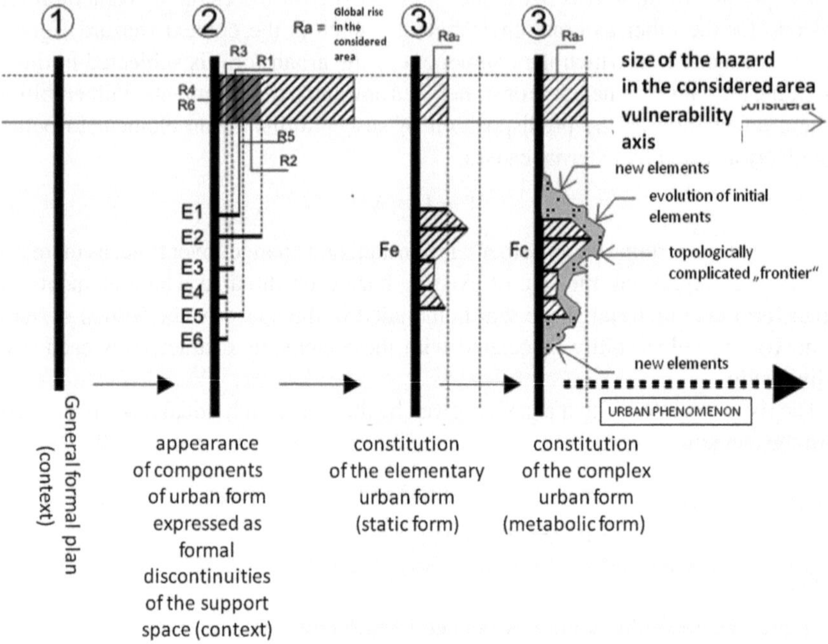

Fig. 10.1 It is illustrated on 4 levels: the uniform formal plan considered of the context; the emergence of some formal elements (E1 ... E6) and of the associated risks (R1 ... R6), the creation of an urban elementary form (fe) and of the risk associated to this (Re < sum of R1 + ... + R6), the constitution of a complex urban form (Fc) and of a new grade of risk Rc (Re < Rc < R1 + ... + R6)

10.2.4 Example

To understand the above demonstration more easily, we will consider a very simple and concrete example. We consider two distinct formal elements that can form a simple architectural form (basic urban form). The first site is an enclosure wall on a square plan, with height equal to the side, while the second element is a square plate with the surface equal to the square enclosure, in the same area (neighborhood).

In the first case we consider these elements as separate entities that are situated in the same area (Fig. 10.2).

We assess the vulnerability associated with these elements in relation to a certain hazard, considering the example of wind, and the protection they offer against it, for example on a scale of 1 to 10. The probability of occurrence of the wind phenomenon is expressed by the hazard H, which has the same value for both elements, the size being characteristic for the area in which they are situated. Assuming $V_{e1} = 3$ and $V_{e2} = 10$ (the board offers no shelter), then if the two elements coexist separately, the risk in that area associated with wind phenomenon is

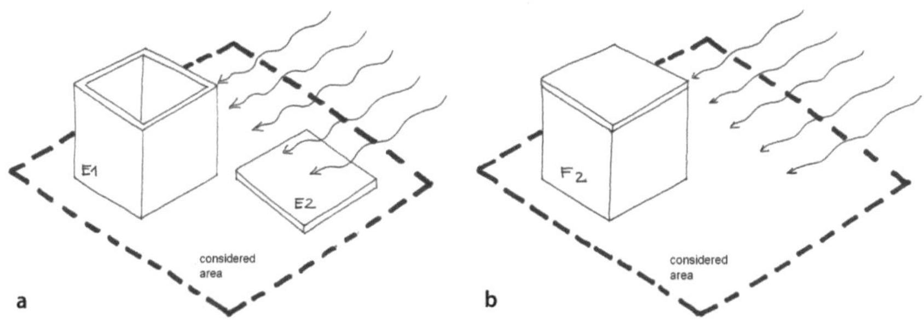

Fig. 10.2 **a** illustrated the risks associated to the component elements considered separately, **b** illustrates the risk associated to the form (elementary urban form) which is lower than the sum of the risks associated to all of the component elements taken separately

$$R_{a1} = R_1 + R_2$$
$$R_{a1} = H \times Ve_1 + H \times Ve_2$$
$$R_{a1} = H \times (Ve_1 + Ve_2)$$
$$R_{a1} = H \times 13.$$

In the second case, if we place the plate over the inner wall, we get a simple form, a cube without a base (as the support is offered by the plan). The share of the two formal elements in the form of incorporation is 80 % for the wall and 20 % for the plate, as we have five sides of a cube.

The vulnerability of the form considered is

$$V = V_{e1} \times 0{,}8 + V_{e2} \times 0.2 = 3 \times 0.8 + 10 \times 0.2 = 2.4 + 2 = 4.4.$$

In this case the risk turns out to be

$$R_{a2} = H \times V$$

$R_{a2} = H \times 4.4$, smaller than R_{a1}.

In other words, in the second case, the respective formal system is characterized by a lower risk related to the hazard considered, offering greater safety when established in that area. So the risk associated to the basic urban form in this case, and, in a general case to the urban form, is lower than the aggregate risks associated with all elements, taken separately.

10.2.5 *Metabolic Urban Forms*

It is worth noting regarding the example given in the previous paragraph that the risk associated considered form (R_{a2}), although smaller in the respective area than the original one, when the elements were a group of separate ones, is higher than that of some of the associated elements forming for example the enclosure wall considered separately (one wall element). This shows that the working together of several elements in building a form (including an urban form) predispose this form to another level of risk than that associated to the elements out of which it consists, and that objectively this is higher than that associated with some of them separately and lower than that associated to others, and that in the frame of the overall risk there also the share these elements have in building up those forms has an influence.

Since, objectively seen, more than one hazard can be met in an area, the attitude towards them, together with their hierarchy, justifies the formal composition as response for the reduction and mitigation of risks. In the example discussed, if we consider other associated hazards such as those connected to rainfall, for example, the risk corresponding to the form made up by the two elements (wall enclosure and cover) is obviously much lower than that of any individual element.

Once the elementary urban form is established as a phase in building up the global urban form which is the urban organism, its evolution over time and inclusion of other elements which emerged as discontinuities of the support plan which is the context, as well as the evolution of the newly created urban form, leads to the emergence of a global limit topologically increasingly more complicated, resulting finally in a complex urban form, which has the attributes of metabolic forms. Metabolism is a characteristic of living organisms, such as the urban organism is in its development along the urban phenomenon. Crăciun (2008) dedicated her work to the metabolism of the urban organism exploring how different disciplines, including music but also astronomy can lead to its understanding.

The necessity of approaching the urban form considering the response this gives to external influence is given by the fact that the definition of the urban form is too general and cannot be precise applied to specific research, like the one considered here about urban risk. For example, considering a subgroup of the "urban frame" elements as one of the expressions of the urban form related to the function of the space, we can talk of a static form. But if we consider the same component of the urban form in interaction with other activity systems, for example the urban modernization, the rehabilitation of urban areas, functional reconversion or urban remodeling, it is clear that these are supposed to transformations which can be included in another formal category. Such activity systems are, as will see, even if rarely, expression of reducing urban vulnerability to hazards. This later form belongs to the category of metabolic forms, which, in a different way related to the static form can have a complicated "frontier" from a topologic point of view: generally they are very sensitive to perturbations and display fluctuations. Subjected to weak perturbations the metabolic form resists through structural stability, but the topology of its support (the context) is affected, and thus changes. When a certain threshold of interaction is passed, the form is solved immediately in a continuum of elementary forms with more simple internal structure, in static forms or in more simple metabolic forms. This process is called generalized catastrophe, according to the Theory of Catastrophes of Thom (1984).

We will give an example for a better understanding of the meaning of metabolic forms. For this purpose we consider the urban form which is the bringing to material shape as presence in a territory, at a certain moment, of a settlement (a town). Depending on the size, the importance and the role which the respective city has in the territory, the "frontier" of this form (its "limit" as existential expression) can be complicated, non-uniform, or in other words topologically complicated. The interaction with external factors, represented by the environmental factors, or the economical, administrative and social factors can lead to the change of its topological support, through the change of its frontiers (administrative limits), characterized for example through the raise (development) phenomenon. Up to a certain threshold, the urban form keeps its internal structure and cohesion. When it exceeds certain development limits (in the economic, population or territory plan), it becomes administratively difficult or inefficient to deal with and it is necessary to divide it into administrative parts (sub-systems) which build separate entities. Such ones are the sectors for the city of Bucharest, the arrondisements for other capitals etc. This way a new urban form, also a metabolic one, is created, which is the sum of more simple metabolic forms. In this case, the limits of the growth (even if it is about a linear growth) mark the passing to another existence stage of the respective urban form, which represent, according to the theory of Thom (1984) an inflexion moment in the evolution, called catastrophic point. We can talk also about the effective risk of dissolving (disappearance) of an urban form in case of hazards and disasters, the object of this chapter and the book it is part of.

In the case of interaction of two metabolic forms we can have as a result fundamental qualitative changes, so that the forms suffer deformations as an effect. Taking into account the characteristics of the metabolic forms, from which we highlight their sensibility to the interaction with external factors or with another form, if this interaction is above a certain threshold, the forms change. In case of the interaction between a static form and a metabolic form, the metabolic form suffers the transformation because its "frontier" is much more sensitive to perturbations. In case of this interaction type we talk mainly of the general effects of the architectural forms (the building) defined as static forms on the global urban form, defined as metabolic form. The effects of these interactions can lead to the growth (development) of the urban form, to its extension in the territory, to an increase of density of the component elements of the urban form, to bringing material shape to some n-functions (continuity) or to dis-functions (discontinuity), as defined previously in our explanation on the theory of systems, in frame of the expression of the existence of the urban form, which is the urban phenomenon.

10.3 Discussion

An example of the interaction of the architectural form as an urban static form with the urban metabolic form is the proposed datascape for the impact of the 1,755 earthquake on Lisbon proposed in frame of the "Semantic enrichment of 3D city models for sustainable urban development" project. The movement suggested by agents is best suitable for it. Another possibility is the representation through fractals, as proposed by Chirvasie (2010), which is however only suitable for organically grown cities, not for those on a pre-established plan.

In case of earthquakes it shall be aimed at the definition of units (urban textures) for element reordering efforts towards an efficient urban structure in case of seismic risks, a seismic vulnerability reduction method for an urban system, through reordering according to Deleuze's (1980) philosophical approach on the flat and striated space, as explained in another chapter of this book (Bostenaru 2014). Urban textures, an element of urban morphology, have been successfully employed in the Earthquake Master Plan of Istanbul (Balamir 2004), one of the few successful practical implementations of the theoretical concept of strategic planning, developed in the 1980s, in the context of urban seismic risk management. It is a systemic approach, as strategic planning is, based on the concept of "risk-sector", which are defined as "sets of causal relations focused on specific urban risks or vulnerabilities". The megacity of Istanbul has the size of a complex metabolic form composed of simple metabolic form as explained above. The simple metabolic forms in this case can be model zones, like (Balamir 2004):

- Extensive development areas,
- Intensive development in seismically safe areas,
- Areas endangered by secondary events,
- Historical-culturally significant areas,
- Rehabilitation areas, highly vulnerable due to seismic or building stock properties, requiring public declaration of their microzonation maps in local action planning.

The later one is the example given by us of a metabolic form.

Bostenaru (2005) identified four historically constituted planning types in disaster management as layers today, and later it was proposed to underlie a grid to these layers, the corner points of which will be analytic elements of morphological subdivision into components. It includes an analysis of the 2003 Earthquake Master Plan of Istanbul. The earthquake vulnerability and the impact of retrofit in the system are evaluated through an analysis of the system functionality. If the Earthquake Master Plan of Istanbul is an example of mitigation planning, an example of resilience planning analyzed there is the one of the "Urban system analysis" work package (Masure and Lutoff 2003), developed in frame of the RISK-UE project. Masure and Lutoff (2003) propose the "urban system exposure" method which comprises the division of elements of the city in strategic and common ones, ranked according to their role in the system during normal, crisis and recovery periods. It is an application of the systemic theory to studies of vulnerability to earthquake hazards. Both strategic and vulnerable elements are mapped. The vulnerability of the "elements at risk" is superposed with their exposure, defined through indicators. In an approach followed by us in teaching "the state of crisis and architectural vulnerability" we mapped such factors for an area in Bucharest (Gociman et al. 2011) and this will be taken into account in the conclusion of this book about the role of urban planning for such vulnerability analyses.

Another approach of involving urban planning into earthquake protection is the Italian one called "Struttura Urbana Minima"—"Minimal Urban Structure"—(Mercuri 2005). The concept was aimed, in line with the Earthquake Master Plan of Istanbul and the RISK-UE project to be developed into the "Strategic Urban Structure". It comprises, like in the Masure and Lutoff (2003) approach the definition of the elements of the city which have to keep their function during the crisis period of a disaster so that the urban organism maintains its integrity. All elements we analysed here regarding the urban organism, urban system and urban structure maintain their applicability. What is new in our approach regards the urban form. However, the Italian example of the city of Crotone in this book (Goretti et al. 2014) follows the approach by Masure and Lutoff (2003) directly, instead of that of Mercuri (2005).

Bostenaru, personal communication, project application (2012) aims at further developing the applicability of the concept of textures in the urban structure is based in the approach in Bostenaru (2005). Aimed is a grid pattern like in computer applications such as the games defined in a step of the project, one example being the simulation of urban development in SimTorino (http://www.laq-tip.polito.it/SimTorino/), or in cities with a pre-established plan, as we mention, and thus opposed to those which can be modeled by fractals considered by Chirvasie (2010). Similarly to the computational morphogenesis of the constructive structure, according to the strategic role of elements in earthquake resilience planning efforts elements will be switched from strategic to common depending on potential vulnerability (result of foreseen retrofit—employing methods from philosophy according to Deleuze's (1980) philosophical approach on the flat and striated space regarding the reordering itself. There are a number of projects dealing with computational representation of morphology, and among them a European Research Council one lead by Debra Laefler.

10.4 Conclusion

The relationship continuity/discontinuity is particularly important for the study of urban form and urban organism in the context of their dynamics. The attempt to define in greater detail the concept of urban form in theory, to interpret its changes through the general dynamics of the world of forms as well as the approach of the urban form related to risk issues, and vulnerability of the urban form are the theoretical and practical contribution of this chapter. From the findings of this chapter we can draw conclusions regarding principles of urban practice regulations regarding the minimum content of the documentation of urban planning for natural hazard prone areas, providing theoretical and practical support for further research in urbanism. A new way of understanding the urban form was done by introducing new parameters in the relationship as urban form—urban phenomenon: risk factors, evolving bifurcations, continuity of urban form, formal development through morphogenesis, the interpretation of urban form in relation to the general laws of the world of forms.

The evolution "urban static (elementary) form"—"urban metabolic (complex) urban" always implies an increase of vulnerability, and therefore of the risk, owing to the qualities of its "border". Increase of formal sensitivity by increasing vulnerability, as an objective process, actually expresses the developing nature of the urban form and its permanent susceptibility to trans-formal and morphogenesis processes, demonstrating its dynamic character, reinforcing the conclusion that there cannot be an "ideal complex urban form", formally established over time and eternal. Hence the deviations in the history of urban planning (ideal cities) and even of architecture, which came to expression in the urban elementary forms which are the architecture objects (the "Falaster", the dwelling unit by Le Corbusier, the cyber home, etc.).

In conclusion, it is interesting to emphasize that regarding the issue of risks to which the urban form is subjected from outside, the context is the determining factor. The context in itself (the physical frame of existence of the urban form) reveals the existence of hazard, while the nature of the context (geographical, economical, social, political, etc.) gives measure of the risk degree, as well as of the vulnerability.

The existence of an urban form that is not subject to any risk factor would be an urban form indifferent to the context, and this, as noted, is not and will not be possible as long as urban form will remain a complex manifestation of human existence in the territory.

References

Alexander C, Ishikawa S, Silverstein M (1977) A pattern language: towns, buildings, construction. Oxford University Press

Aou T, Ohmori H (2005) Computational morphogenesis of structures by extended ESO method ~ application to dynamic problems. Proceedings of international symposium on shell and spatial structures. Bucharest and Poiana Braşov, Romania, p. 175–178

Balamir M (2004) Urban seismic risk management: the earthquake master plan of Istanbul (EMPI). Proceedings of the 13th World Conference on Earthquake Engineering, Vancouver, B.C., Canada, August 1–6, 2004, Paper No. 9005

Bostenaru Dan M (2004) Review of retrofit strategies decision system in historic perspective. Nat Hazards Earth Syst Sci 4:449–462

Bostenaru Dan M (2005) Multidisciplinary co-operation in building design according to urbanistic zonification and seismic microzonation. Nat Hazards Earth Syst Sci 5(3):397–411

Bostenaru Dan M (2014) Aspects of Architecture and Urbanism in the Reconstruction of Disaster: Comparison of L' Aquila (Italy) with Kolontar/Devecser (Hungary) and Corbeni (Romania) in the Context of Participative Reconstruction

Chirvasie C (2010) Simulari urbane. Urbanometrica. Presentation at the ARA Symposium, for abstract see http://www.simpara.ro/ara11/a11_01_09.htm retrived 5 december 2013

Crăciun C (2008) Metabolismul urban. O Abordare Neconventionala a Organismului Urban. Editura Universitara "Ion Mincu", Bucharest

Deleuze G (1980) Mille Plateaux. Minuit, Paris, p. 592–625

Florescu T, Ionita S (1999) Ghid privind exigenţele minime de conţinut al documentaţiilor de urbanism şi amenajarea teritoriului pentru zonele cu riscuri naturale, UAUIM şi Urban Proiect, Bucureşti

Gociman CO, Bostenaru Dan M, Florescu T (2011) *Urban strategies for risk reduction in central protected area – study case Bucharest, in* Proceedings of the International Emergency Management Society (TIEMS) 18th Annual Conference Edited by: Snjezana Knezic, Meen Poudyal Chhetri, Alexandru Ozunu, p. 608–617

Goretti A, Vetta S, Palmieri F, Adamo F, Berlingeri M, Palmieri L (2014) The urban system of Crotone, Italy, facing the earthquake impact, Chapter 8 in this book.

Ioan A (2012) Marcel iancu şi alfabetul său formal: un exerciţiu didactic în derulare (I), Arhitectura 3/2012

Masure P and Lutoff C (2003) Methodology of urban system exposure (USE). Assessment to natural disasters, report, European commission project report RISK-UE

Mercuri C (2005) La "struttura urbana minima" (SUM) come strumento di inquadramento delle azioni di mitigazione nella prevenzione e nell'emergenza, Convegno AssoDiMa

Sandu A (1988) Teoria structurilor urbane. Editura IAIM, Bucharest

Thom R (1984) Stabilite structurele et morphogenese. Intereditions, Paris

Xie YM, Steven GP (1997) Evolutionary Structural Optimization. Springer Verlag

Chapter 11
Timber Frame Historic Structures and the Local Seismic Culture—An Argumentation

Maria Bostenaru Dan

Abstract In this argumentation we will present different viewpoints on the matter that timber skeleton structures are constructive types that are characteristic for zones with high seismicity. Within this scope, we have done research on structures with timber skeleton through the published literature and in the field. Some of these constructive types are vernacular, other resulted from following seismic codes elaborated by the political authorities in periods as far back as the Baroque. Although these structures sustain the hypothesis of local seismic culture, the type is also spread in coastal cities, especially in zones with adequate wood as a resource, which is the "winning" argument.

Keywords Local seismic culture · Architecture resource · Construction materials · Diagonals · Spatial distribution · Traditional architecture · Vernacular · Timber

11.1 Introduction

This argumentation is about the employment of timber frame structures and the local seismic culture. We have chosen this theme for the simple reason that this type of construction, described in the literature as being particularly resistant to earthquakes, is also frequent in countries with low seismicity, such as Germany or England. However, a series of studies, which will be described in the argumentation, attributed the occurrence of this type of buildings to the so-called local seismic culture, a form of vernacular architecture adapted against threats to life and property posed by earthquakes. Some known examples mentioned in this literature do not belong to vernacular architecture, but were assigned to this category by decree of respective forms of government after a strong earthquake. The rationale is open and a counter-argument to satisfy us has not yet been found, that is, one that explains the appearance of this type of construction in both the vernacular architecture of

---

M. Bostenaru Dan (✉)
Department for Urban and Landscape Design, "Ion Mincu" University of Architecture and Urbanism, Academiei 18–20, 010014 Bucharest, Romania
e-mail: maria.bostenaru@iaim.ro

M. Bostenaru Dan et al. (eds.), *Earthquake Hazard Impact and Urban Planning*, Environmental Hazards, DOI 10.1007/978-94-007-7981-5_11,

© Springer Science+Business Media Dordrecht 2014

Fig. 11.1 Architecture of the National Romantic, inspired from vernacular architecture. **a** Szekler Museum, Sfântu-Gheorghe, Romania, architect Károly Kós (1911–1912). Photo: M. Bostenaru, 2011. **b** entrance to the Zoo, architect Károly Kós designed together with Dezső Zrumeczky, Budapest, Hungary (1910). Photo: M. Bostenaru, 1998. **c** Church with the Cock, Cluj-Napoca, Romania, architect Károly Kós (1912–1913), Photo: M. Bostenaru, 2009. **d** Former reformed school, today pupil club, Sfântu-Gheorghe, Romania, architect Károly Kós (1930). **e** House Keresztes, Sfântu-Gheorghe, Romania, architect Károly Kós, (1914). **f** Mortuar Chapel, Sfântu-Gheorghe, Romania, architect Károly Kós (), Photos: M. Bostenaru, 2011. **g** Restoration of the King Mathias House, Cluj-Napoca, Romania, architect Károly Kós (1944). **h** House in the Bretfü street, Cluj-Napoca, Romania, architect Károly Kós, Photos: M. Bostenaru, 2009

countries with low seismicity and those with high seismicity. The subject is topical, as lessons learned from traditional architecture can be useful for designing modern buildings of reinforced concrete, which often have poor performance in earthquakes.

11.2 Definition of Terms

Vernacular architecture is a term used to refer to anonymous architecture or architecture without architects. It uses construction methods based on the use of resources and local traditions. However it is not to be confused with traditional architecture, which may include cult architecture. Many modern architects were inspired by vernacular architecture, especially in the early twentieth century when there was a style called "national romanticism" (Fig. 11.1), creating cult architecture. Today also, some architects are oriented towards lessons learned from vernacular architecture, as star architect Imre Makovecz is, and we see this in our other contributions in the book. These lessons lead to an adaptation of the construction to the environment in the so-called organic architecture. The statics of buildings in vernacular architecture is based on knowledge passed down from generation to generation, the experience of disposal of construction and not on strict calculations as in educated architecture. Therefore, vernacular architecture adapts to local conditions such as climate, culture and construction materials. However the argument is relevant that vernacular architecture also adapts to earthquakes that affect a certain area.

Local seismic culture (Ferrigni 1990; Ferrigni et al. 1993; Helly 1995; Niglio and Ulivieri 2005, a concept promoted by EUCCH in 1991, see publication in 1993) means that traditional architecture meets the earthquake safety requirements of buildings. It is stressed that it is traditional architecture, and not vernacular architecture, because, as it will be demonstrated in this argument, often the constructive adaptation of the system was a result of official orders.

The German language differentiates between two types of building structures, "Massivbau" or massive building load bearing structure, where the function cannot be separated from the spatial conformation of the building and "Skelettbau"/"Fachwerksbau" or skeleton structure in which the two functions can be well differentiated as skeleton and filling it with separate functional spaces. "Fachwerk" however, is the name used for a particular type of skeleton structure, namely timber frame. In English these buildings are called "half-timbered" ("half-timber') and the structure is made of timber and infill.

11.3 Timber Frame Construction Types in Areas Affected By Strong Earthquakes

This type of building with timber structure and stone or brick infill from the vernacular architecture currently performed well in earthquakes in different parts of the earth.

In Europe this type can be found (Fig. 11.2):

- in Greece, on the island of Lefkada (which behaved well in the 2003 earthquake, Ionian Islands, see Karababa 2007; Makarios and Demosthenous 2006),
- in Bulgaria, for example in the Bansko (see Marinov Tchavdar, personal communication), spread through the rest of the Balkans such as Macedonia (Piotr Dimov, personal communication: http://photo.net/photodb/photo?photo_id=9026418, http://photo.net/photodb/photo?photo_id=9026415); on the Black Sea coast there are other types of timber construction (Fig. 11.3),
- in Turkey ("himiş" type, frequent in Safranbolu, which has withstood the earthquake in 1999 well, see Gülkan and Langenbach 2004, http://www.tandfonline.com/action/showPopup?citid=citart1&id=F0011&doi=10.1080/15583050601125998 and http://www.tandfonline.com/action/showPopup?citid=citart1&id=F0002&doi=10.1080/15583050601125998 in Langenbach 2007; while the first type is related to full timber-infill construction such as the typical "Fachwerk", the second is related to the Balkan architecture from Bulgaria and Macedonia which were long under Turkish rule),
- in Italy, in the north, in the Dolomitelor area (see Cecotti et al. 2006),
- in the triangle Germany-France-Switzerland, the southern type of Fachwerk (see Bostenaru 2004, Figs. 11.4 –11.6),
- in Romania (the so-called construction in "paiantă" in the Danube Plain, see ICCPDC 1989).

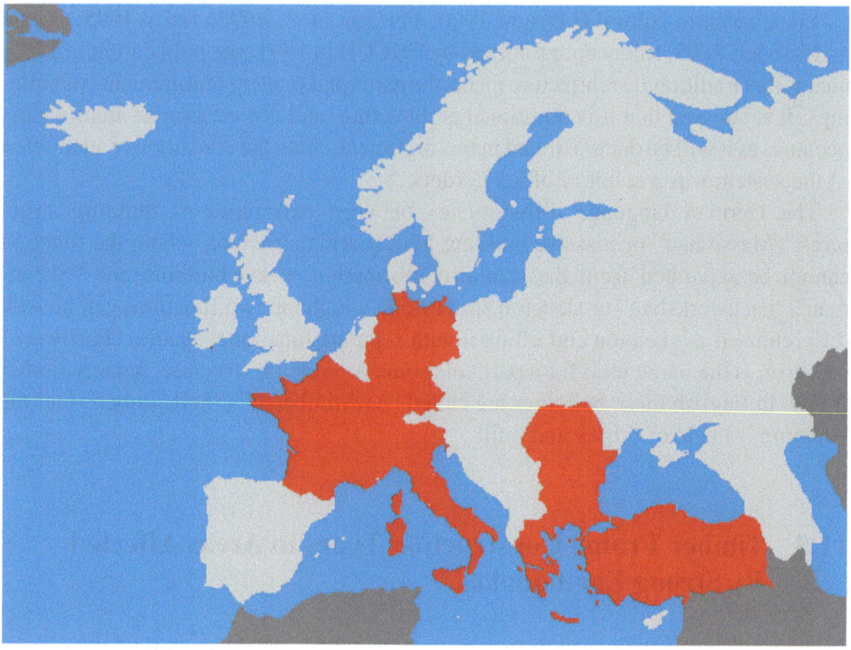

Fig. 11.2 Map of Europe having highlighted the countries analysed in the paper where vernacular timber architecture reflects the local seismic culture

Fig. 11.3 Timber architecture in Bulgaria, Nessebar. Observe the timber lacing in the stone masonry of the ground floor and the entirely timber structure with consoles in the floor above. Photos: M. Bostenaru, 2011

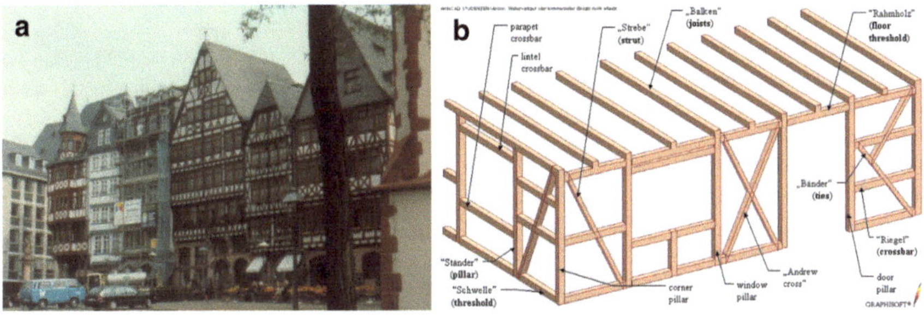

Fig. 11.4 Southern German "Fachwerk" type, **a** Reconstructed houses in Frankfurt on the Maine, Frankfurt is situated in Hessen, in the land where in the city of Wetzlar we find one of the oldest Fachwerk buildings, from 1356, the year of the Basel earthquake, and which is subject of tourist tours. Photos: M. Bostenaru, 2004. **b** Structural skeleton

Fig. 11.5 "Colombage" type, **a** an example relevant for the history of architecture and **b** city-scape with this typology, from Strassbourg, France. Photos: M. Bostenaru, 1997

In Asia constructions belonging to this typology can be found (Fig. 11.7):

- in Nepal (the "dhaji Dewar" type, see Langenbach 2009, http://www.tandfonline.com/action/showPopup?citid=citart1&id=F0005&doi=10.1080/15583050601125998 in Langenbach, 2007);
- in Pakistan (the "bhatar" type, which resisted better the earthquake of 2005, see Schacher 2008), in Kashmir (the "Taq" type, Langenbach 2007, http://www.tandfonline.com/action/showPopup?citid=citart1&id=F0004&doi=10.1080/15583050601125998).

In Southern America the constructions belonging to this typology can be found (Fig. 11.8):

- in Haiti (the "gingerbread" type, see Phillips 1984, which has withstood the earthquake of 2010 well, Langenbach et al. 2010, http://haiti-patrimoine.org/),
- Peru (the "quincha" type, see Carbajal et al. (2005), which stood in the Baroque period),

Fig. 11.6 Southern "Fachwerk" houses in Switzerland, close to the dome, but not the ones surviving the 1356 earthquake. Photo: M. Bostenaru, 2006

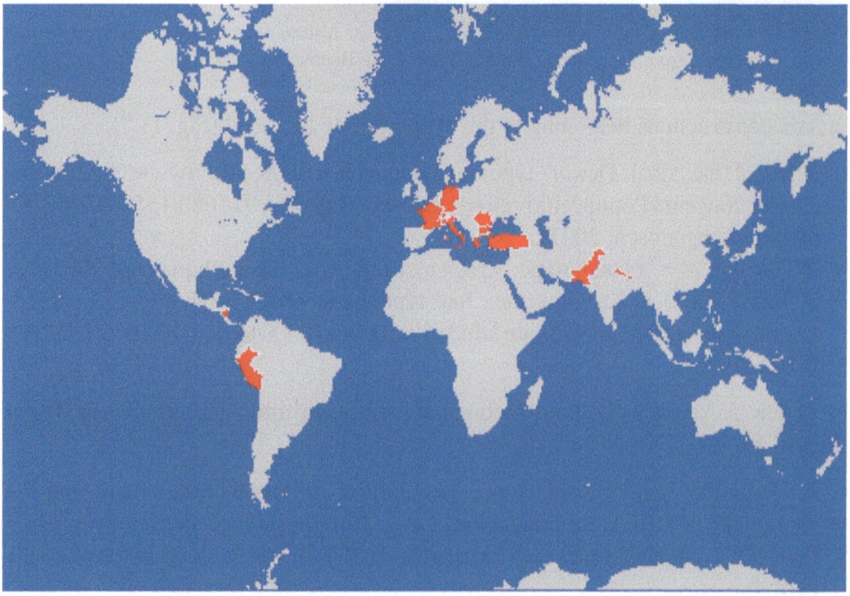

Fig. 11.7 Map of the world having highlighted the countries analysed in this work where vernacular timber architecture reflects the local seismic culture

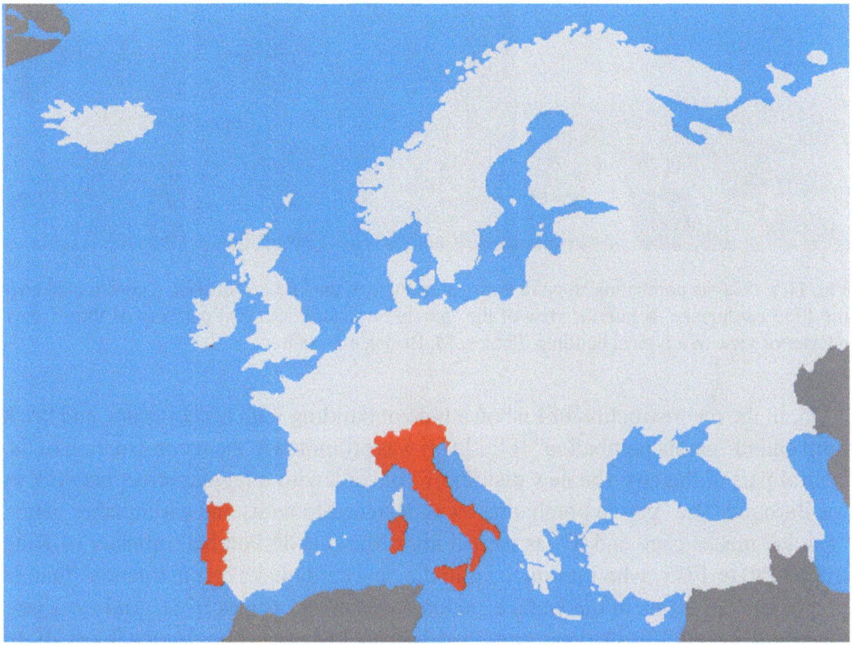

Fig. 11.8 Map of Europe having marked countries analysed in this work where the local seismic culture has been revived by a decree

- El Salvador (the "bahareque" type, see Lang et al. 2007, which withstood well the 2001 El Salvador earthquakes, to name the most recent),
- Nicaragua (the "taquezal" type, see Langenbach 2007).

Some of these types of buildings were investigated by Randolph Langenbach (http://www.conservationtech.com and http://www.traditional-is-modern.net). Langenbach found the existence of this type of construction in ancient times and has documented that of such a construction in Herculaneum, Italy up to this day (Langenbach 2007 http://www.tandfonline.com/action/showPopup?citid=citart1&id=F0007&doi=10.1080/15583050601125998). In the triangle Germany-France-Switzerland the timber skeleton type has performed well after the 1,356 earthquake in Basel (in German South-type "Fachwerk" in France "colombage"), thus preventing from collapse the housing of the then poor population and exposing that of stone of the rich population. An example which survived is to be found close to the dome (It is protected as a monument and was built between two sandstone struts of the church St. Martin, the back part being of stone http://www.meteoriten.ch/www/laden1.html). The "gingerbread" type in Haiti is derived from the French type "colombage" and was introduced by the colonists.

In some areas, the seismic culture was lost, as local seismic culture depends on the frequency of recurrence of earthquakes (EUCCH 1993; Helly 1995). Therefore earthquake resistant techniques had to be revived by decree. After the earthquake of

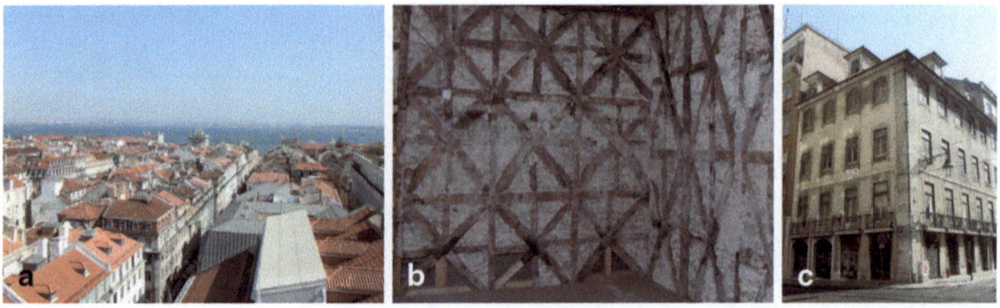

Fig. 11.9 "Gaiola pombalina" type in Portugal. **a** View of the "Baixa" district, reconstructed after the 1755 earthquake. **b** Interior view of the "gaiola" structure, site visit courtesy of Vitor Coias. **c** Exterior view of a typical building. Photos: M. Bostenaru, 2008

1755, in the reconstruction of Lisbon a type of building with timber frame and brick infill called "gaiola pombalina" (Fig. 11.9) was promoted in the reconstruction of the central part of the city, the new district of "Baixa", with a regular street network as for Baroque. The "gaiola pombalina" was designed to withstand earthquakes better. "gaiola" means cage and it was named after Marquis de Pombal, minister of King from 1750 to 1777, who introduced it in the post earthquake reconstruction. "gaiola pombalina" presents in the walls a skeleton of timber crosses in St. Andrew shape with infill and thick-walled masonry load-bearing losing thickness to the floors above in the exterior walls. In the exterior the buildings have no Baroque style features, but are simply covered, mostly with ceramic tiles of "azulejos", but sometimes just plastered. The masonry may fall out during earthquakes, but the building remains standing because of the timber skeleton inside. This type of construction is considered the world's first building designed to be resistant to earthquakes, with experimental research on the models. In Italy as well, a type called "casa baraccata" was promoted in a building code by the Bourbon government in a similar period following the 1783 earthquake in southern Italy (Tobriner 1983), as the "Gaiola Pombalina" during the Baroque. Different from the "Gaiola Pombalina" in the "casa baraccata" the timber frame is in the exterior walls, but plastered and difficult to be recognised, not like the timber uncovered following the romantic movement in "Fachwerk"; thus, this is an addition to traditional architecture. Both "gaiola pombalina" and "casa baraccata" are subject of investigation today (Duţu 2011; Ceradini and Bianco 2008). A theme suited to further study is the relationship between vernacular architecture with timber frame in northern Italy (as described in Cecotti et al. 2006) and the introduction of the standard type in the south, where there was such vernacular culture, as the ruins of Pompeii documented (Langenbach 2007). A similar situation we enter in case of the "gaiola pombalina", as in the northern part of Portugal the timber frame typology is present (for example in the historic city of Guimaraes, Fig. 11.10), but this is not an earthquake prone zone. Unlike Lisbon, in Nothern Portugal the outside walls are also made with timber skeleton, and present consoles like the "Fachwerk" type, or recesses as we will later see for reinforced concrete in early twentieth century urban planning. The basement is out of a heavier material as in the Balkan area. The vernacular influences might be due to the fact that commerce was done between the

Fig. 11.10 Buildings in the historic centre of Guimaraes, a city in Northern Portugal. **a** a typology characteristic for Portugal, with Mediteranean balconies. **b** a typology related to the Balcan timber typology. **c** Typical buildings. **d** Detail. Note the stone basement and timber frame upper floor, as in Lefkada and other Balkan areas. Photos: M. Bostenaru, 2010

Balkan countries and Portugal, commerce proved by the origin of names (for example for oranges), a supporting issue for our later argument regarding the occurrence of the type in areas close to the sea.

In the seventeenth and eighteenth centuries, not only Lisbon and Sicilian cities (Sicilian Baroque) but also many other areas were affected by earthquakes and rebuilt in Baroque style, after the earthquake of 1693 like some cities in the Philippines and Guatemala (http://en.wikipedia.org/wiki/Earthquake_Baroque). In the framework of the conversion after an urban earthquake, changes were easier to implement than, for example, the plan of reconstruction of London after the great fire of 1666.

Therefore it appears that timber frame structures have proven to be resistant to earthquakes in the following two situations:

- Emerging vernacular in countries where earthquakes are common and thus may have been an attempt to adapt the structures to be more earthquake resistant over many earthquakes;
- Published by decree in times when urban regulations received another statute, the Baroque, following a major earthquake.

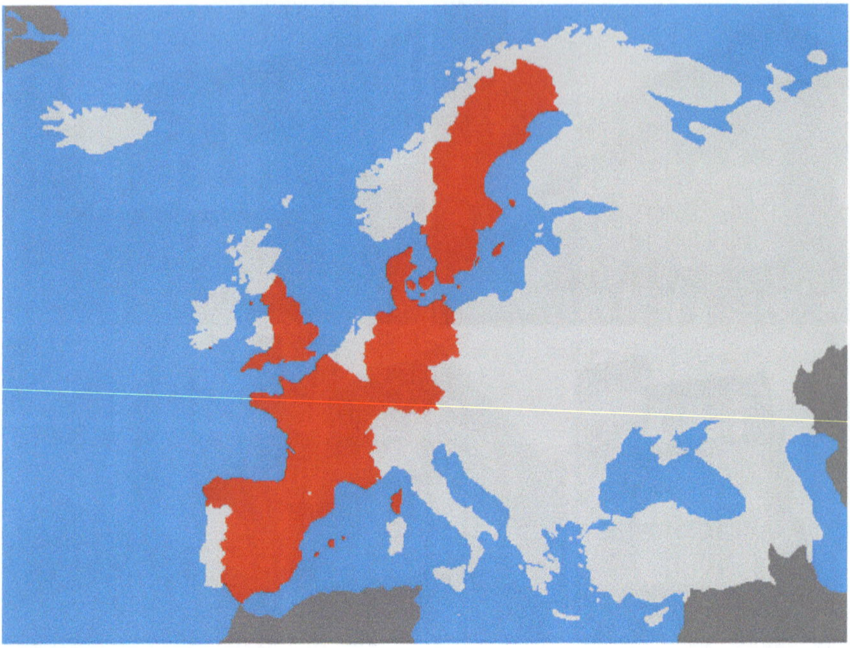

Fig. 11.11 Map of Europe having marked the countries analysed in this work with vernacular half timbered houses but low seismicity

However, this distinction is rarely made in the literature. Because of the reasons stated above this type of building is considered to be an expression of the local seismic culture.

11.4 Timber Frame Construction Types in Low Seismicity Areas

These are common buildings found also in areas where seismic risk is not a problem (Fig. 11.11), such as the "half-timbered" type in England and "Fachwerk" in northern Germany, or "colombage" in France. The same typology can be found in Skandinavia (for example Denmark and Sweden, see Copani 2007). It was also spread, at the time of Baroque in the Madrid area of Spain (Gonzales Ridondo and Aroca Hernandez-Ros 2003), which is less seismic than other areas of the country of moderate seismicity.

Uhde (1903) documents the existence of such buildings in France in the regions of Normandie, Bretagne and Alsace (Dreux, Laval, Annonay, Bayeux/stone in-filled), Morlaix, Dol, Yville, Compiegne/stone infilled, Rouen, Rheims, Abbeville, Boulogne, Beauvais, Angers, Lisieux, St. Brieux, Caen, Strassbourg).

"Fachwerk" buildings are a complete timber framework, with irregular bracing, both in the exterior and interior walls. The construction details are very similar to those of "dhaji Dewari" specific for Nepal-Kashmir. Although the details and the

Fig. 11.12 "Half-timbered" house in England (Hyde Park, London). Photo: M. Bostenaru, 2002

logic of design are similar, "Fachwerk" may be an example of vernacular architecture of the local seismic culture. The infill is made with adobe on a mat, masonry or wooden boards. The southern type was subject to detailed analysis in the World Housing Encyclopedia (Bostenaru 2004, 2010a), but the northern type (Lachner 1887) is now interesting for the argumentative discourse. These two sub-types are one placed in a seismic risk area, while the other is not. The northern "Fachwerk" type is more structurally coherent than the southern type. In the Northern type the consoles are subordinated to the overall concept of a ship. Characteristic of this type of construction in southern Germany are outbuildings and annexes, like "Erker" (balcony), "Chörlein", "Ecktürmchen" (corner little tower), "Lugaus" and "Dacherkertürmchen" (combinations of balconies and towers). "Lugaus" are rectangular front buildings spanning more stories, starting either on ground floor level or in a console/cantilever over the stone ground floor. At the upper side it ends with an independent little tower (Bostenaru 2008).

The "half-timbered" buildings are different, consisting mainly of vertical frame elements at low distances, without horizontal ones. It is found not only in England (Fig. 11.12), but also in the British crown territories, for example in Canada. In England they are in high number also in the area that was affected by the 1884 Colchester earthquake.

11.5 Common Features of Timber Frame Construction Types in Areas with Low and High Seismicity

We sought an alternative explanation for the origins of this design. Two expressions of this type have been examined more closely: "gaiola pombalina" (Portugal) and "Fachwerk" (Germany). Common to the "gaiola pombalina" and the northern and sub-type "Fachwerk" is the inspiration from structures like ships, as some authors have noted (Cardoso and others 2004; Lachner 1887). In their purest form, sub-type

North German buildings “Fachwerk” have a similar structure in the way ships are made with consoles on the upper floors. “Gaiola pombalina” was also generated with knowledge of building ships. And sub-type “himiş” added to the sea proximity of the earlier two, might suggest that the origin of this construction type can be searched in urban culture and skills related to the urban occupations of port areas, including the exchange of knowledge (in Bulgaria this type is found in both sea proximity and in the mountains, but in Romania as well as in the Greece islands it is only in the sea proximity) rather than in local seismic culture, as we argued in previous articles (Bostenaru 2005).

11.6 Architecture Resources

One should also look to other aspects of domestic architecture, such as availability of timber resources. The “Fachwerk” type and its regional varieties may be a result of efficient use of wood resources. In Romania, where wood resources were rich, there was no need of economy in both types of wood and as such historical timber houses have massive wood walls. As it can be demonstrated by a visit to the Village Museum, in Romania the timber skeleton is not typical of vernacular housing, and often the wood material is limited to use in one of the upper floors, while the lower floor is made of more resistant material, as is the case of the neighbouring Bulgarian architecture. As well as in other countries where this type of construction is found, like Denmark, Sweden and Germany, the accent lays on timber of quality for massive construction, as these countries are all rich in forests. A study of the vegetation is imposed to draw further conclusions.

It is true that forms of deliberate use of this structural type which prevent the devastating effects of earthquakes were signs of cult architecture: in Portugal and in Italy they came as a decree from the government after a strong earthquake. A topic for further research is also the division between north and south of the different typologies across countries, as in urban planning we differentiate between development in the north and the south of the city in the Northern Hemisphere.

11.7 Lessons from the Timber Frame House for Modern Concrete Buildings

In the 1999 Izmit earthquake (Turkey), the 2005 Kashmir earthquake (Pakistan) and the 2010 Haiti earthquake timber frame buildings have performed well, and many of the modern reinforced concrete frame buildings collapsed. For this reason Langenbach (2004, 2007) promote the lessons learned from timber structures to modern concrete, namely the existence of “reinforcement” (bracing) cross.

In Romania local seismic culture exists. It was shown, for example, in adapting Saxon fortified churches in Transylvania towers taking into account the distance to the epicenter in Vrancea (personal communication Sever Georgescu, based on presentations at the European University Centre for Cultural Heritage 1995, 1997, at the University of Karlsruhe 2002 and at the "Ion Mincu" University of Architecture and Urbanism 2011). The typology of timber framed homes can be also found in the Danube Plain in the popular form of paiantă, but in 1830–1840 the use of wood was reduced by royal decree (personal communication Sever Georgescu, presentation at the University of Architecture and Urbanism "Ion Mincu", Georgescu 1994, 1995, 1997). They held up well and we can use as an example the 1977 earthquake in Zimnicea (ICCPDC 1989). ICCPDC (1989) published an extensive study of traditional architecture in Romania, which included aspects of construction materials and earthquake resistance. A map of Romania displays this interdependence. The typology of village homes in Romania includes timber frame construction in one component or multi-material form (rigid floor, wooden floor).

In Romania after the 1940 earthquake there was a special type of building with concrete frame structure but diagonal bracing (Bostenaru 2004b, Fig. 11.13a), making it different from the more spread vulnerable type with reinforced concrete skeleton designed for gravity loads only (Fig. 11.13b). Such buildings resisted better in the 1977 earthquake (ex. the building staying near the collapsed one at "Colonade"). It is an example of local seismic culture today, which in fact would support the theory that such constructions as the ones analyzed in this argument are an expression of local seismic culture, but not only that.

From the local seismic culture of timber framed buildings are also drawn lessons for the retrofit of reinforced concrete skeleton or frame structures with:

- Steel diagonals (Higashi et al. 1984; Penelis and Kappos 1997; Dritsos 2004; Bostenaru 2006, Fig. 11.13c),
- Energy dissipaters (Antonucci et al 2001, http://www.fip-group.it/fip_ind_eng/prodotti/dispositivi_antisismici/frame_dissipatori_viscoelastici.html),
- Fibre reinforced concrete diagonals to bind the infill to the frame (see Borgogno 2001; Öztürk and Bostenaru 2007).

Steel diagonals in reinforced concrete frames performed well in subsequent earthquakes, as they were applied in the 1980s in Mexic (Del Valle 1980; Del Valle et al. 1988; post-tensioned diagonals Miranda and Bertero 1990). Charleson (2001) studied the architectural expression of seismic strengthening with steel diagonals on reinforced concrete buildings of the 1960s, which are similar to those in the Mexican case. Bostenaru (2006) included a study on the cost efficiency of adding steel diagonals compared to other measures, as well as of steel diagonals in different positions in a building.

Energy dissipaters were first applied by "fip industriale" in 2000 in Italy after the 1997 Umbria and Marche earthquake at the retrofit of the heritage school building Gentile Fermi from 1950s, in Fabriano, Ancona, Italy, following the pre-war style of Rationalism; this was also the case for the 1950s rationalist Liceo Classico Perticari, in Senigallia, provincia Ancona, central Italy, which was affected by the Molise earthquake. In the first case the diagonals are visible at the exterior, penetrating the walls, with special attention to the windows; in the second they are hidden in the

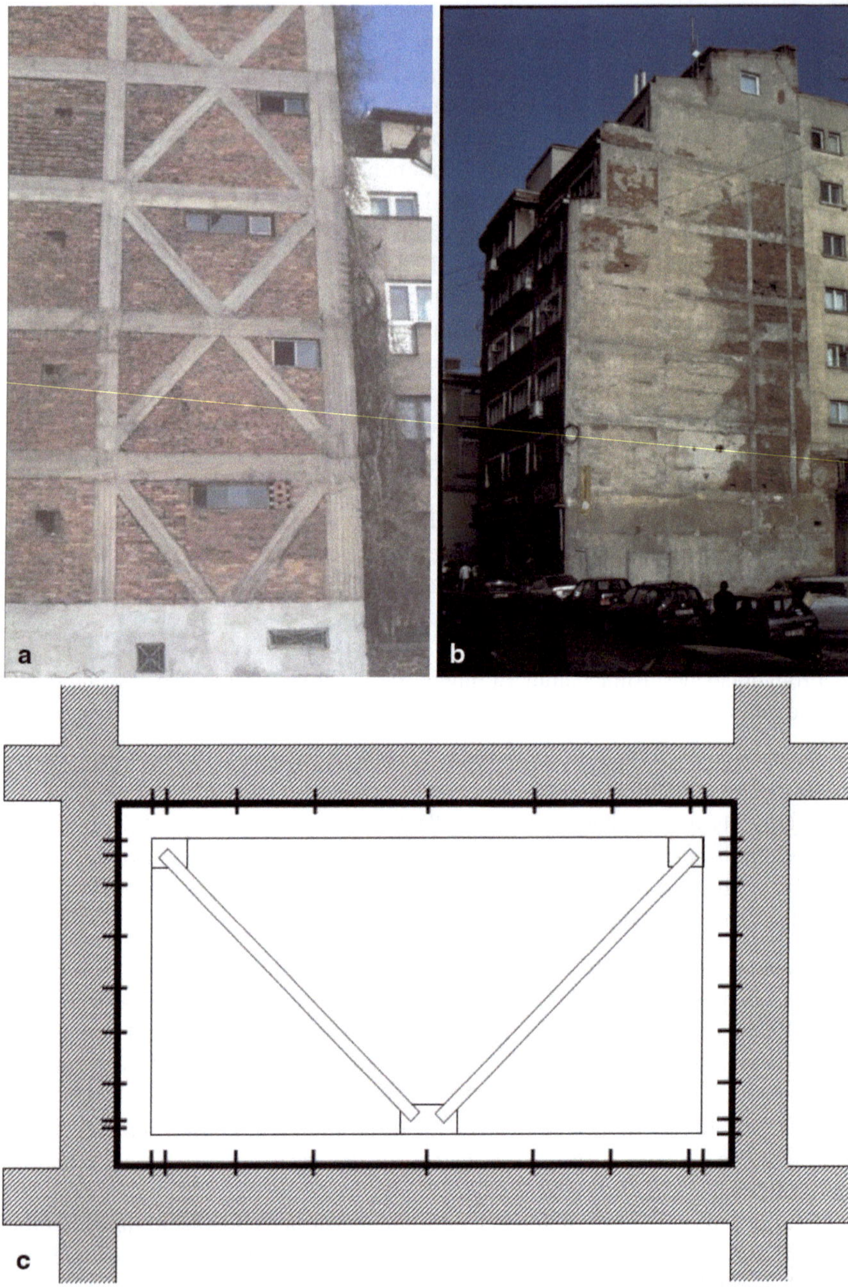

Fig. 11.13 Example of lessons learned from the local seismic culture for reinforced concrete buildings: **a** reinforced concrete diagonals in a post 1940 Vrancea earthquake building, Bucharest, Romania. **b** the vulnerable reinforced concrete skeleton building with recesses. Photos: M. Bostenaru, 2002. **c** steel diagonals in a reinforced concrete frame

interior and the brick façade is intact, using different dissipative mechanisms (viscoelastic elastometric dissipators at Gentile Fermi and hysteretic metal dissipators at Petricari). Two schools in Potenza: Scuola Domiziano Viola and Luigi la Vista were retrofitted in 2010, in the first case with visible colored diagonals inserted in the exterior of the facade and windows adequate to these (Ponzo et al. 2010). Potenza is situated in one of the most seismically endangered areas in Southern Italy, where also the casa baraccata was introduced. Retrofit of school buildings is a priority in Italy (Grant et al. 2007), taking also into account the fact that in the 2002 Molise earthquake most of the victims were school pupils in a school in San Giuliano di Puglia. Adding dissipation elements is according to Fardis (1988) not as effective as it might seem because the displacement demand is reduced by increase of stiffness. An approach related to that of the Romanian post-war typology with diagonals. Also, Bostenaru (2010b) showed the modeling of the space-structure approach in historic reinforced concrete buildings outgoing from Neoplasticist painting and the 45° rotations of Mondrian as in the De Stijl movement, which led to triangle shaped windows, so that the intervention is feasable.

Also Tobriner (2006) remarks on the historic role of bracing against earthquakes affecting San Francisco. Bracing should be more often and more creatively employed in case of reinforced concrete, learning not only from timber, but also from steel.

11.8 Conclusions

Examples of timber frame architecture have been investigated from around the world in order to determine whether development of this design is or it is not the seismic expression of a local culture. Local seismic culture was defined as a characteristic of vernacular architecture, or architecture without architects, in which over several generations adaptation of architecture to the environment has been practiced in case of earthquake resistance. The literature considers timber frame construction, as the type of "Fachwerk" as an expression of local seismic culture and promotes the lessons to be learned from these examples which withstand earthquakes better than advanced technical forms such as concrete construction. There are indeed examples of timber frame construction which proved to be resistant to earthquakes in places where they have a tight recurrence period. Many were also deliberately promoted in the reconstruction of cities during the seventeenth and eighteenth century. At the same time, this type of construction is prevalent also in areas with low seismicity, such as Germany, England and France. It is our theory that the use of timber in the form of "Fachwerk" can be assigned to more than the local seismic culture. We can say that the use of urban development codes promulgated in Portugal and Italy might have been inspired by the vernacular type that survives well in earthquakes, but the reasons for the development of the vernacular type are different. Although there are explanations for this type being able to better withstand earthquakes there, the reason for the spreading of this type can be a general vernacular architecture reason, such as availability of local building materials. Especially if the local culture and more often occurring fires suggested the employment of the stone house, which was more expensive (Bostenaru 2007).

References

Antonucci R, Balducci F, Castellano MG, Ahmadi H, Goodchild I, Fuller K (2001) Viscoelastic dampers for seismic protection of buildings: an application to existing buildings. In: Proceedings of the 5th world congress on joints, bearings and seismic systems for concrete structures, Rome

Borgogno W (2001) Erdbebenertüchtigung eines Hochhauses, CFK-Lamellen und – Schubwinkel zur Verstärkung von Mauerwerk [Seismic retrofit of a high rise building, Carbon Reinforced Polymer laminates and angles to strengthen masonry], SIA: tec21, 36 ("Erdbebensicheres Bauen" [Earthquake safe building]): 13–16 (in German)

Bostenaru Dan M (2004a) Half-timbered house in the "border triangle" (Fachwerkhaus in Dreiländereck). In: Brzev S, Greene M (eds) World housing encyclopedia—summary publication 2004, EERI, Oakland, report # 108/Switzerland

Bostenaru Dan M (2004b) Reinforced concrete frame structure with brick infill walls and diagonal bracings. In: Brzev S, Greene M (eds) World housing encyclopedia—summary publication 2004, EERI, Oakland, Report # 71/Romania

Bostenaru Dan M (2005) Reassessment of lessons learned from multi-storey shells for space structures. In: Mihailescu M, Mircea C (eds) Proceedings of the International Symposium on shell and spatial structures, Mediamira, Cluj Napoca, vol II, p 503–510

Bostenaru Dan M (2006) Wirtschaftlichkeit und Umsetzbarkeit von Gebäudeverstärkungsmaßnahmen zur Erdbebenertüchtigung. Grundlagen und Lösungsansatz unter besonderer Berücksichtigung der Situation in Bukarest, Rumänien. [Applicability and economic efficiency of building strengthening measures for seismic retrofit. Basics and solution approach under special consideration for the situation of Bucharest, Romania]. Shaker Verlag, Aachen (in German)

Bostenaru Dan M (2007) Építési kőanyagok használata és a történelmi vázszerkezetű kőépületek előnyös viselkedése földrengés alatt [Using materials of the building stone and advantageous behavior of the historical stone buildings skeleton structures during earthquake], Anyagvizsgálok lapja. J Mat Testers Mag 17(1):36–42. http://www.anyagvizsgaloklapja.hu/en/old.php?cid=1069 (in Hungarian)

Bostenaru Dan M (2008) "Gaiola pombalina" buildings in Portugal and Fachwerk buildings in Germany. In: Proceedings of the Historic Structures International Conference: Historic Timber Structures, 16–18 october 2008, Cluj-Napoca

Bostenaru Dan M (2010a) Vernacular and modernist housing in Germany and Romania. An analysis of vulnerability to Earthquakes, Cuvillier

Bostenaru Dan M (2010b) Historical use of concrete and innovation in architecture, in: Paulo J. S. Cruz (ed.): Structures and Architecture, CRC Press, Leiden, p 433–434. Full paper on attached CD ROM

Carbajal F, Ruiz G, Schexnayder CJ (2005) Quincha Construction in Perú. Pract Period Struct Des Constr 10(1):56–62. http://ascelibrary.org/doi/abs/10.1061/(ASCE)1084-0680(2005)10:1(56)

Cardoso R, Lopes M, Bento R (2004) Earthquake resistant structures of Portuguese old 'pombalino' buildings. In: Proceedings of the 13th World Conference on Earthquake Engineering, IAEE, Vancouver, BC, Paper # 918

Ceccotti A, Faccio P, NART M; Sandhaas C, Simeone P (2006) Seismic behaviour of historic timber-frame buildings in the Italian dolomites. Proceedings of the ICOMOS International Wood Committee 15th International Symposium, Istanbul and Rize (Turkey), September 18–23. http://www.icomos.org/iiwc/15/ceccotti.pdf

Ceradini V, Bianco (2008) A the Casa Baraccata after the earthquake of 1908 in Reggio Calabria: methods of investigation for knowledge techniques and assessment of the conservative conditions. In: Binda L, di Prisco M, Felicetti R (eds) Proceedings of the International RILEM Conference on Site Assessment of Concrete, Masonry and Timber Structures SACoMaTiS 2008, 1–2 September 2008, Varenna, Como Lake, Italy, p 1209–1218. http://www.rilem.net/gene/main.php?base=500218&id_publication=63&id_papier=2410

Charleson A, Preston J, Taylor M (2001) Architectural Expression of Seismic Strengthening. Earthquake Spectra 17:417–426. doi:http://dx.doi.org/10.1193/1.1586182

Copani P (2007) Timber-frame buildings in Scandinavia: high deformation prevent the system from collapse, from material to structure—mechanical behaviour and failures of the timber structures. In: Proceedings of the ICOMOS IWC—XVI international symposium—Florence, Venice and Vicenza 11–16 November 2007. http://www.icomos.org/iiwc/16/copani.pdf

Del Valle Calderon E, Foutch DA, Hjelmstad KD, Figueroa-Gutierrez E, Tena-Clunga A (1988) Seismic retrofit of a RC building: a case study, Proceedings of the Ninth World Conference on Earthquake Engineering, August 2–9, 1988, Tokyo-Kyoto, Japan, Vol. VII, pp 451–456

Del Valle, E. (1980) Some lessons from the March 14, 1979 earthquake in Mexico city, Proceedings of 7th world conference on earthquake engineering, vol 4, pp 545–552

Dritsos S (2004) Retrofit of reinforced concrete buildings. The University of Patras, Greece (in Greek)

Duţu A (2011) Contribuţii privind analiza comportării la seism a structurilor cu schelet din lemn şi umplutură din zidărie [Contributions to the analysis of the seismic behaviour of structures with timber skeleton and masonry infill], PhD thesis, UTCB Bucharest (in Romanian).

European University Centre for Cultural Heritage (EUCCH: Ferrigni F, Helly B, Rideaud A.) (1993) Atlas of Local Seismic Cultures, Supplement to Stop Disasters. The United Nations International Decade for Natural Disaster Reduction Newsletter, 12 (March–April).

Fardis MN (1988) Seismic assessment and retrofit of RC structures. Proceedings of the eleventh European conference on earthquake engineering—Invited lectures, Paris, France, pp 53–64

Ferrigni F (ed) (1990) San Lorenzello A qui recherche des Anomalies protegent. CUEBC, Ravello

Ferrigni F, Helly B, Rideaud A, Valiante M (1993) Atlas of Local Seismic Cultures, European University Centre for Cultural Heritage

Georgescu ES (1994) Characteristics of local seismic culture in Romania. The 4th Intensive Course on seismic vulnerability of old buildings and local cultures. European University Centre for Cultural Heritage, CUEBC, Ravello

Georgescu ES (1995) Local seismic culture in Romania—Transylvania Investigation of medieval buildings. The 5th Intensive Course on seismic vulnerability of old buildings and local cultures. CUEBC, Ravello

Georgescu ES (1997) Aspects of local seismic culture in Romania compared to other Mediterranean countries. The 7th Intensive course on reducing vulnerability of century existing buildings fund by rediscovering local seismic culture. CUEBC, Ravello

Gonzalez Ridondo E, Aroca Hernández-Ros R (2003) Wooden framed structures in Madrid domestic architecture of the 17th to 18th centuries. In: Proceedings of the first international congress on construction history. Instituto Juan de Herrera, Madrid

Grant DN, Bommer JJ, Pinho R, Calvi GM, Goretti A, Meroni F (2007) A prioritization scheme for seismic intervention in school buildings in Italy. Earthquake Spectra 23:291–314. doi:10.1193/1.2722784

Gülkan P, Langenbach R (2004) The earthquake resistance of masonry and timber buildings in Turkey. In: Proceedings of the 13th World Conference on Earthquake Engineering. Vancouver, BC, Paper # 2297

Helly B (1995) Local seismic cultures: a European research program for the protection of traditional housing stock. Annali di Geofisica, vol XXXVIII, (5–6):791–794

Higashi Y, Endo T, Shimizu Y (1984) Experimental studies on retrofitting of reinforced concrete buildings. Proceedings of Eighth World Conferencce on Earthquake Engineering. San Francisco, United States, vol. 4, pp 477–484

Institutul Central de Cercetare, Proiectare şi Directivare în Construcţii [Central Institute of Research, Design and Guidelines in Construction] (ICCPDC) (1989) Locuinţa sătească din România. Studii de arhitectură tradiţională în vederea conservării şi valorificării prin tipizare [Village housing in Romania. Studies of traditional architecture for the conservation and valorisation through typisation] (in Romanian)

Karababa F (2007) Local practices construction seismic vulnerability reduction as a means to sustainable development and, PhD Thesis. University of Cambridge. http://www.island vulnerability.org/europeanunion.html and http://www.researchgate.net/profile/Faye_Karababa/publication/

Lachner C (1887) Geschickte der Holzbaukunst in Deutschland [History of the art of construction in timber in Germany], EA Seemann, Leipzig, 1887. (Reprinted in "Libri rari", Hannover, Germany, 1983), 1–132 and 1–130 (first part the South-German subtype, the second part the North-German subtype) (in German)
Lang D, Merlos R, Holliday L, Lopez Menjivar MA (2007) Vivienda de Bahareque, World Housing Encyclopedia Report, available online at http://www.world-housing.net/whereport1view.php?id=100159. Accessed 6 Oct 2011
Langenbach R (2007) From "Opus Craticium" to the "Chicago Frame": earthquake-resistant traditional construction. Int J Archit Heritage 1(1):29–59. doi:10.1080/15583050601125998, free access at the online version of the journal as indicated in the article
Langenbach R (2009) Do not tear it down! Preserving the earthquake resistant vernacular architecture of Kashmir. UNESCO, New Delhi
Langenbach R, Kelley S, Sparks P, Rowell K, Hammer M, Olsen JJ, Ed. Avrami E (2010) Preserving Haiti's Gingerbread houses: 2010 earthquake mission report, World Monuments Fund. http://www.wmf.org/dig-deeper/publication/preserving-haiti%E2%80%99s-gingerbread-houses-2010-earthquake-mission-report
Makarios T, Demosthenous M (2006) Seismic response of traditional buildings of Lefkas Island, Greece. Eng Struct 28:264–278
Miranda E, Bertero VV (1990) Post-tensioning techniques for seismic upgrading of existing low-rise buildings. Proceedings of the fourth US national conference on earthquake engineering, Palm Springs, California, Vol. 3, pp 393–402
Niglio O, Ulivieri D (2005) Vernacular architecture and "historical seismography" yearly research experience. In Modena C, Lourenço and Roca P (eds) Structural analysis of historical constructions, Taylor & Francis, London, p 203–212
Öztürk G, Bostenaru Dan M (2007) Comparative study of FRP seismic retrofit of existing URM infilled RC structures from a construction management point of view. In: Triantafillou TC (ed) Proceedings of the 8th international symposium on fiber-reinforced polymer reinforcement for concrete structures, 16–18 July, 2007. University of Patras, p 434–435
Penelis G, Kappos AJ (1997) Earthquake resistant concrete structures. E & FN Spon, London
Phillips AA (1984) Gingerbread houses: Haiti's endangered species. H. Deschamps, Port-au-Prince (first edition 1975)
Ponzo FC, Di Cesare A, Arleo G, Totaro P (2010) Protezione sismica di edifici esistenti con controventi dissipativi di tipo isteretico: aspetti progettuali ed esecutivi. Progettazione Sismica 4:50–75
Schacher T (2008) Timber reinforced stone masonry in northern Pakistan in the context of the post-earthquake reconstruction efforts. In: Proceedings of the International Seminar on Seismic Risk and Rehabilitation. Faial, Azores
Tobriner S (1983) La Casa Baracca: Earthquake-Resistant Construction in 18th-Century Calabria. J Soc Archit Hist 42(2):131–138
Tobriner S (2006) Bracing for disaster: earthquake-resistant architecture and engineering in San Francisco, 1838–1933. Heyday Books, Berkeley
Uhde C (1903) Die Konstruktionen und die Kunstformen der Architektur [The Constructions and the Art Forms of Architecture. Their Genesis and Historical Development at Different Nations] Vol II Der Holzbau: seine künstlerische und geschichtliche geographische Entwickelung, sowie sein Einfluss auf die Steinarchitektur, [The Timber Construction: its artistic and historical geographical development, as well as its influence on the stone architecture]. Ernst Wasmuth: Berlin (in German)

Chapter 12
Design in Time of Crisis

Sergiu Cătălin Petrea

Abstract Romania's regional structure includes different types of hazard-prone areas, most of them very active in recent years. Even though the extreme events that occurred here lately lasted for only short periods of time, they inflicted important damages on local communities and the built environment. The impact on people and the whole context of their existence is very complex and had multiple consequences in the long run: increased death rate, various diseases and posttrau-matic stress caused by delay of intervention by local authorities, failure to provide humanitarian and/or medical support, social and economic decline of the entire area and radical changes of the natural environment. In this paper we will analyze a possible model of intervention and reconstruction for hazard-prone areas, focusing on the role of a sustainable design process and durable planning of new dwellings/communities and we will suggest some strategies for a multidisciplinary approach that integrates community structure, local tradition, cultural diversity and disaster management planning.

Keywords Shelter · Open architecture · Vague space · Heterotopies · Transarchitecture

12.1 Models of Intervention—from Global Problems to Local Solutions

Until recently, the issue of social activism in civilized states of the developed world addressed only marginal aspects of the problem: temporary housing for the disadvantaged, local interventions in the ghetto areas or in mobile home communities or intervention policies to integrate the homeless, etc. However, today there is an area that has seen a general expansion in the world: the emergency housing program, necessary in the wake of a disaster or a war. The world has seen earthquakes in Turkey, Iran, Greece, Japan and Latin America, tsunamis and typhoons from Southeast Asia,

S. C. Petrea (✉)
Department of Basics of Architectural Design, "Ion Mincu" University of Architecture and Urbanism, Academiei 18–20, 010014 Bucharest, Romania
e-mail: sergiu.petrea@visdesign.ro

M. Bostenaru Dan et al. (eds.), *Earthquake Hazard Impact and Urban Planning*, Environmental Hazards, DOI 10.1007/978-94-007-7981-5_12,
© Springer Science+Business Media Dordrecht 2014

hurricanes in the U.S. and Australia, newly recognized extreme climate changes such as desertification, or the alarmingly current spectrum of nuclear disaster, such as occurred at the Fukushima nuclear power plant in Japan that followed an earthquake of magnitude 9.0 on the Richter scale, and the tsunami that hit the northern Japanese archipelago on March 11, 2011. All these are engines that can, at any time and in the blink of an eye, brutally transform the ordinary daily drama of life into a major global crisis. The current answers to such catastrophic events, usually managed by local authorities or trans-state entities (such as the Red Cross or the World Health Organization, for example) are limited to providing temporary shelters, spread over vast territories, without the responsibility or ability to provide anything other than minimum conditions of subsistence.

Architecture, which has for centuries been recognized as a true image of society, and its most well-known representatives—local housing programs—must find the resources to become actively involved in responding and resolving, even if only partially, these new contemporary issues. Creation of buildings as a contribution to the self-satisfaction of the world's population is no longer an effective goal on which to center contemporary architectural discourse, which is now called upon to attack various problems regarding the chaotic growth of cities, climate changes, extreme poverty and the slow response capacity of individual states.

Japanese architecture is in the forefront of this shift of goals in contemporary architecture through a series of exponents who carry on the tradition of the background matrix generated by the impressive vernacular architecture and the Metabolist theories. They are now the vanguard of a new way of working with materials and technology, by seeking rationality in each discourse and focusing on current societal needs. On the other hand, the American architectural landscape is a fertile ground for the emergence of an impressive number of voluntary forms of labor for the community, particularly those affected in different ways by extreme weather, poverty and military conflicts. Centered on the assistance of specialists and the availability of firms to contribute to programs of volunteer work, getting involved in this kind of activities finds a powerful example in the pro-bono design program—1 % Solution, whose appearance was influenced by successful similar experiences in the medical field of organizations such as Doctors without Borders. The 1 % Solution Program aims to attract and train non-governmental organizations that adhere to a common cause and that are seeking theoretical or practical support for a better involvement. The purpose of the program is to increase and improve the work dedicated to the community by obtaining the broadest coverage of all possible needs for intervention, while perfecting their level of expertise on all levels of society.

One of the main goals of this ideological movement is to incorporate sustainable principles in all the projects of the 1 % Solution program[1], whether they relate to building design, or to object design, urban development programs or landscape planning. The ultimate goal of this movement is to synchronize all forces involved in the

[1] The 1 % Solution relies on volunteer and pro bono involvement of architects and related professionals and aims to catalog and organize the taxonomic work on the organization's website, the main engine of publicity and promotion for the movement: www.theonepercent.org/.

process to strengthen contacts between professionals and those whom they address, and to provide a common database, thoroughly organized, which will constitute a possible source of development for future activities.

There are still more innovative ways of interventions such as the projects of the Research Division of the Massachusetts Institute of Technology called MIT SENSEable City Laboratory—a program that studies and implements strategies to anticipate the urgent needs of a society that is in a more accelerated development and to respond in a feasible manner. Centered on the anticipation, participatory design, on the use of "strong concepts, low technology" and addressing problems through new perspectives, the members often promote an extreme development vision. This is influenced by the fact that the research team consists of an interdisciplinary collective animated by the idea that our way of relating to everyday life continually undergoes radical changes and, therefore, it needs to investigate new ways to approach it critically.

Regarding Romania, harmonization between local shelters and the globalized society is beyond any goal, a necessity, especially in hazard-prone areas. This involves both the economic and the geo-political context and demands that the network of separate territories must be changed. For Romania it is very important to demonstrate that there can exist local facilities that can deliver firm and prompt responses, surpassing the delayed reactions of the authorities. In order to create such facilities, a structural reform at the conceptual level is required, starting perhaps from a better understanding of emergency housing needs, and by linking these needs to the conditions of unfortunate global events. Creating a coherent theoretical framework should be the first step for controlling the chaotic housing development, whose independence our society is still reluctant to integrate.

After the impact of a natural catastrophe, the built environment is abolished and architecture becomes silence, darkness and void. Our study investigates how a recursive matrix could be created, one that would be able to generate new living spaces and trans-architectures by using information taken from the affected sites and extending the local knowledge in order to develop generally valid methods. The result must be able to be modified according to different requirements in two main aspects: the dynamic-topological component and the functional-substantial component. On the other side, any further intervention will be reflected in the social space of the community, with great impact on the daily life of its inhabitants. In this case, the meaning of the intervention collides with the abstract elements of the physical and geometrical relations that rule the contemporary science of construction and the rigorous disaster management planning.

Human migration, which is often a result of un-settlement and lack of territorialisation of a certain population due to a natural hazard that has occurred in a region, often accompanied by extreme poverty, social problems regarding membership in a no longer available community and mental illness for those who have suffered heavy losses. For those persons living on the edge, in a constant tragic reality, the design approach must focus on the object (minimal dwelling, emergency shelter, prefabricated buildings) as well as on the built context (rural or urban areas) in terms of cost, technical efficiency and durable development. On a deeper level, it is necessary to encourage active involvement of the population in order to facilitate the integration of built intervention in the collective mentality, maintaining local identity by using traditional construction techniques—reinterpreted and perfected by modern technology to become environmentally friendly methods—and promoting low-tech architecture.

The relentless movement and the constant feeling of physical and mental uprooting lead to a fundamental dialectic of the concept of minimal shelter—which represents the most common intervention in a hazard affected area—that operates with the terms of *dependence/independence*. The first term is the accumulation of cost efficiency and the technical limits imposed by organization of the habitat with limited resources and manpower, at a given place and in a short time, while the second refers to the amount of specific attributes such as utility, aesthetics, ideology and tradition.

We will follow a *meta-modern*[2] perspective over the phases of a possible intervention in a hazard affected area, focusing on the main aspects that usually influence the planning process, synthesized in the image below (Fig. 12.1).

An emergency dwelling that comes in a hazard affected area must be developed at the same time as the life processes of the inhabitants that it protects, with self-completing and rewriting layouts, similar to the vernacular tradition. The Emergency Housing that we promote is intended to be a phenomenon open to perpetual changes, additions and transformations, dictated by environmental conditions of insertion, but also by bio-spiritual aspects of the life of those it receives for housing, as Jean Nouvel asserted in his conversations with Jean Baudrillard over the current aspects of architecture.

> And we feel that the architectural act revolves around settling into a repurposed architecture. This could involve something that's built inside or on the roof or even on a terrace. Nonetheless this process of sedimentation is a form of creation and a complete qualification of the space. It's not only a modification; it's a mutation. The space is no longer experienced the same way, there are different things inside; we play with scale differently, change the meaning, and starting with what was a large, poorly defined, purely functional volume, we've gradually managed to produce a regenerative recreation that no one would have thought possible (Baudrillard 2002, p. 41–42).

Minimal shelter, as an archetypal form of living, faces a complex situation, becoming a daily pattern of appropriating a space. It is an open and suitable space for the stranger, the refugee, the hazard victim, *the other* who has his own vision of territoriality. The alternative emergency housing is perceived as a phenomenon directly interconnected with the persons it fosters, therefore it is always open to changes and directly influenced by them. The key words that phrase the phenomenon are *multifunctionality* and *transformability*, and we will permanently operate with these concepts in order to delineate emergency housing in that vague area of the

---

[2] The term was introduced by Timotheus Vermeulen and Robin van den Akker, two Dutch philosophers, in 2009 in the article ***Notes on metamodernism***, published in the ***Journal of Aesthetics & Culture, Vol 2***, 2010. doi:10.3402/jac.v2i0.5677, p. 1–14, as a specific phenomenon of the 2000s and a reaction to a post-modern phenomenon that had reached, in their view, its ideological limits. Defined as an unstable, dichotomous relationship between modernism and postmodernism principles, the term carries a semantic load, assuming its theoretical descent in the issues regarding climate changes, contemporary financial and energy crises and geopolitical instability.

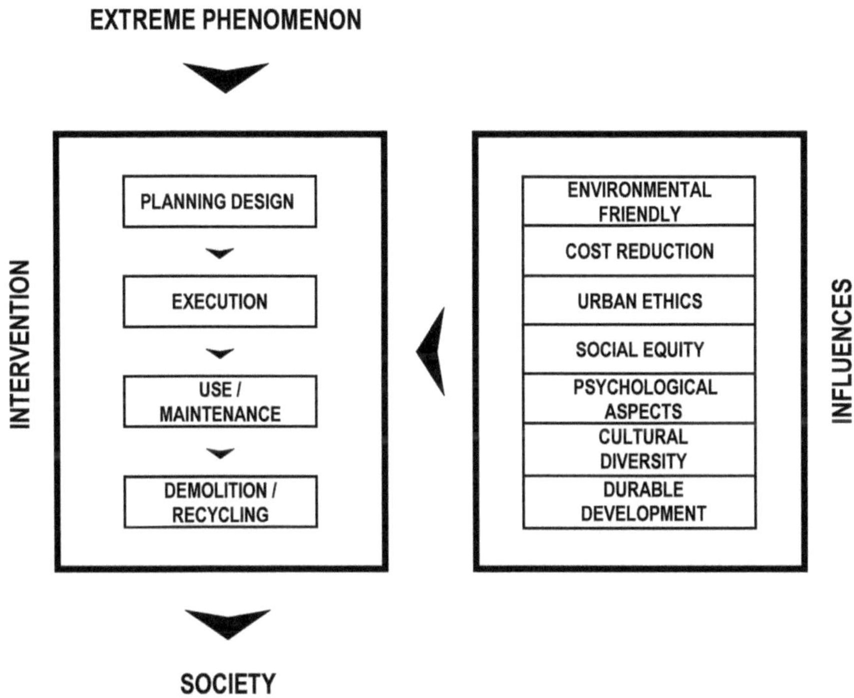

Fig. 12.1 The phases of intervention in hazard affected areas and their contemporary influences

shelter. It must be permeable for antagonistic activities with a minimal effort for redevelopment/reconformation.

This paper speaks about the confluence of multiple disciplines which manage the phenomenon of contemporary housing, preserving the architectural and the urban concerns regarding the program of emergency living spaces. The thesis identifies the spatial and temporal context in which these phenomena occur, by studying the well-irrigated ideological theories they rely on and it tries to develop further intervention methods.

We consider that it is imperative to use the principles of sustainable architecture during the design process, in a way that minimizes the consumption of materials by optimizing their performance and shortening the execution time by using prefabricated and modular elements. This method also resorts to locally produced materials and local resources for construction, using constructive systems that were tested by traditional use and that could easily be put into practice. The constructive approach should be adapted to the restrictions and features of the site, in an attempt to find an *open architecture* focused on process, economy of resources and technological means rather than only on the finished product and its poetry. Therefore, emergency architecture assumes as basic goals the recovery/ reuse of waste materials and self-sustainability, working to reduce resource consumption in construction and operation and making use of the supplies and technologies that could give them energy independence, incorporating the principles of passive houses and reducing their carbon footprint to zero. We adhere to the terms of design stated by Werner Sobek in his quest for an architecture that has sustainability as a natural daily practice, especially in the situations where the interventions in the hazard affected areas will become permanent.

> Frankly, I can no longer stand the word sustainability. Architects brag about making sustainable buildings as if this was something special, whereas by now sustainability should be as essential as fire safety and structural stability. I hope that in five years it will no longer be necessary to speak explicitly of sustainable construction. Throughout my career as an architect and engineer, I have always treated sustainability as an integrated aspect of architecture and construction, not as an additional or optional gimmick (Sobek 2010, p. 34).

As it is shown in the survey ***Guidelines for Evaluation of Post Disaster Programmes of UNCHS***[3]—the need for a standard emergency housing program is often overrated, as the attempt to apply pre-given, local-indifferent solutions for the current needs of survivors ignores long-term consequences. Therefore, architecture for emergency programs must relate to a number of issues and requirements that could be drawn from the studies made in different regions where the hazard phenomena occur periodically:

1. *Emergency sustainable architecture must respond to the local and specific needs of the people affected by natural disasters*. It must relate to considerations dictated by the insertion site and the local community with all its characteristics (religion, tradition, background culture, customs etc.). Our own studies and investigations showed that, in many cases, recovered elements from previous houses, destroyed by the natural catastrophe, were used or incorporated in the new homes: old furniture, pictures, decorative objects belonging to the families for generations, religious icons. Often, the modern layout of the house is modified in order to make a better room for these very emotionally-valuable artifacts.
2. *The dynamic component of the phenomenon is essential.* Understanding the dynamics of the situation and the territory in which a project is implemented gives the extent to which the intervention could be a success or not. Also, by mapping the migration paths, called by some experts *inhabited trajectories*, the process of sustainable spatial planning and land management provides various tools to prevent further natural hazards in the living areas.
3. *The intervention must focus on the approach to a specific context.* The projects of intervention and post-disaster reconstruction have a specific character for a certain place (construction of temporary or permanent housing, reconstruction and rehabilitation of the damaged infrastructure). The specificity of each place should be integrated in the general context of the solutions, especially in the case of those inserted constructions which are planned to remain permanently in that location.

[3] Available on-line on: www.unhabitat.org/content.asp?cid=1264&catid=286&typeid=16.

4. *Emergency architecture and urbanism must be considered with the participatory involvement of the community.* Most post-disaster reconstruction projects develop processes with important implications for human and material resources, which create a powerful bond between the local communities and their members, focusing on mutual support and ecumenical spirit. We believe that it is recommended to encourage the active participation of the affected persons in all the stages of the reconstruction so that the new intervention will be more easily accepted in the collective mentality once it is finished.
5. *Fighting mental alienation must be a basic policy of the intervention strategies in the disaster affected areas.* Global research studies have concluded that, often, the people that suffer from a natural disaster prefer to move in with relatives or to build themselves houses instead of living in a prefabricated housing unit, whose design denies the traditional and cultural peculiarities of the area. Therefore reconstruction should be performed together with the community, in multidisciplinary teams, focusing on critical regionalism principles and psycho-social aspects.
6. *It is required to find independent funds for the management of certain types of intervention that could go together with those developed by the local authorities or the United Nations Organization and other transnational organizations.* The projects must be developed on stages of prevention, mitigation, preparedness and intervention, each with their own budgets. Issues such as storage of prefabricated construction materials or emergency housing kits and the quick transport of large quantities to distant areas could be solved if it would be possible to appeal to the logistics of the global organizations. The solutions for design could be given by nonprofit organizations opened to professionals who are willing to help others. Labor for projects is made up of volunteers, trained to coordinate activities and help people who already received minimal previous training. Following these strategies serves as a basic approach that we intend to develop in the future, exploring at least theoretically, its latent possibilities, so that in the next phase we would be able to detail possible housing projects for crisis conditions, applicable to the specific case of certain hazard-prone areas from Romania.

12.2 Spaces of Intervention

Emergency shelter is a form of resistance to the drama, chaos and poverty that works with the frugality of resources and calls for ways to alter the absence of minimal means of building. It works with immediately available items in order to achieve a protective living space. At this time, a shelter model is less than a house and, simultaneously, more imperatively needed than a house, in its current sense. If we add to this the fact that the phenomenon of living in such conditions is radical, we can question the sense of a classic way of living when it is faced with a crisis. The result, although apparently limited to certain harsh circumstances, contains a high degree of complexity, given the many issues that it is called to respond to.

The place generically called *home* is actually a point of intersection between man and society, between the individual, subjective, living space, and the social, objective, community place. The house is a place of refuge from daily meanders, where we carry out our personal lives and we intimately show our true face without being forced to wear masks or play roles. Romanian archetypal dwelling space, represented by traditional architecture, was designed as *telescopic space*, a succession of spaces of various degrees of intimacy on the path between outside and inside. From the road, gate, porch, guests living room, all are areas designed to receive strangers who are facing the household, while the privacy is hiding in the last room, the family area, being oriented to the backyard.

We will consider these aspects and we will place them in the dimensions of the possible shelters, to experiment how we can design multifunctional spaces according to four categories of attributes (as shown in the figure below). Also, we will try to see how the boundaries of the shelter interact with the natural environment and what the use of the tectonics is in the materials' conformation.

There are some gestures in the professional field of our current issue (the work of Japanese architect Shigeru Ban, Instant Architects from Germany or Rural Studio from U.S.A.), that send an oblique reference to a return to the origins of architecture, at the primary topos of the constructive and social practice of the building, from which all other facts descend. The result is an architecture that can be self sufficient, affordable, contemporary as an expression of mostly natural materials, proper to essentially different communities, so, with great potential for universality of the phenomenon (Fig. 12.2).

a. *Living space.* The relationship between the individual (affected by a crisis or by a displacement forced by a natural disaster) and the living space of his own habitat is usually unstable and therefore needs to be improved through consistent practice and careful building. Contemporary living spaces, often containing a wide range of activities, can be packed in a multilayer surface that may suffer adaptations due to various conditions of insertions. The appropriate spaces for contemporary living are versatile, characterized by lightness and adaptability and influenced by avant-garde. They are defined by new keywords: matrix, relationship, possibility, probability, dematerialization, virtualness, metamorphosis, transarchitecture. Inside—outside relationship represents the primary aspect of any living space and determines the variable degree of *openness* and *closeness* it possesses.

   Usually, any enclosure of an area becomes manifest in relation to the land that surrounds the marked place. The way it is realized contains information that outlines the closing gesture and gives value to the space created inside. This living space is the unit resulted from a concrete sealing operation of protection that becomes a *center* that is able to preserve human life and activity and can function as a focal point for all their hypostasis. Paraphrasing Norberg Schulz we can say that living space is a *space system* which would imply the fact that the idea of living space has its origins and is intimately linked to concrete situations of life, although, in general, "spaces can be described and defined by abstract mathematical terms" (Norberg Schulz 1980, p. 10).

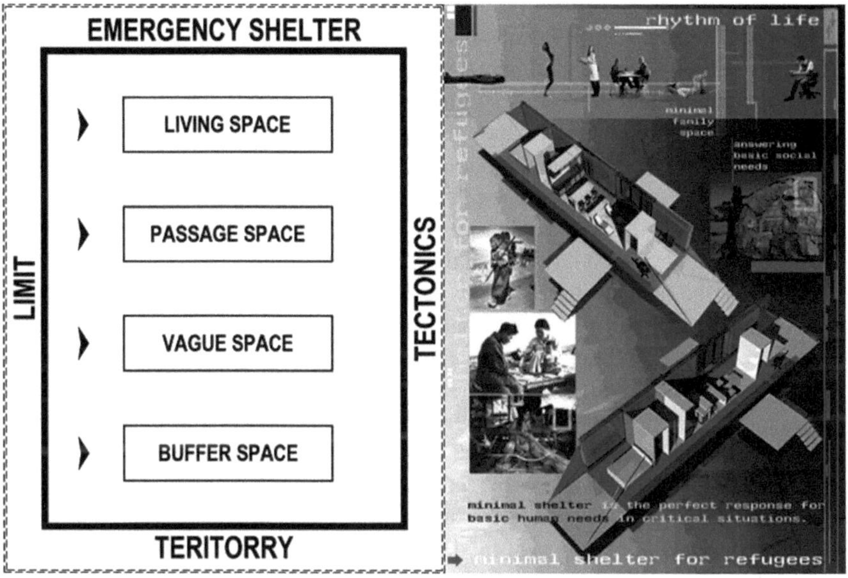

Fig. 12.2 The categories of spaces that may occur in the conformation of an emergency shelter and the practical use of the concept: a working model for OSAKA International Competition in 2005 that explores the concept of living machine of Le Corbusier, trying to create a new vision for a typology of intervention in crisis affected areas

For contemporary living, which is influenced by the emerging problems of our society, generated by different types of crisis, the living space for occupancy and use, can no longer contain promises of modernist utopian theories, but it is rather perceived as a potential breeding ground for temporary employment, slightly supervised, prepared for a possible, discreet withdrawal and always open to changes. Moreover, emergency housing space is open to the stranger in need, in the sense that Georg Simmel has foreseen it: "not the wanderer who comes today and goes tomorrow but the person who comes today and stays tomorrow."[4] By extending its meaning, we could observe that the traditional wanderer has evolved, updating and being multiplied in lots of avatars: the homeless, the vagabond, the squatter, the refugee, the natural disaster victim, terms that apparently refer to the same topos of existence, but which contain inherent differences in how each of them interacts with the living space.

b. *Passage space*. The living space of an emergency shelter is a passage space in the sense that Anthony Vidler gives to this type of location: *estrangement*, in its multiple forms of existence: "the estrangement of the inhabitant of a city too rapidly changing and enlarging to comprehend in traditional terms; the estrangement of classes from each other, of individual from individual, of individual from self, of workers from work" (Vidler 2000, p. 64).

[4] Simmel (1971).

These spaces are generated by what Georg Lukacs calls, in Heidegger's affiliation, the "transcendental homelessness of the modern world" (Idem, p. 65) and are influenced by the spatial paradigms of modernity that are direct tributaries of the materiality and the excessive sensitivity of contemporary life. According to the American theorist, there are a series of landmark locations that are specific for the nomadism of the contemporary urban life, for consumerism and for excessive mobility: the portico, the specific commercial passage of nineteenth century architecture, the hotel reception areas, the galleries and the metro stations. All these locations, which have stated their existence in the core of transition itself, influence the habitat of emergency because they are characterized by a certain type of versatile permissiveness that is open to various uses. Their stake is even greater as we talk about the space generated by emergency architecture that addresses the disaster affected communities, which can be either common spaces or areas for group activities. Moreover, they contain inherent availability and are open to changes according to a certain use that makes the emergency shelter turn from a simple passage space to a temporary resting place.

c. *Vague space.* Vague as an architectural term was used by the French philosopher Jacques Derrida, the father of modern deconstruction in collaboration with the architect Peter Eisenman for the project Choral Works. At the inauguration of their joint work, Derrida apologized to Eisenman for his absence[5], through a text written in his characteristic spirit, which debated the presence of absence, vague as architectural manifestation and the invisible as visible in the architect's work. The French philosopher challenged Eisenman to participate in a discussion in which he wanted to point out the relationship between architecture, poverty, capitalism, war, the need for shelter and housing scandal, as he repeatedly tried to anticipate the ways of solving future problems.

   Vagueness of the emergency shelter space must allow any needed function to accommodate, accurately and efficiently, by being always open to new transformations, no matter the period of time. Vague area prefigures a subtle blend of living and working spaces, meditation space, hospital, school or asylum, the ambiguity of the functional and the stylistic rewriting being always based on the ways in which it is used by its tenants. Being located between the multifunctionality postulated by the postmodernists, (where versatile space was addressed in a systemic perspective as a complex system of complementary elements), vernacular tradition (where we find the ability to expand with additional new spaces according to the needs of the human subject, without harming the whole) and the simultaneous existence of Cubism and, later, of Deconstruction (where the synchronous approach of an area is essentially influenced by temporary factors), emergency shelter spaces inherently carry the full range of features and characteristics of the derridian approach without vacuuming its ideological status.

d. *Buffer space*. In computing science, a buffer space is a common area of data, shared by hardware devices or software processes operating at different speeds

---

[5] As mentioned in Peter Eisenman, ***Eisenmanesie***, in ***Architecture + Urbanism***, Extra ed., August 1988, p. 137.

> or with different sets of priorities that allow each device or process to operate independently of the others. Such space is a point of balance between different activities that allow them to perform optimally. By extension, an architectural buffer space is a particular form of vague space that can perform various complementary activities, facilitating their independent operation, without favoring any of them. Emergency Housing is a buffer space in the sense described above, as an exponent of a new system, using a living space by different categories of people who need similar responses to a certain issue, independent of their personal aspects (sex, age, religion, social or cultural background).

The regionalized world is a place of infinite new possible territories, either physical or abstract, scientific or cultural, overlapping, flexible and mobile. It is characterized by a perpetual motion, in which individuals always have to learn in each new place how to live together and to respect the otherness. Then we have the positive or negative psychological tension of each person who comes to a new place, near someone else or something else and this, usually called "first impression", can easily become chronical. Starting with the volume, the various scales, the spatial relations, textures and colors, sensations and sounds, inhabited spaces must give the stranger the chance to feel free. In this approach, we think that the buffer space must be sufficiently permeable and flexible enough to allow a plurality of protective measures for a human being for a relatively short time period, until the person will be integrated in a new form of living or will return to the previous state of existence.

Seeking a theoretical location for the phenomenon of emergency housing in a paradigmatic approach that would support us, we will try to show that the idea of *heterotopy* is operational for our study and that the practices involved in this phenomenon can enroll in a complex meaning, applicable to this kind of architecture. Emergency shelter aims to be a compensatory space (even for a specified period) to correct local unnatural order, instituted by abuse and hazard (natural disaster), by offering an alternative to improve the situation, extracting the being from the real daily life and trying to put it into a *different space* until the natural order of things is restored.

> Heterotopies have the power to juxtapose in a single place several spaces, several sites which, by themselves, are incompatible (Foucault 1984, p. 47).

We referred to Michel Foucault and the concept of heterotopic because it is likely to be circumscribed, thanks to its features, to our research problems. We know that space is generated by the relations between various locations, and is also defined by neighborly relations between points or elements; discovering and making use of living space gives birth to complex algorithms for interconnection and networking, depending on these determinants. All the concepts of possible spaces of intervention may rely on this concept and we may find in the semantic area of the idea a very fertile ground for further theoretical investigations connected to emergency housing phenomenon.

Following these typologies of spaces, the objectives of an efficient intervention strategy in a hazard-prone area, after the disaster stroke and it is required to rebuild the affected communities, can be ordered as follows:

- It is wise to explore the space types that recurrently develop, that are investigated by monitoring the direct use of the existing shelters for additions or changes made by the inhabitants, in order to improve future typologies.
- By operating with clear concepts about the used space typologies—vague transitional space, buffer space or living space, their birth places and spatial limits, certain preventive intervention directions by which to influence the design of new spatial structure may be predicted.
- The study of the human types associated with these concepts of space, features of each distinct human typology, must be secured in the equation as unpredictable factors that interact with the shelter, especially due to post-traumatic stress. The sudden collapse of an average person to an affected one or a refugee is a problem one finds very hard to face. For this reason, a psychological component must be added to the design process, in an attempt to counter the effects this type of shock has upon the victim's state of mind. It is important to rely on the personal identity of the community (ethnic, religious, linguistic) to prevent further drama after the emergency even though the shelter must accommodate in a given, generally valid space refugees from different areas and cultures.

12.3 Case Study—Houses for the Flood Victims in Suceava County, Romania

Credits:

Authors:	Arch. Sergiu C. Petrea, Arch. Cristina Petrea
Design planning:	S.C. Vis Design Media S.R.L.
Investor:	S.C. Holzindustrie Schweighofer S.R.L.
Client:	Grăniceşti City Hall
Objective location:	Dumbrava Village, Grăniceşti, Suceava County
Execution Period:	4 months (September—December 2010)

As the floods in 2010 affected hundreds of households, the investors—honoring the concept of active involvement in the support of poverty-struck communities—immediately came up with a solution. They financed the construction of eight dwellings for those families who were most in need (with either the largest number of members or with people with disabilities). The houses—built with a low budget, in the area where the company operates—were to be finished in less than a few months, before winter came.

The Austrian investor wanted the image that promotes the new housing to capture the atmosphere of Romanian vernacular architecture. Thus, wood gained a major role in the process of construction—this decision was also facilitated by the fact that it is the main component of Schweighofer's production activities. Although the houses were destined for the underprivileged, the aesthetical aspect has not been neglected for a minute. The architectural vocabulary gravitates around minimalism, keeping a

Fig. 12.3 Map of Suceava County, situated in the northern part of Romania. In June 2010, the floods caused by heavy rains affected 72 villages, forced 2,390 people to be evacuated, temporarily isolated 501 people, destroyed 47 houses and damaged another 4,228. (Image source: map extracted from the Romania Atlas and processed by the author)

formal and volumetric austerity, warmed only by the use of wood on the facades. The houses were built in less than four months, so that the end of the year already found them occupied by their new families. Even though the budget did not rise above 300 € /sqm, the homes managed to obtain the desired high comfort and quality of living (Fig. 12.3).

The use of local materials, the saving of resources and the close association and dialog between what was built and the natural environment, were all important goals that had been established by the investors and attained one by one. Generous glazing surfaces, along with the recurrence of material tectonics, assured the subtle transition between the interior and the exterior spaces.

The team of architects used the game of light and shadows as recursive means of composition so that the geometry of the interior is nuanced in different depths, just like in the traditional architecture, which is still present in the area. The use of black and white dialectic in shaping the exterior and interior finishes, punctuated by wood accents, allowed indoor and outdoor views to become unified and the living space became more generous, enhanced by the ample windows that open picturesque perspectives to the surrounding hills. A subtle play of reinterpreted traditional wooden

Fig. 12.4 The two types of layouts that were used: ground floor type and ground and one level floor type, chosen according to the number of the members of the families. (Photos: Sabrina Ene)

elements, arising from the use of local craftsmen, help in alternating the lights and the shadows, giving rhythm to the elevations. The minimalist approach is backed by the subtle use of nuanced color wooden surfaces for interiors and exteriors (Fig. 12.4).

In Romania there are frequent cases in which the crisis is self-written in the human activities or vernacular planning of the rural houses (households rebuilt in beds or above major flood dams that have proved to be unable to fulfill their role in extreme conditions, settlements and infrastructure on the slopes exposed to landslides etc.). The frequency and the universality of this reason is related to a particular attachment to a certain territoriality which prefers taking risks better than facing the house-losing spectrum. Once the disaster occurred, people are excluded from participating in a given territory, they are brutally snatched from a certain affiliation, from a location between familiar limits. One of the major challenges of the project was to find a way to put the new houses in a deserted landscape and to connect them with the existing community. It was also difficult to find a functional layout that would be suitable for the daily needs of its inhabitants in a minimal image which to maintain, still, some of the traditional image of the local house-types.

We explored two distinct possibilities: the use of prefabricated houses, made of minimal features that can be transported and quickly assembled in the affected area or manufactured on site houses, with local resources and the help of the people in need, according to principles previously studied. We finally chose the second option in order to help strengthen relationships between community members.

We can conclude by saying that the intervention was built as a private space, whose understanding is not definitively reduced to its physical evidence, but is constantly rewritten, by the events and the persons that enter into dialog. Half a year after the families moved inside the new locations, the living spaces of the houses and the surrounding areas were full of life, each of the inhabitant families customizing them according to their specific needs (Fig. 12.5).

Fig. 12.5 External views of the houses showing the relationships that arise between buildings and the related landscape. Due to property reasons, the in-between spaces of the houses were all different and, therefore, used by each family according to its own needs. (Photos: Sabrina Ene)

12.4 Conclusions

In a factual sense, emergency shelter aims to be a compensatory space (even if it is only for a limited period) that corrects local unnatural order, instituted by natural hazard, by offering an alternative to improve the situation. The intervention appeals to the fundamental component of Heidegger's living, *the care for the Being*, which is the basic quality of post-drama shelter and manages the entire methodology of intervention in such situations. This is the main foundation on which the construction process is based in this sense, beyond any other implications (economic, aesthetic, formal). This implies the affiliation of the human beings together, which means having an ontological advantage in the concrete existential sense. Reaching the limits of non-living puts man in the fundamental problem, that of passing and retrieving lost archetypal space. This is where the protective phenomenon of emergency architecture appears: from the need to bind the individual, recovered from an anarchic living trajectory, in a hospitable space, for a specific period of time, until he is able to retrieve a new existential order.

References

Baudrillard J, Nouvel J (2004) Obiectele singulare. Arhitecturăşi filosofie. Ed. Paideia, Bucharest

Foucault M (1984) Des espaces autres, in Architecture, Mouvement, Continuité, no. 5, octomber. Published in Michel Foucault, Theatrum philosophicum (eseuri, stdii, interviuri), (2001), Cluj, Casa Cărţii de ştiinţă

Gociman CO (2001) Tipologia hazarduluişi dezvoltarea durabilă. Editura Universitară Ion Mincu, Bucharest

Leach N (1997) Rethinking architecture- a reader in cultural theory. Routledge, London

Merleau-Ponty M (2002) Phenomenology of perception. Routledge Classics, New York

Meyhoefer D (2008) Touch wood. The rediscovery of a building material. Verlaghaus Braun, Berlin

Norberg Schulz C (1980) Genius Loci. Towards a phenomenology of architecture. Ed. Rizolli, New York

Simmel G (1971) On Individuality and Social Forms, ed. D. Levine, University of Chicago Press, Chicago, p. 143

Sobek W (2010) Architecture isn't here to stay. Towards a reversibility of construction. In: Ruby I, Ruby A (eds) Re-inventing construction. Ruby Press, Berlin

Vidler A (2000) Warped space—art, architecture and anxiety in modern culture. MIT Press, Cambridge

Zumthor P (1999) Thinking architecture. Birkhauser, Basel

Part IV
Impact and Decision Making Tools

Chapter 13
A Proposal to Improve the Disaster Management Cycle Model: the Importance of Community Participation

Alessia Cerqua and Silvia Rapicetta

Abstract Earthquakes, hurricanes, severe flooding and other events that we call natural disasters are not simply natural at all. They are also the result of the erroneous way we have planned and built our cities, buildings and infrastructures in hazard-prone areas. In this paper we do not analyze the contributing factors of human activities that cause extreme events, such as those resulting from climate change, but we will analyze the problems arising from an incorrect planning of the disaster management cycle in hazard-prone areas, and suggest improvements in the disaster management cycle through a community participation approach to better integrate urban planning and disaster management planning.

Keywords Participative planning · Disaster management · Community · Strategic choice approach

13.1 Models of Disaster Risk Analysis

In the area of disaster research for disaster risks according to the traditional model by Alexander (1993), risk is expressed as:

$$\text{Risk} = (\text{Hazard})*(\text{Vulnerability}) \tag{13.1}$$

According to Wisner et al. (2004), vulnerability is defined as 'the degree to which someone's life, properties and other assets are put at risk by events in nature and in society'.

It implicitly includes preparedness, which was explicitly included in the definition by Smith (2001) of risk as (Hazard) * (Loss)/(Preparedness).

S. Rapicetta (✉)
Adjunct Professor at University of Perugia—Civil Protection Degree Course,
c/o Centro Studi Foligno, 123–16034 Via Oberdan, Foligno (PG), Italy
e-mail: silviarap@gmail.com

A. Cerqua
Independent researcher, Environmental Management Advisor, Rome, Italy
e-mail: alessia.cerqua@gmail.com

M. Bostenaru Dan et al. (eds.), *Earthquake Hazard Impact and Urban Planning*, Environmental Hazards, DOI 10.1007/978-94-007-7981-5_13,
© Springer Science+Business Media Dordrecht 2014

At the same time, the word resilience was introduced as the capability of a socio-economic system to recover after a disaster. An integrated risk management framework should then aim at decreasing vulnerability and increasing resilience (Ikeda et al. 2008).

On the other hand, in the environmental or technological fields of risk analysis, an interdisciplinary expression, such as 'The potential for the realization of unwanted, adverse consequences to human life, health, property, or the environment' is used.

This definition is originated from the discussions on defining an interdisciplinary concept of risk at the Society for Risk Analysis (SRA; http://www.sra.org), and Kaplan and Garrick proposed the expression in terms of a risk triplet vector (Kaplan and Garrick 1981) as:

$$\text{Risk} = \text{R}\{ < \text{Si,Pi,Di}\},(\text{i} = 1,2,\ldots) \tag{13.2}$$

where Si is a set of scenarios concerning the nature of possible disaster events, Pi is a set of likelihoods of hazard in terms of the frequency or probability, and Di is a set of consequences in terms of unwanted damages to the concerned objects.

The risk triplet can be flexible enough to incorporate value choices linked to human and socio-cultural factors (Ikeda et al. 2008).

13.2 The Disaster Management Cycle

Traditional emergency management programs and planning are essential based on the so-called 'disaster management cycle' which is a core concept within environmental health management in disasters and emergency. The objectives of such an approach are to reduce, or avoid, losses from hazards, to assure prompt assistance to victims and to achieve rapid and effective recovery (Borrego and Amorim 2007). It consists of four phases: Ex-Ante disaster planning and Preparedness in the pre-disaster stage, and Response and Recovery in the post-disaster stage (Fig. 13.1).

The 'Ex-Ante disaster planning' phase consists of two steps: prevention and mitigation.

The 'Response' phase includes activities such as rescue, first aid, fire fighting, activation of alarms, evacuation and shelter for people, activation of emergency communications, closing schools, offices, etc.

This phase is managed by an Emergency Response Coordinator and conducted by an Emergency Response Team who are appointed following the activation of a specific Emergency Command Centre.

The 'Recovery' phase concerns the activities of 'Rehabilitation' and 'Reconstruction', which are the activities necessary to return the community to normal, such as temporary housing, medical care, livelihood support, etc.

The 'Ex-ante disaster planning' includes 'Prevention' and 'Mitigation' whose purpose is to minimize the effects of the disaster. In these stages there are activities such as public education, vulnerability analyses, building of active defenses and implementation of passive defenses, depending on the type of hazard, etc.

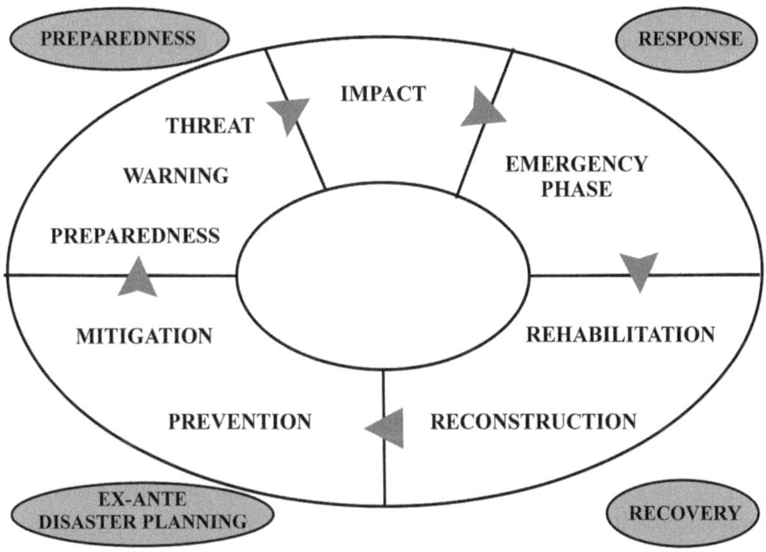

Fig. 13.1 The disaster management cycle

The 'Preparedness' phase, whose purpose is planning how to respond, includes the 'preparedness', the 'warning' and the 'threat'. It goes from the preparedness plans, with preparation of hazard maps, construction of early warning systems, preparation of emergency kits, emergency training, etc., to management of the warning and threat phases.

Technologies and knowledge for natural hazard reduction are crucial in the disaster management cycle, but they do not reduce disaster impacts unless institutions that have responsibility for emergency management and hazard response and recovery exist (Hills 2000). From a technical point of view, in the disaster management cycle, it would also be very important to utilize a 'direct' method that envisages the continuous and punctual supervision of technicians, who can monitor, check and control the method during its application (Rapicetta and Zanon 2008) , and allowing at the same time its critical interdisciplinary elaboration.

Of crucial importance are also that communities and individuals know how to translate hazard knowledge into actions, use and adapt technology effectively and organise socially and politically to structure risk reduction (King 2008).

In the disaster management cycle according to King's (2008) point of view, planners, as key partners in the local government response to risk management and mitigation, are not yet engaged and still regard hazard mitigation as a low priority.

Furthermore 'management is a fundamentally top-down approach. In a post-modernist world, debates about top-down and bottom-up tend to argue a case for the relevance and empowerment of the bottom-up, community-based view, to the detriment of the top-down approach, which is easily labeled an old paradigm of command and control. The reality for emergency managers is that they are necessarily at the top and have the responsibility to manage and plan for diverse and complex crises, through a layered and hierarchical system. The effectiveness of top-down management is not necessarily its complete alternative, but the extent to which it percolates and engages all levels of community and stakeholders' (King 2008).

Another problem is that the recovery phase of the disaster management cycle, seems completely extraneous to the urban planning process. The recovery phase should be more integrated into the planning phase of ex ante-disaster planning.

In other words, urban planning and in particular the reconstruction phase should be fully integrated with the phases of mitigation and prevention in the disaster management cycle.

According to several authors, the only real way to minimize the growing human and property losses from extreme natural events is rooted first and foremost in how we design and build our communities in the first place in these hazard-prone areas.

The DRC (disaster resistant community) concept was explicitly created to provide this vision, the 'overarching guidance that informs development in hazard-prone areas', that Mileti (1999) has found to be so wanting.

A DRC is a means to assist communities in minimizing their vulnerability to natural hazards by maximizing the application of the principles and techniques of mitigation to their development and/or redevelopment decision-making process.

Therefore, considering the importance of promotion, participation and involvement of community and stakeholders in the entire cycle of planning, design, implementation, and recovery processes in the disaster risk management (Ikeda et al. 2008, King 2008) and at the same time considering the need to better integrate the planning phase and the reconstruction phase with the disaster management cycle, a deep variation of the traditional disaster management cycle model is here proposed (Fig. 13.2).

This proposal aims to decrease the vulnerability of the community. It is known in fact that a better adaptation of the community to the environment helps to create a community that is hazard resilient and disaster resistant (Geis 2000).

13.3 Importance of Community Participation

In the traditional disaster management cycle, community participation is not explained and it is not adequately taken into account. The abovementioned hypothesis of modification aims to fill this gap.

Participation in planning plays an important role in empowering community members to become part of the general political process and to have a voice in decisions that shape the community: the requirement for community involvement is recognized by major stakeholders in post disaster re-construction programs. The idea of 'community participation' has been widely expressed: the term 'Community' has been used to refer to a neighborhood, a slum, a group of local NGOs, residents of a small towns, etc.; in this sense, the term neither denotes what this group of people really have in common, nor their differences. The term 'Participation' is used to denote civil debate and communication consultation, partnership, delegation of activities, self-help construction, communal meetings, etc. (Davidson et al. 2006).

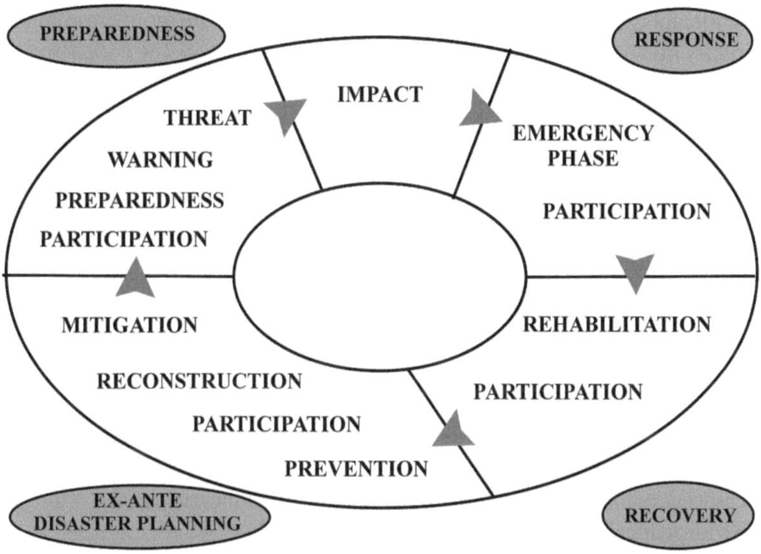

Fig. 13.2 Hypothesis of disaster management cycle change

In a participation recovery plan, the community can assume the role of active participants, rather than helpless victims. Local people can define goals, control resources, and direct redevelopment initiatives with long term economics and social benefits. The underlying theorem in post disaster reconstruction is that the more the recovery relies upon local resources, the quicker the community will be able to move to self—sustainability, and thus from recovery to normalcy (Lawther 2009). Another important opportunity in community participation is to alter physical development patterns to reduce future hazard vulnerability (Berke et al. 1993). Support for hazard mitigation is typically strongest immediately following a disaster (Rubin et al. 1985): with appropriate construction, repair and land use standards, a rebuilt community can be at lower risk to future disaster, compared to a pre-disaster condition; moreover longstanding community problems can be resolved through reconstruction (Berke and Smith 2009).

Literature has identified distinct opportunities associated with community involvement in the recovery and mitigation process (Lawther 2009):

- Involvement of the community should lead to more sustainable outcomes of projects. It stands to reason that the more people are engaged in the process, the greater the level of stakeholder engagement is, the more they are able to influence and take ownership of the outcomes.
- Capacity building programs can be developed and implemented to enhance future employment prospects.
- Significant relationship building opportunities can be created between implementing Agencies and the local community.

- Significant economic stimulus for the local economy. The more the community is directly involved, the more likely that the financial resources provided to undertake the project will directly benefit that community.
- Community involvement in planning can be used as a key indicator of project success. Types of involvement can be clearly identified with measurable indicators of achievement.

Some recent research identified advantages of community driven projects (as opposed to contractor driven approaches) as:

- Being more cost effective.
- Providing a potentially better product quality where technical and supervision skills are available.
- Being more empowering.
- Allowing for incremental re-construction thereby permitting occupancy before a house is fully completed.
- Restoring confidence in those traumatized through the experience of disaster.
- Providing local capacity building and employment.
- Preserving of local cultural heritage through land use planning.

13.4 A Participative Approach for a Community Involved in a Recovery Phase

In the following paragraphs we describe a participative approach for involving a community in a mitigation and recovery phase. The suggested approach is cyclic and incremental, and combines two methods of planning/ design: dynamic construction of future scenarios (DCFS) and Strategic Choice Approach (SCA). The incremental nature of this approach is a direct consequence of the fact that the development of 'vision' of local actors inevitably leads to change, in time, of the future scenario envisioned by the actors themselves, and choosing the most appropriate actions to realize it.

The Strategic Choice Approach (SCA) originated in the late 1960's in research projects concerned with understanding the processes of strategic decision making in city government. SCA differed from that of the rational comprehensive approach in (Friend and Hickling 2005):

- An emphasis on decisions, rather than on systems;
- An emphasis on the planned management of uncertainty, rather than on methodical collection and analysis of information;
- An emphasis on the making of incremental progress, rather than the production of a comprehensive end-state plan;
- An emphasis on collaboration between people with many different perspectives, rather than reliance on professional expertise.

In SCA, it was the emphasis on the planning management of uncertainty which marked the most radical departure from established planning theory. Three primary categories of uncertainly were identified, each of them calling for a different type of response: (1) *Uncertainties on the working environment* (uncertainties of this kind lead to calls for various form of investigation, such as research, survey, analysis, estimation of costs, mathematical modeling, consultations with experts, etc.); (2) *Uncertainties about guiding values* (uncertainties of this kind lead to calls for some policy-clarifying activities, such as meetings to reconcile opposing interests, formulation of clearer goals or objectives, consultation with politicians or representatives of other stakeholder interests, etc.); and (3) *Uncertainties about related choices* (outside the boundaries of the problem that is currently being addressed).

In SCA the planning process is divided into four modes: *shaping*, *designing*, *comparing* and *choosing*. These modes can be operated in a cyclic process: each mode consists of a number of steps that are carried out using special techniques. In the *shaping* mode, the decision areas and problem focus is decided upon. This means that the group of participants outlines the decision areas of their planning problem, looks at their linkages and decides which ones are more urgent to focus on. In the *designing* mode, the most urgent decision areas are analyzed in terms of different decision options and their interconnectedness. A special technique is used to limit the decision options by looking at their incompatibility. Decision schemes are constructed to outline the different feasible combinations of decision options to work with for the remainder of the workshop. In *comparing* mode, different criteria or comparison areas are now discussed to determine the requirements for the strategies to construct. Assessments of the various combinations of decision options and comparisons are made. In *choosing* mode, for the combinations of decision options that look most promising, considerations of uncertainties of different types are made. Additionally, it is decided how these uncertainties can be dealt with, for example by taking stepwise decisions; action schemes and commitment packages are constructed to outline the different decisions that are made now and in the future.

According to SCA, these are valuable methods that help actors, institutional or not, to plan, choose and coordinate interventions aimed at changing the territory, whereas all those autocratic forms—designed plan, project-plan, etc.—that purport to prefigure once and for all the shape and uses of the territory, as in a sort of 'one-act play', are doomed to failure.

Scenario making is the first step of a learning process that continues through the SCA. The construction of a scenario in the distant future plays a very important role because it directs the planning process in the initial phase and helps to identify the projects to be implemented immediately. The scenario is not generally a coherent 'vision' shared by all territorial actors, but rather a set of wants that can be fragmentary and inconsistent, sometimes conflicting with each other. SCA can gradually bring out a vision based on the values shared by the majority of actors and help the planner to design and choose interventions mutually compatible and ultimately consistent with this vision.

In a Visioning exercise, all the interested parties—such as local authorities and other public bodies, entrepreneurs, professional, cultural and environmental associations and neighborhood committees—are invited to attend and help develop a participatory scenario.

A participatory scenario is a useful tool for identifying problems and exploring possible actions to enable a community to move towards a preferable future. In our proposed version of Visioning, the scenario takes the appearance of a 'chronicle from the future'. The community interacts, under the guidance of facilitators, to write a short story set in the distant future, 15 or 20 years later. Every person involved in scenario building identify themselves with the protagonists of a story and describe what they see, say or do in order to build a 'vision', that is a representation of a desired future.

The scenario provides the main input to the Strategic Planning. We identify issues as decision areas, options, uncertainty areas or comparison areas, updating them—for instance, by opening a new Visioning session –whenever the decision situation changes.

In the designing mode the working group identifies some alternative proposals (options) for each problems individuated (decision area). This planning/design process is cyclic. It is not important always to start with Visioning: for some decision areas, options and sources of uncertainty can be identified from the existing decision situation, before developing any scenario. The scenario making can also be interrupted to allow the participants to use diagnostic tools and deepest analysis to understand the problems of the area and access some suggestions that can help to solve them.

13.5 Conclusions

Considering the importance of involvement of community and stakeholders in the entire cycle of disaster risk management and the need to better integrate the planning phase and the reconstruction phase with the disaster management cycle, we proposed a variation of the traditional disaster management cycle model.

We also presented a participative approach for involving community in mitigation and recovery phase: the suggested approach is cyclic and incremental, and combines two methods of planning/design: dynamic construction of future scenarios and Strategic Choice Approach. This participatory nature of the approach provides a basis for building a resilient community and helps urban planners to be more informed about the decision contexts in which their scientific knowledge is used.

References

Alexander D (1993) Natural disasters. UCL Press

Berke P, Kartez J, Wenger D (1993) Recovery after disaster: achieving sustainable development, mitigation and equity. Disaster 17(2):93–109

Berke P, Smith GP (2009) Hazard mitigation planning and disaster resiliency: challenges and strategic choices for the 21st century in planning the risk: Spatial Planning as a Strategy for Mitigation and Adaptation to Natural Hazards. ISO Press, Amsterdam

Borrego C, Amorim JH (2007) Risk and emergency modelling for environmental security: general aspects. In: A. Ebel, T. Davitashvili (eds) Air, Water and Soil Quality Modelling for Risk and Impact Assessment 1–13. Springer

Davidson CH, Johnson C, Lizarralde G, Dikmen N, Sliwinski A (2006) Truths and myths about community participation in post-disaster housing projects. Habitat International. doi:10.1016/j.habitatint.2006.08.003

Friend J, Hickling A (2005) Planning under Pression. The Strategic Choice Approach. Elsevier, Oxford

Geis DE (2000) By design: the disaster resistant and quality-of-life community. Nat Hazard rev 151–160

Hills A (2000) Revisiting institutional resilience as tool in crisis management. J.Conting Crisis Manage 8(2):109–118

Ikeda S, Sato T, Fukuzono T (2008) Towards an integrated management framework for emerging disaster risks in Japan. Nat. Hazards 44:267–280

Kaplan S, Garrick B (1981) On the quantitative definition of risk. Risk Anal 1(1):11–18

King D (2008) Reducing hazard vulnerability through local government engagement and action. Nat. Hazards 47:497–508

Lawther PM (2009) Community involvement in post disaster re-construction—case study of the british red cross maldives recovery program. Inter J Strategic Property Management 13:153–169

Mileti DS (1999) Disasters by design: a reassessment of natural hazards in the United States. Joseph Henry Press, Washington, D.C.

Rapicetta S, Zanon V (2008) GIS-based method for the environmental vulnerability assessment to volcanic ashfall at Etna Volcano (Springer). GeoInformatica 13:267–276

Rubin CB, Saperstein MD, Barbee DG (1985) Community Recovery from a Major Natural Disaster, Monograph No. 41, Program on Environment and Behavior, Institute of Behavioral Science, University of Colorado, Boulder

Smith K (2001) Environmental Hazards: Assessing risk and reducing disaster. New York, New York, USA: Routledge

Wisner BP, Cannon ST, Davis IN (2004) At risk: natural hazards, people's vulnerability and disasters. Routledge, New York. 472 p.

Chapter 14
Aspects of Architecture and Urbanism in the Reconstruction of Disaster: Comparison of L' Aquila (Italy) with Kolontar/Devecser (Hungary) and Corbeni (Romania) in the Context of Participative Reconstruction

Maria Bostenaru Dan

Abstract This paper proposes to analyse the architectural and urbanism aspects in two different recent cases and one historic one related to the second recent case of reconstruction after disastrous events: 1. the progetto C.A.S.E., which created a series of new neighbourhoods around the city of L' Aquila, Italy, affected by the earthquake from 6th of April 2009. The old city was temporarily protected by means of propping, with the inhabitants evacuated, while new buildings, following the highest standards of seismic protection with base isolation, were built at a certain distance from the city, lacking services and public transport. The new architecture of the progetto C.A.S.E. is put in the context of similar developments of "Zeilenbau". "Zeilenbau" is an urbanism approach typical for the edges of the city, rejecting tradition. 2. Reconstruction in Kolontar and Devecser after the chemical catastrophe from Ajka, Hungary, in 2010 as well as in Corbeni after floods on the Argeş river in Romania, in 1942. Unlike in L' Aquila, here the houses damaged by the red scum had to be demolished, and completely new buildings were erected in their place. Much of the design was done for free by star architect Imre Makovecz. Unlike in L' Aquila, it follows the vernacular tradition in architecture and in an urban setting. Building the village of Antoneşti in the settlement Corbeni not far from the highest mountains of Romania an architect of the style of Functionalism, trained in Italy, designed peasant houses. In both cases of traditional architecture the new architecture will be put in context as well. Common to all projects is that the building times were short, in about half a year the inhabitants were able to occupy the new houses. In L' Aquila prefabrication was used for this purpose. Timber was employed in all cases often along with reinforced concrete. All developments will be judged from a point of view in which a disaster is a force flattening a city, and the reconstruction produces a new striation, if we were to follow the philosophy of Deleuze as a research method of "the flat and the striated space". This again will be put in connection with historic examples of restructuring a city after a disaster, investigating when striation followed the old

M. Bostenaru Dan (✉)
Department of Urban and Landscape Design, "Ion Mincu" University of Architecture and Urbanism, Academiei 18–20, sector 1, 010014 Bucharest, Romania
e-mail: maria.bostenaru@iaim.ro

M. Bostenaru Dan et al. (eds.), *Earthquake Hazard Impact and Urban Planning*, Environmental Hazards, DOI 10.1007/978-94-007-7981-5_14,
© Springer Science+Business Media Dordrecht 2014

pattern and when not. Another research method from the philosophy employed here will be deconstruction of the term “modern” and “tradition”. The article features images taken during the site visits at the analysed settings one year after the disasters for the recent ones, archive images and contemporary views for the historic one as well as 3D modeling approaches, along with images of the comparison objects.

Keywords Methods of philosophy in architecture · Participative planning · Tradition · Innovation · Functionalism · Reconstruction · Rural housing typology · 3D modeling

14.1 Introduction

Participation means co-decision in planning processes designed by others. Central elements of the concept are influence, interaction and information exchange. In the 1960s’ and 1970s’ theory a closed model of planning was employed, in which the planning subject disposed of all information and power means to create a new world out of a “tabula-rasa”-situation (Häußermann and Siebel 1993). The model suited the theme of the time: urban extension through satellite towns on the town edge. Intervention-administration and adaptation describe the image of the planning. Participation attempts were isolated cases (Byker in Newcastle or the theoretical approach of Christopher Alexander 1977). At the beginning of the 1970s the planning focus shifted towards conversion of city zones, where the planner had to mind numerous actors, uses, constructions. A new, “open”, planning model was looked for, with somehow the same enthusiasm with which the architects of previous centuries were looking for a new style. The application of participatism in Germany began, as in any other country later on, with forms of protest: what the citizens did not want, but ended in a positive form with the association of citizens implementing what they did want. It began on the small scale of architectural objects and spread to the large scale of the region (Bostenaru 2007). It spread from the pilot projects of the 1970s to a routine in the days of today, when it is embedded in urban legislation.

Bostenaru (2005) identifies four layers of planning, historically constituted: recovery/reconstruction planning, emergency planning, mitigation planning and resilience planning. “At post-disaster reconstruction stage, land-use and physical planning activities are employed.” First examples in this sense date from the Baroque urban plans; after the 1693 Noto-earthquake (Tobriner 1982), 1746 Lima-earthquake (Garcia 1997), 1755 Lisbon-earthquake (Cardoso et al. 2004), 1766 Istanbul-earthquake (Ambraseys and Finkel 1987), which included the development of safe individual building structure types. (Coburn and Spence 2002) identified some twentieth century examples giving insights into the “long term nature of earthquake protection from the decisions implemented in the aftermath”:

- early: Pakistan 1935 (Coburn and Spence 2002),
- mid: Skopje 1963 (Greene 1987), featuring the plan by star architect Kenzo Tange (in Skopje, like in Lisbon, an example ruin was kept as witness of the earthquake: the clock of the train station stopped at the time, in the ruin),
- late: Mexico City 1985 (Coburn and Spence 2002)”.

If the Lisbon earthquake reconstruction from the eighteenth century is a successful example, although the only ruin kept is the Carmo Convent, in the twentieth century requirements are higher in what regards keeping the substance. The reconstruction of Gibellina, a Sicilian locality, in the 1960s (film by Jörg Burger 2007) was done with means of functionalist architecture, as in L' Aquila more recently. The new city is a magnet for tourists with its brutalist concrete reminding the one of Rouen, "ville moderne", but deserted by the inhabitants. The same phenomenon is seen in the reconstruction of San Giuliano de Puglia after the 2002 Molise earthquake (Langenbach and Dusi 2004), where citizens could not attach themselves to the new locality, the demolishment of the old buildings leading to sociologic problems. Also in Romania the city of Zimnicea, almost completely destroyed in the 1977 earthquake, was rebuilt following the functionalist principles by star architect Cezar Lăzărescu, an award winning project.

This paper proposes to analyse the architectural and urbanism aspects of the progetto C.A.S.E., the neighbourhoods built to the city of L' Aquila after the 2009 earthquake compared to the new developments in Kolontar and Devecser after the 2010 chemical catastrophe. The research has been extended by research on the Antoneş village in the Argeş county in Romania, constructed after floods in 1941 by Richard Bodernache, an architect trained at the Romanian School in Rome, using traditional architecture elements of "New Romanian" style, and not the Modernism he was aware of. Starting point of the argumentation was the fact that in L'Aquila it was chosen to be built in the periphery, in "Zeilenbau" type, while the historic centre, with typical Italian block housing was left as a "ghost city" with the walls propped, but not restored. Architecture is an expression of the cultural setting, and we try to analyse this approach of post-modernity in an Italian socio-architectural context, in order to see to what extent the "heritage habitat" (Gociman 2006) was kept.

14.2 Methodology

Methods from philosophy, according to different schools of thought, were applied, if suitable, to the investigated architecture subject, as it will be further explained in detail. It included also insights from philology, concerning the act of creation in writing (novels) compared to that of architecture works, and social sciences (sociology of architecture, more specifically participatory planning case studies).

The philosophical analysis was applied to case studies, using the method of regression. How regression, a method of Joedicke (1976), is applied to decision processes, including participatism was presented in Bostenaru (2004): "It is an approach where characteristics and criteria are deducted based on analysis of existing projects, while know-how is derived by means of scientific disciplines. In the first step of the deductive phase a general hypothesis had to be drawn out from the one formulated after induction. In a second step, statements for a particular case can be derived." Data sets are obtained from case studies and a hypothesis is induced. From these multiple individual hypotheses a single hypothesis is obtained. Such individual hypothesis concern: the mental map a community has of its environment based on the Lynch (1960) analysis, the characteristics of the traditional and respectively modern building typology, and also the lessons from participative case studies. From these individual hypotheses a single hypothesis is obtained. The mean for obtaining this was deconstruction of the term "modernity" in the dialogue with the opposite term "tradition". It is an idea generating from the school of thought of the French philosopher Jacques Derrida.

This concerns the relationship between traditional and modern in seismic prone areas in shaping the mental map which provides psychic security after a catastrophic event. Another school of thought from which methods were borrowed as a philosophy appropriate to this research was the one around Deleuze (who refused to be assigned to a school of thought, but follows the principles of deconstruction). Research on the potential of participation has its roots in phenomenology, from the first theoretical approach of Alexander (1977). The same applies for the "heritage habitat", by learning lessons from the spirit of the place. The single hypothesis is to be integrated into the mission/recommendation in a strategic planning in the process of decision making. Then hypotheses for the individual elements are derived and finally statements about them deducted. These statements are feedback compared with the induced hypothesis in order to regressively reformulate the later. (Bostenaru 2004) The individual elements for which statements are derived are the three case studies of reconstruction, the subject of the paper: L' Aquila, Kolontar/Devecser and Antoneşti village in Corbeni. For these two case studies a comparative analysis was performed, with criteria derived from the traditional-modern dialogue investigation.

The analysis was preceded by field studies at all locations of the case studies to which the hypothesis is applied, while for the case studies from which the hypotheses were derived were analysed on the basis of literature review. In the literature study three lines were followed: investigation of the architecture concept, investigation of the participatory concept and lessons learned and influence on other similar approaches. In the field, study methods such as photography, but also reproduction through sketch drawing by hand and computer 3D modeling, were employed. Archive research was conducted to retrieve the plans of the ancient buildings, while the new ones became available through literature research, in books or through internet resources. We aim in a further research to apply to these data methods of the analysis of plans developed during previous research projects such as PIANO (http://bostenaru.natkat.org/) such as ontology, space syntax or building information modeling, for which our 3D modeling is the begin. The space parcour was investigated through direct perception. Following the field study, research by design was applied. This did not encompass design of whole buildings but of scenographic installations aimed at a space of memory after disasters. For designing the installations the same investigation methods as in field work were employed. Then these installations were compared to endeavours at the site of L' Aquila to model not only the memory, but also the current state of the site. The interpretation of time in a line of space and time comes from Heidegger and thus phenomenology, but also from Nietzsche (Bostenaru 2011a). Thus the methods from philosophy of space were applied to space design.

14.2.1 The First Phase of the Regression: Research Framework for the Induction

A concept of philosophy was applied to the analysis of urban reconstruction after natural and man-made disasters. The concept was launched from the philosophy of Gilles Deleuze (1980) and refers to smooth and striated space. The hypothesis is that after a catastrophe, the city, which is a striated space par excellence, is smoothed, and then striated. Striation is similar to that before the catastrophe in examples such as reconstruction after World War II or London after the Great Fire of 1666. Also, examples of different striation in accordance with the principles of Art Nouveau and Baroque are given: these examples are Ljubljana 1895 and, respectively, Lisbon 1755 after large earthquakes, which affected them in different centuries. Smoothing and striation may occur also in the absence of disasters, but how to apply striation depends on political will, such as in case of Le Corbusier and the Voisin Plan for Paris (Le Corbusier 2007) or interventions in Bucharest, when the Athens Charter was applied or in the time of Ceauşescu. The earthquakes in Bucharest are related because they led to increased vulnerability, physical and mental respectively. These issues are dealt with more extensively in Bostenaru (2010a).

Data can be transformed into information for decision making intervention. Although the decision is the element to be ultimately reached, it has a number of supporting elements. The participatory aspect has to be taken into account in decision-making systems through user involvement. This is foreshadowed in the documents underlying urban conservation or environmental intervention, such as the Declaration and Charter of Amsterdam 1975, in the European Year of Architecture, which promoted integrated conservation, at conservation level, or in the Texts to HABITAT II (OECD) "Final report on the ecological city", at urban planning level. A participatory approach will take account of the so-called "heritage habitat" (Gociman 2006) defined by a study of the typology in that area. For this purpose we studied the interwar typology of buildings in Italy, from both Rationalism and Novecento, and related them to the International Style, to Avant-Garde (Gibson 2003) and Modernism (Harrison 2003). It is to be noted that Gruppo 7, the promoters of Rationalism, proclaimed another five principles (Etlin 1991, p. 250) than the five principles of Le Corbusier (2007), and that contextualism, in Italy called "ambientismo", as well as the expression of the constructive structure, belonged to them. We will return in the analysis to what results gave the employment of this principle and of that of free pilotis in the ground floor of Le Corbusier in L' Aquila. Another important aspect of conservation is to preserve the object memory architecture, which is why there was an incursion into a multimedia art project, such as the door to the "re-discovered space" (Bostenaru 2008a), multimedia itself playing a role in conservation interventions for the relationship between old and new. The door to the "re-discovered space" is a scenographic installation of us symbolising a unique encounter. The door to the rediscovered space has long since merged with the environment. So we have the door, which is old, which belongs to the past, to the time when that space was still known, and we have the lighting, the light of today which tries to rediscover itself, which tries to show us a way, or to show us even what's behind the door, but we don't see this. The door is laying horizontally on the ground covered with sand, and half tilted, so the light coming from a box can come thus from below, coloured by the transparent elements in its way. When archeologists try to find some signs of a culture that no longer exists, or at least in the form it was, they collect some objects, and put them in a museum and then say: 'there you can get the feeling of the culture'. As I think of it, it is very seldom that a museum of objects of a culture really reveals a culture. On this idea I built my memories. Boxes in the sand, like the door I once built. Just boxes in the sand, needing to be dug out. In a computer transposition of the installation of boxes in the sand, which contain objects addressing the memory, are transformed into cubes playing with composition of linear and curved walls shaping thus spaces which give a feeling. This way they transpose the book that preceded the door, in which instead of walls, lines in different shapes separate black and white surfaces in order to give the feeling of different spaces—the drawing in the plan. Such multimedia installations are possible to revive old sites, as Xenakis did in Cluny and Persepolis, or as it was done in film, for example "Dust Devil" on the sands of Namibia, to play with the same metaphor of covering with sand, although other materials, such as vegetation, water or even the propping as we will see, are thinkable. In vegetation such as rose bushes in "Sleeping Beauty" even the cut of such spaces is possible. In Bostenaru (2010b) we designed a pavillion to symbolize the employment of reinforced concrete which influenced this "heritage habitat" and which could well be one of these boxes. The role of photography was stressed, and how to organize it (Bostenaru 2011b) in the cybernetic systemic organization of an ontology. Finally seismic strengthening, which can be applied to both existing and new buildings, is a particular intervention to be considered for conservation. In this case we return to the relationship between traditional and modern lessons applied to modern construction.

Seismic risk management involves measures for pre-disaster vulnerability reduction, retrofit, but also even to prepare for action in case of emergency, as well as post-disaster response measures in reconstruction. The impact of seismic events is divided between keeping the memory and project opportunity, namely urban remodeling. Urban remodeling was analyzed by means of philosophy as a research method in architecture. We then went out to preserve memory in intervention, considering how so-called "heritage habitat" which is appropriated by the community and the new philosophy as a research method in architecture, by means of deconstruction, creates the relationship between tradition and modernity. Applying these principles in a case study, reconstruction of L' Aquila, Italy, helps us understand to what extent the Italian "heritage habitat", as identified in Bostenaru (2011c), the subject of our research, was employed in the reconstruction. In the European context, we relate it to a reconstruction which respects Hungarian (Danube Swabian) tradition, that of Kolontar/Devecser, and to Romanian peasant tradition, that of Corbeni. From post-disaster reconstruction lessons can be learned for future disaster-resilient construction, an expression of local seismic culture, as presented in an earlier paper in this book. This is why we examined to what extent such lessons can be learned from timber as material for construction in a reinforced concrete frame system, not least because timber was first an inspiration for the language material of concrete, both being tensile materials.

The aim of the study of the typology of historical buildings of reinforced concrete structure is, as mentioned, to define so-called "heritage habitat" (Gociman 2006). Under this concept, the reconstruction should call upon the memory of the affected population to rebuild using urban and architectural structure compatible with their memory map. The idea of "heritage habitat" is based on the approach of Lynch (1960) regarding the perception of the urban image, which we previously used for urban analysis in decision for retrofit (Bostenaru 2004). Rather than considering all buildings equally, differences according to the importance in collective memory should be considered. As mentioned, and presented in Bostenaru (2008a), memory can be accessed by means of art, to enable people to express their perception of "heritage habitat", considering that not every citizen is able to speak in this language, as they were considered by Alexander et al. (1977). We will see that recent approaches in L' Aquila address this research by artistic project into the memory of inhabitants, not only through photography, but also, as we propose, through artistic installation. The study of the "heritage habitat" is important, both in the design of new buildings and reconstruction as in retrofit of existing buildings, if relocation of site occupants is necessary for a period, in order to minimize disturbance related to their displacement. This would increase acceptance retrofit measures among the inhabitants, and increase resilience in facing future earthquakes. We considered thus first the theoretical basis underlying the need for "heritage habitat" in the reconstruction, then its absence in a case study: reconstruction of L' Aquila, Italy, reported the approach of contemporary European reconstruction, namely in Hungary, and to a historic European reconstruction, in Romania, by an architect trained in Italy. In Italy and Hungary there are, among others, some models in reinforced concrete approach in dialogue with traditional materials, for example the birth of the idea of transparent ground floor on columns, which lead to the flexible ground floor in seismic areas (Béla Lajta, Rózsvölgyi building, Budapest, 1911–1912), but also essays on the connection between reinforced concrete and timber (Benko-Medgyaszay 1909).

We made also a brief incursion into the cultural nationalism theme, a theme of modernity, and so the historical period we are dealing with in this paper. From a cultural nationalism (Hobsbawm 1994) point of view, reconstruction becomes a milestone for a nation, in cases when the remodeling after an urban disaster coincides with a historical event (in which case we speak of a historical accident). Such cases were reconstructions after World War II, but as such the reconstruction of Ljubljana after the earthquake of 1895 coincided with separation from the Austro-Hungarian Empire. The reconstruction of Corbeni took place during World War II. The combination of traditional and modern architecture in the twentieth century can be seen also as a form of cultural nationalism. The relationship between literature and architecture has several facets, not only on architecture representation in literature, but also the use of literature as a means of expression in participation (e.g. the approach of "A Place for Words" http://www.urbanwords.org.uk/aplaceforwords/index.shtml in the UK). Relevant for our research is the fact that modern historical novels treat fear and anxiety. Most often they are related to disease, and conjure up a natural disaster such as the plague. This is the case of Ken Follett's novel "World without End", which is used in promotion of architecture in literature, even in connection with construction management activities, especially in the games developed afterwards. Traditional architecture is modern, according to Langenbach (2009) since it meets the requirements of contemporary resistance to earthquakes. Because modernity was born in 1755, with the Lisbon earthquake and shaking that it provoked in all disciplines of the Enlightenment, and earthquake resistant buildings of the Marquis of Pombal. In our analysis we made a distinction between Avant-Garde, Modernism, and Modernity. The Italian interwar typology distances itself from the "classic" Modernism (Avant-garde) by the socio-economic aspects to be addressed and therefore the space-functional solutions adopted and the materials used, and even the urban solution, the building in context, instead of so-called "Zeilenbau".

Italian architecture uses also traditional materials, such as stone in which the country is rich (ex. Casa del Fascio by Giuseppe Terragni), and which influenced also Romanian architects who spent a great deal time there (ex. the ASIT building 1935 by Richard Bordenache in Bucharest), a characteristic which came to other architects from the influence of vernacular architecture, such as in Estonia or by the Romanian Henriette Delavrancea Gibory villas in Balchik, use of natural stone in a language for "other modernisms". The use of timber has led to a different language, the National Romanticism of Károly Kós in Romania/Transylvania (for Károly Kós architecture, see the other paper in this volume), inspiring the architecture of Imre Makovecz, including the reconstruction of Kolontar/Devecser discussed in this paper.

14.2.2 Review of Participative Aspects in Disaster Management

In the twent-first century a new concept arose, reflecting computer support for participative approaches: public participation GIS. Despite the name, a number of practices associated to this concept are low tech and realized through hard copy models, such as participative 3D mapping. However, aspects of computer support are still included, for example in mapping, using voluntary geographic information or so-called crowd sourcing, as we will see also in the creation of the Google Earth3D model of L' Aquila. Remarkable however is, that most (P)PGIS practices are related to locations in the developing world (P3DM world wide locations at http://maps.google.com/maps/ms?ie=UTF8&hl=en&msa=0&msid=10355550579-3015427581.00043e70e56245f7c03b2&t=h&om=1&source=embed&ll=44.37098-7,7.624512&spn=11.119462,27.46582&z=6&iwloc=lyrftr:msid:10355550579301-5427581.00043e70e56245f7c03b2,00046cdeb2fb91f7bfd3f,4 accessed 31.01.2011). So are the participative approaches to post-disaster reconstruction presented at the i-rec conferences. I-rec is a research centre, based at the University of Montreal, Canada, and specialising in information and research for reconstruction.

Locations to which participative approaches in reconstruction relate are India (reconstruction after the 2004 tsunami or 2001 earthquake), other areas affected by the 2004 tsunami (such as Sri Lanka), Turkey (reconstruction after the 1999 earthquakes), Bam, Iran (reconstruction after the 2004 earthquake), Central America (hurricane Mitch 1998 and El Salvador earthquakes 2001). But also some areas in the industrialized countries are included, such as the UK and Australia (affected by floods) and the US (hurricane Katrina).

Jayantha and Gunasekera (2006) wrote about participatory post-tsunami reconstruction. One of the author's first observations is that community involvement contributes to sustainable development. Top-down approaches are short term, but this is the case in most post-disaster reconstruction. Participatory approaches include pre-disaster preparation for a possible disaster that would affect the community again. Jayantha and Gunasekera (2006) wrote about the involvement in such a reconstruction of the Practical Action Association, namely the construction of community-based boats, road construction, waste management approaches, the design of decentralized disaster preparedness, disaster-resistant housing and environmental sustainability work. In emergency disaster recovery these activities are neglected and the participation of particularly vulnerable groups is minimal. One of the instruments adopted was the law. Another method was to invite community representatives, both men and women in strategic planning meetings. Another method was to identify forms of communication that can meet their requirements. Another method was the workshop "Building back better" to define guide lines. Also post-tsunami rehabilitation in Sri Lanka is discussed in Harvey's (2006) contribution, this time focusing on gardens. In fact, housing and green space are areas of participatism par excellence, as we found from several analyses of disaster. In this contribution, the rehabilitation of agriculture is treated as a component of disaster recovery. Some of the methods used were interviews with governmental agents, NGO participation in meetings, and discussions over the Internet (EMail and web group). CBOS (Community Based Organisations for Selfhelp) were formed. The role of gardens in the developing world has been recognized by adapting to studies of improvement possibilities with exercises based on the Green Revolution Game at the University of Karlsruhe, Germany. Lang (2008) investigated the response to community needs in post-disaster housing on Nias Island, Indonesia. Although the amount of destruction after the 2004 Indian Ocean tsunami was here not so big, it was followed by a destructive earthquake in 2005. Dwellings built by contractors have been a failure, contracts were not honored, and therefore a pilot project run by community building was developed. Residents participated in the decision making process. Such a construction leads to knowledge construction management and construction methods by the community. There were no standard methods to involve the community, but the group discussions and partnership in design and construction are among the most successful. With the contract extension the incremental construction of houses was difficult. Lang (2008) describes the exact phases of design and execution and how the community was involved in them. Sivaja (2002) wrote about the unique experience of training in community post-disaster scenarios with cyclones in India.

Arslan and Ünlü (2008) investigated the role of participatism in reconstruction after the earthquakes in Turkey in 1999. Participatism was seen in their paper as mediated by NGOs. To measure the success of NGO involvement in mediating between the affected people and the government, which built temporary and permanent housing after the earthquake, a SWOT analysis was used. It was found that those most affected were former tenants who lost housing and property in the earthquake so they remained forced to live in temporary housing. The conclusion is that such participatism must be supported by appropriate legislation in order to be successful. In another paper (Arslan and Unlu 2006), the authors compared permanent housing built with government funding, the World Bank and two NGOs in terms of participation. In the latter case participation exists, but the result was a small district contribution to the total reconstruction, even if it is a good example. The bottom line is that if rebuilding is participatory, the community psychosocial rehabilitation is faster. Dikmen (2006) investigated the acceptance or refusal of housing relocation after the earthquake disasters in Turkey in 1999. The finding was that the refusal was caused by much too rapid decision making, lack of participation of those affected in decision making, the site selection criteria and poor lack of interdisciplinary work for this action, not taking into account the lifestyle of inadequate guidance to those affected and the beneficiaries during the construction phase of the houses. The lack of architects and designers of site selection appears as a negative factor. To avoid these negative effects a strategic pre-disaster design should be done. Some phases of post-disaster housing design should be completed in advance before the disaster. Unlike previous contributions which are case studies, Özden (2006) developed a model for community involvement in post-disaster housing programs, focusing on Turkey, namely the housing after the 1995 flood and the 1999 earthquake in Marmara Senirkent. The employed method of research was site observation as well as questionnaires on satisfaction of residents. Both districts face the same socio-cultural problems post disaster as those of L' Aquila: breaking the memory and social, economic problems by the featuring both distance from work, and problems of transport/infrastructure. In addition to other issues such as quality of construction, a key issue is the lack of tenant participation in decision making. The strategy of the Government was to take full responsibility for design and construction. Neighbors are separated from each other in post-disaster buildings. In Turkey there are disaster management systems, without a component for the active participation of communities at risk. An example is seen in New Orleans involving the active participation of all, provided by the Internet and public communication means. The author proposes to involve NGOs or community representatives and local architects and engineers who have knowledge of vernacular building in the disaster management cycle.

Asgary et al. (2006) examined the role of actors in post-disaster reconstruction after the 2003 earthquake in Bam, Iran. Citizens of the community were identified collectively as one of the actors, and the NGOs as another actor. Gharaati and Davidson (2008) investigated the role of participation in the reconstruction. Non-traditional construction methods were promoted, namely the use of steel frames, but these methods continued to be used after the employment of technical specialists was suppressed, without knowing whether or not lessons had been learned. A combination of a top-down approach based on technology and a bottom-up community-based approach was used in the reconstruction in Bam: those affected could choose the material and the reconstruction technology they preferred. There were three studies on site, one immediately after the earthquake to examine the causes of failure, one during and one after the formal reconstruction. The last study was not conducted through questionnaires, but by investigating the tacit knowledge with the researcher admitted in the working group under study. A surprising finding was that people tend to want over-size building blocks, even those in engineering projects, to ensure that buildings are earthquake resistant. These are lessons learned. The construction with metal structure proposed in the reconstruction in Bam is presented in detail in a report in the World Housing Encyclopedia (http://www.world-housing.net/). In another contribution on this Bam 2003 topic, Gharaati (2008) presents alternative techniques of reconstruction such as confined masonry. As noted in the current case of Haiti, the spread of this technique is based on the educating of local craftsmen in the use of technology. This contribution is critical for the seismic design of the steel model adopted; it underlines the lack of participation and the lack of flexibility of the plan, which will be difficult to be extended for a larger dwelling. Confined masonry is a construction technique promoted for non-engineered housing by the World Housing Encyclopedia through a network.

Duyne Barenstein (2008) investigated the post-disaster housing in India after the 2001 Gujarat earthquake and after the tsunami that followed the 2004 Sumatra earthquake. The focus fell on research conducted to compare the reconstruction with the owners led top-down by the contractors. It was found that participation leads to a more effective reconstruction regarding all aspects, including costs. Future residents participated in the project only after the whole project cycle had been decided, including housing projects and selection of materials, and thus their role may be reduced to participation in the construction process. The employment rate of post-disaster building with community input tends to be high and the homes are more durable. An approach from the owners tends to maintain the cultural heritage and the vernacular architecture, elements of cultural identity of the community. This is an important psychological factor after the devastating experience of a disaster. It should be taken into account and the application for construction would not be vulnerable. Duyne Barenstein (2008) conducted empirical research in India to compare these two types of construction. In Tamil Nadu, along with the relocation, damaged homes were demolished and trees were felled for the new construction, resulting in adverse environments. Unfortunately, the experience of Gujarat did not lead to lessons learned and applied in Tamil Nadu. The lesson to be learned from this comparison is that reconstruction is more than building houses resistant to multiple hazards.

Forwood (2004) wrote about community participation in post-disaster housing projects in Honduras after Hurricane Mitch, 1998. It was an NGO project of Irish Goal, which has made three projects. GOAL opted for a community participation model. Setting up a community was desirable for reasons of lack of land available for building more communities that should be grouped in a single development. Its neighboring communities were hostile to refugee resettlement in this place. Weekly meetings with the designated project engineer were held. In the long run integration into the wider community should be monitored. Building models used were: (i) Build your own home under supervision, (ii) use a traditional (contractor), (iii) form a building community (the urban plan is prescribed by the NGO and the beneficiaries organize a building teamwhich goes from house to house to build them, including houses that are not allocated to the beneficiary until the project is complete) and (iv) choice of material only. Sliwinski (2006) wrote about social dynamics in participatory reconstruction—a review of El Salvador after the earthquakes of 2001. Participatory methods often used by NGOs in developing countries to create a sense of a community reconstruction project are identified by author. These methods can be successful or not according to their adaptation to local context, namely the local values and social structure. The research method was direct observation of the site. The first case investigated was the initiative to create a neighborhood (pilot) model, which is rare in post-disaster projects. The work-for-food scheme was adopted, those involved giving up their services for a year and working at home. The project followed a very top-down approach. It was prevented from forming a community development association.

Glavovic (2008) wrote about the need to build sustainable communities, resilient to hazards, with the example of recovery after Hurricane Katrina. In the process of rebuilding after the hurricane, the existing management programs failed, which stimulated the initial design activism and community participation. But community activism must be complemented by institutional leadership in order to translate into projects. Resilience is the reverse of vulnerability. The principles set out by Glavovic are in line with the ones set by Dela and Arambepola (2000) on the creation of resilient communities: encouraging participatism and the holistic approach to disaster. Worth mentioning is the contribution of Rebecca Solnit "Paradise Built in Hell" (Solnit 2009) who argues that a hell left by a disaster develops humanitarian aid and initiatives worthy of paradise.

Buckle (2004) studied disaster recovery management by comparing the situations in England and Australia. Unlike the developing countries referred to by the previous case studies presented, in both England and Australia community participation in disaster management is rooted in politics that are far from exclusive basic emergency services. Involvement in politics is similar, but there the formal design is less evident in England. Case studies were Lewes, a town in England affected by floods in 2000 and the village of Leonard Stanley, damaged by storm in 2002. Local commitment had design support. Marsh, Buck and Smale (2004) have written about the concept of community capacity and community disaster management in emergency management research in Australia. Disaster management agencies should adopt a strategic approach when working with communities. The research is based on response rather than disaster risk assessment and vulnerability of the community. It was found that there is a discrepancy between the perception of risk by the agencies and the community, focusing on natural hazards, while the community feels vulnerable because of problems like unemployment or the departure of youth. A better term is a community of interests.

Findlay (2002) wrote about the design of sustainable communities in relation to resilience in three sites located in different parts of America and investigated by the author, an architect. A sustainable community must have a healthy social network.

In a table it summarizes the criteria for sustainable communities in different phases of disaster response.

Fox (2004) has written about design for increased resilience. It introduces the new term "disaster planning" different from the design development, an established term. This term stems from research in sociology, to develop multidisciplinary contributions from geosciences, health and engineering. The contribution analyzes the 2002 contributions of i-rec to the participative approach.

14.2.3 The Second Phase of the Regression: Deduction

14.2.3.1 The Reconstruction in L' Aquila

On April 6, 2009 an earthquake of magnitude Mw = 6.3 (USGS)/ML = 5.8 (INGV, on the Richter scale) hit the Abruzzo region of central Italy, west of Rome, causing significant damage in the city of L' Aquila and surrounding villages. The earthquake occurred along the fault of the Eurasian and African plate along the Adriatic Sea which led to the formation of the Apennines. 10,000 to 15,000 buildings were damaged, with 305 dead, the highest number since an earthquake in Italy in 1980, the Irpinia earthquake, with 1500 injured and 60,000 to 80,000 victims who needed new homes. L' Aquila is built on a former lake, with alluvial soil such as in Bucharest (EERI 2009), but has a medieval urban tissue with old buildings.

Causes and effects of the earthquake are not yet sufficiently documented, but what is even less documented is the provision of housing reconstruction quickly before the cold winter in the mountains, for the homeless. After publishing a special issue in the Italian magazine "Progettazione Sismica" (3/2009), including chapters on the hazard, the impact on structures, and reconstruction, which was later translated into English and distributed for free, the IUSS Press published in November 2010 a book produced by the EUCENTRE exclusively on reconstruction, "L' Aquila. Il Progetto CASE"(Turino, 2010), only in Italian. The book bears the subtitle "a reconstruction project unique in the world to succeed in providing housing to 50,000 in just 9 months". C.A.S.E. stands for "Complessi Antisismici Sostenabili Ed Ecocompatibili" which means" seismically resistant, sustainable and eco-compatible complexes". On the other hand, C.A.S.E. is plural for "casa" which means "house" in Italian, and makes a reference to the provision of collective housing. The buildings designed around L' Aquila are corresponding to a housing typology promoted by the Avantgarde: "Siedlung"—midrise and so-called "Zeilenbau" (linear design), composed of blocks of flats. This typology was used in the interwar avant-garde in countries such as Germany (Fig. 14.1), but not in Italy. C.A.S.E. itself is a true renaissance of interwar housing types, a goal of our research project CA' REDIVIVUS performed at the IUSS 2005–2007. The prototype used in the "Siedlung" neighborhoods of the C.A.S.E. project were defined as urban cores in an earthquake. It should be noted that the idea of urban cores that respond to earthquakes is not a new idea in this project, as Gociman (2006) defined security zones, or the Italian approach of

Fig. 14.1 "Zeilenbau" of the Modernist Avantgarde: **a** Student dormitory Fondation Suisse in Cite Universitaire, Paris, Le Corbusier (1931–1933). Photo: Bostenaru 2010. **b** German Siedlung of Dammerstock, Karlsruhe (1928–1929). (Photo: Bostenaru 2004)

minimum urban structure—struttura urbana minima (SUM)—that is strategic, but this core is not safe against hazards and emergency response. Such a nucleus could be part of SUM.

In the C.A.S.E. project (Fig. 14.2) renewable energy was used, from, among others, solar panels. The prototype includes a house on piles, free round columns on the ground floor, which can be separated from the rest of the house by seismic isolation. The implementation has given interesting architectural solutions in the mountain landscape of L' Aquila, with free drivers for only a part of the building and land, behind. It is a combination to build on sloped land providing seismic isolation solutions. The ground floors serve as semi-garage. The areas where Siedlungen were built are located on the outskirts of L' Aquila, in so far as they were outside of historic places nearby, but also the places themselves were located in the existing tissue. Turino (2010) documents the choice of these sites using satellite images and also the planning solution, by placing the "Zeilenbau" of Le Corbusier to reverse urban tissue (Fig. 14.1). The neighborhoods have worked with a limited number of "Zeilen", which were combined in various different neighborhoods by repeating from one neighborhood to another, with the result that they can be recognized based not on the type of blocks, but on the size and urban distribution. Eight types of houses are in wooden structure, six in concrete and two in steel, all about the same price of 2 million € each. Execution speed was a consequence of pre-fabrication. Today, prefabrication has lost the fascination exerted on the architects to build mass housing for preference for individual homes, but it is a legacy of the interwar period, when there was a search for economic efficiency. However, pre-fabrication is still used in some European countries for social housing. In a study of ours (Bostenaru 2008b) the finding is that in earthquake-prone countries in Eastern Europe, prefabricated concrete housing performed better because of better supervision when put into practice. All types of prefabrication are used by Progetto C.A.S.E.: from columns and beams to precast concrete cells in the size of a room space. Seismically isolated base columns were cast in situ concrete. In giving over the residential units management was done with a Raumbuch. Energy efficiency was verified by thermographic analysis. We visited the project site of C.A.S.E. in May 2010, which was after the homes were completed but not yet documented. Unlike the housing itself, vegetation needs time to grow. But the need to grow together is near. The neighborhoods themselves do not stipulate

Fig. 14.2 The reconstruction in L' Aquila part 1: Progetto C.A.S.E.: **a** protests of citizens for the lack of participation, **b** "Zeilenbau" in the new neighbourhoods: here an industrially constructed concrete building, **c** newest seismic technique: the use of base isolation in the new neighbourhoods (the prototype), **d** also newest technique: solar panels in the new neighbour-hoods. (Photos: Bostenaru 2010)

and Turino (2010) does not document community centers—the urban analysis should have been made not only for free available sites, but also for the available facilities. Thus, the main technical achievement is the solution of environmental energy, seismic structure, and rapid construction management, which were outstanding.

Participation started at a small-scale of architectural objects, and then spread widely in the region. It spread from pilot projects in 1970 that became a routine the day when participation was incorporated into planning legislation. The situation in L' Aquila is a special case vis-à-vis the post-disaster participatory approach. Although reconstruction initiatives in post-disaster participation are reported, as noted, following the earthquakes in Turkey in 1999, in L' Aquila housing was built during the 6 months to provide residential space for those whose homes had be-come uninhabitable after the earthquake of April 2009. One year after earthquake the historic center of L' Aquila was still marked by a ban on access, and almost three years after still not rebuilt. The residents were resettled in new neighbor-hoods on the outskirts of neighboring villages and attending L' Aquila to socialize, since distribution of new districts was not based on neighborhoods (Fig. 14.3). New homes are isolated by the latest seismic engineering techniques, but lacks infrastructure: transport, trade. On the bordering of propped building walls, to prevent further structural deterioration,

Fig. 14.3 The reconstruction in L' Aquila part 2: the historic city: **a** The centre of L' Aquila before the earthquake. Photo: Bostenaru 2003, **b** The same place in the centre of L' Aquila one year after the earthquake, **c** propping of buildings have made streets unusable, one year after the earthquake—inspiration for Michele de Lucchi's installation (propping in an arch), Photos: Bostenaru 2010. **d** Fragment from "Comme facciamo" (Google Earth) showing the place in **a** and (**b**), **e** Building maker model contributed by us to "Comme facciamo?" **f** Integration of the building maker model in the 3D Google Earth project "Comme facciamo?" **g** Position of the park at Piazza d'Armi **h** relationship of the park in Piazza d'Armi with the historic centre (both Google Earth)

citizens protested against the cost of the new housing where they had not been involved through participation. However, it should be noted that following the protest participation a constructive attitude was born, one year after, using participation options of the twenty-first century. The volunteering Project "Comme facciamo" (http://barnabygunning.com/comefacciamo/), to which we participated with some models, practically rebuilt the old town of L' Aquila, using photography and Google Earth for 3D modeling. Such a geographical information system was used rarely in industrialized countries to make a contribution to methods for P (P) GIS (Public Participation GIS) and P3DM (Public 3D Modeling), it is more common in developing countries. The difference between the two tissue types, traditional and "Zeilenbau" can be used to classify models into a GIS analysis of architectural and urban scale seismic vulnerability of the city. Documentation, usually a step in restoration but not always for a new building, carries very important lessons from reconstruction to other sites that are facing such problems as in Haiti, for example. Documentation for new projects has been applied to "Wohnen morgen" in Hollabrunn, Austria, by Ottokar Uhl, and this project is in the participation tradition. As indicated in the discussions "between temporary and permanent" these homes are designed to be temporary (Calvi and Spaziante 2009) during the time the historic center is being rebuilt, in fact, not only on Google Earth. C.A.S.E. are to become student dormitories, in the true tradition of Le Corbusier. The situation in L' Aquila reconstruction led to increasing involvement of professionals in architecture and urbanism, especially in generating ideas for L' Aquila. The International Union of Architects became involved with a workshop in Copenhagen in autumn 2009 followed by a visit to the site, Istituto Nazionale Urbanistica (INU) with a series of meetings and workshops conducted by the University IUAV of Venice. But architects should be involved in advance at selected sites, as demonstrated in the study by Dikmen (2006).

Relocation after an earthquake is not a new phenomenon. A known example in the Baroque period is the famous city of Noto in Sicily (Tobriner 1982). In modern times, apart from L' Aquila such a project was conducted at Gibellina (Burger 2007), which was also a failure in relocation. The city of Gibellina is a tourist attraction, a modern postwar city like Rouen in France after World War II, but it wasn't appropriated by the tenants. Remarkable is the presence in Gibellina of concrete ruins that cover an area covered by an artistic approach/such as the re-discovered space that we talked about. Another example in the same situation is the village of San Giuliano di Puglia, with inhabitants displaced after the earthquake in Molise 2002. But in this approach the seismic technique and participation is new. Relocation due to a disaster can be seen as a turning point to determine traditional revival in the call to "heritage habitat" (which requires prior study to be prepared in case of a event), something to which we return to analyze the elements of Modernism in this reconstruction.

Philosopher Deleuze launched a series of concepts and processes that are based on smoothing opposed to striation (Deleuze 1980). The flat space is that of a nomad, which is grooved for sedentary, and this is reflected in the way Deleuze explains the two concepts in different opposite models/processes. Although the models are different: technology, marine, can be found in the opposition between contiguity and smooth/striated. Sedentary and nomadic space is taken once owned by settlement; i.e., people tended to keep the place. Therefore, although the concept of space granular changed, the site was maintained. Sometimes striations appeared after smoothing. This smoothing was sometimes desired, sometimes utopian, sometimes still applied, but sometimes occurred because of a sudden calamity. New concepts have appeared not only to prevent a streak of disasters, but to further develop urban planning and politics. We analyzed some examples of new or lifting striations in new cities. Striations were applied or were given for reasons dictated by the state, as in Deleuze' s concept of policy, shown in the argumentative discourse. The functionalist architecture of the city proposed a different order than that of the traditional city. Often the edge of the city was developing, in the emergence of interwar and postwar modernism neighborhoods. Le Corbusier' s proposal, the Plan Voisin was more radical. It proposed a smoothing of the historical center of the north of the Seine in Paris and then again striation according to functionalist principles as cross-shaped towers of 60 floors in a rectangular grid of green spaces in park type.

We propose to analyze aspects of architecture and urbanism in the housing project C.A.S.E. The name for the project gives a link between seismic and ecological approach that referred to in texts about Habitat II—Final report on the ecological city (1996). (UN-)HABITAT is a UNO (United Nations Human Settlements Programme) for housing. There were two preliminary conferences before it became a program itself. The second one in 1996 in Istanbul concluded with a report which promotes several important elements for the analysed problem. Crucial is a systemic approach. The introduction is made by promoting innovation, innovation that must take place in urban design, involving the participatory approach, for example through strategic planning rather than regulation, to include education flow, but also tools, ex. the then new Internet and GIS. The immediate goal is ecology, and, more broadly, sustainability. The report consists of the following parts: The first part—ecological city, Part

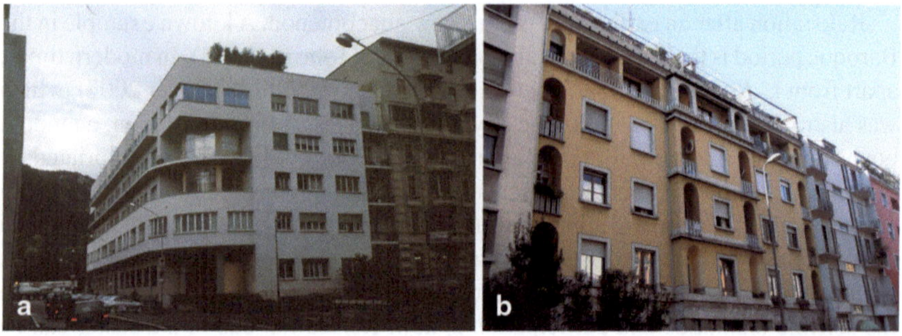

Fig. 14.4 "Ambientismo", the Italian contextualism of the interwar time: **a** building of the Rationalism (Novocomum by Giuseppe Terragni in Como 1927–1928), **b** buildings of the Novecento (House by Giovanni Muzio 1933, 11 Via Longhi in Milano). (Photos: Bostenaru 2005 and 2007)

Two—Innovation as a tool to achieve ecological city, Part Three—the relationship between economy and urban environment, Part Four—Conclusion. We worked on an example application of the report HABITAT II interwar for a neighbourhood in Bucharest (Bostenaru 2002), the same one as addressed by the urban scale retrofit decision in Bostenaru (2004). There was an exercise of urban strategy for a modernist district in the north of the Romanian capital. The district includes historical plots that overall result in several protected areas, and interventions from the communist period, leading to heterogeneity in value. This heterogeneity value was answered by using Kevin Lynch (1960) type analysis, as it allows adequate strategic approach to addressing the level of detail in different areas. A diagnosis was made and a mission formulated, and a mission point was intervention with preservation of the character of the area, which is how specifically protected areas have been defined later on. The district was a subject exercise of a conference in 2011 (http://eaae2011.uauim.ro/), further studies are therefore expected.

The new architecture of the project C.A.S.E. is placed in the context of similar developments, namely:

- Twenty-first century Italian buildings that do not show classic modernism, such as Rationalism and Novecento (Fig. 14.4), in conjunction with other traditional buildings of modernity, namely the Hof-s in Vienna;
- Social construction of the twenty-first and twentieth centuries, taking as examples buildings of the Kramm et Strigl office in the 1990s in Germany (Durth 1999) (Fig. 14.5). Especially the Bürgerparkviertel example of Darmstadt and Wiesbaden prefabricated buildings,
- Construction of dormitories, as housing development under the "Progetto C.A.S.E." are used only temporarily so, and will subsequently become a campus.

First we refer to the dialogue of "classic modernism" and "other modernism" as we met in Italy. We return to this idea talking about traditional-modern relationship in disaster response in the following paragraphs. Adoption of "other modernism" was Italian both in terms of architectural object (due to the 5 principles of Gruppo 7)

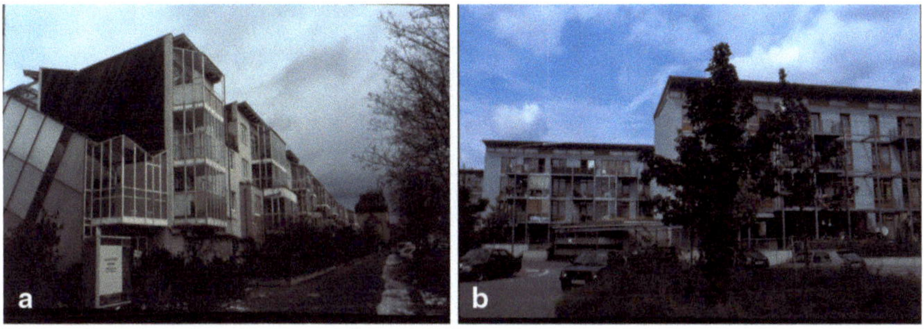

Fig. 14.5 "Zeilenbau" in contemporary building in Germany. Works of the office Kramm + Strigl: **a** Siedlung Bessunger Street, Darmstadt (1988), **b** ecologic Siedlung, Frankfurt Bonames (1995). (Photos: Bostenaru 1998 and 2004)

and urban (island). Other countries, however, have fully adopted the model of Le Corbusier, in the interwar period and later. Bauhaus is the case in Germany. Uhlig (1997) writes that "Zeilenbau" is an option for building yesterday and today. The term "Siedlung" itself, which comes from Germany, and "Neues Bauen", describe these buildings in the suburbs, rows separated by air and light, and green. An example of "Zeilenbau" between the two world wars in Germany is Dammerstock Siedlung in Karlsruhe, a book containing essays by Uhlig. An example of "Zeilenbau" today may be the winning entry in the contest Bessunger Street development, Darmstadt, by Kramm. Kramm et Strigl have developed a set of high standards for social housing Siedlungen (Durth 1999), including one in Wiesbaden, which used prefabricated building techniques, as in L'Aquila, a technique that ensures an effective response to earthquake in countries like Romania (Bostenaru 2008b; Bostenaru and Sandu 2004).

At the same time, it appears that at certain times for nations an earthquake striking, like the L' Aquila earthquake reconstruction after the April 6, 2009 earthquake, was used for reconstruction. The "C.A.S.E. project" approach in "Zeilenbau", did not address urban problems in a typical and traditional Italian way. In fact, the new settlements have new site in the suburbs, many of them "Siedlung"s, although this seems to be a viable solution. The focus of the argument is that it was chosen to build in the suburbs, according to "Zeilenbau", while in the old town with typical Italian housing units in island restoration has not begun, roads are blocked, buildings are temporarily propped not to collapse. L' Aquila is in a state of a "ghost town". Assuming architecture as an expression of cultural background, we consider this approach as one to post-modernity in social and Italian architecture. The so-called "Zeilenbau" is the opposite of traditional tissue islands. This interpretation is helpful for the social construction investigated if it was always above the building on the periphery in L'Aquila. When you get social housing in city centers, like the Viennaise Hof-s, it is building the islands, as these are different from the Siedlungs of "Neues Bauen". Italian examples are useful to see how in Italy the housing for middle class is interpreted, not just the social housing. Construction periods are also relevant, and two examples have been investigated as such, as for Progetto C.A.S.E. the Civil Protection currently documented.

Fig. 14.6 Student dormitory in Germany (HaDiKo): **a** the prize winning building from 1966, **b** one of the later buildings with the 2006 extension. The prize winning wing was extended again in 2012 with a rooftop construction. (Photos: Bostenaru 1996 **a** and 2006 **b**)

The proposed new order was that of the first Avant-garde cities interpreting the so-called "Zeilenbau"—the opposite of traditional urban islands. An example of "Zeilenbau" by Le Corbusier are two dormitories at the Cité Universitaire in Paris: an example of the interwar time (Fondation Suisse) and an example of postwar (Casa do Brasil). We found these to be the most adequate examples of this analysis, as in L' Aquila, because the expected development is that residential units will form later a student campus. Le Corbusier's dormitories meet his principles having a flexible ground floor and the prototype promoted in L' Aquila also has free ground floor columns as in these principles. This is not one among other seismically vulnerable constructions derived from principle, because it is base isolated. In terms of architectural language it is not an expression of public space such as in Loos/Lajta, and even Le Corbusier's housing, but only an expression of the structure. Buildings were made of timber structure 50 %, 30 % of concrete, as the Italian interwar typology and 20 % in metal like the German Modernism, and earthquake resistant housing in Iran (Turino 2010). Another example of student residence considered is HaDiKo (Hans Dickmann Kolleg) in Karlsruhe, Germany, which was also characterized by a reduced time of construction (Fig. 14.6). The building K1 (Büchner and Baréz 1966) won a architecture award, but other buildings were built in parallel wings in accordance with the principles of Le Corbusier ("Zeilenbau"), even with partially open ground floor, with parking function as in L' Aquila but for bicycles, not cars, a difference in addressing environmental problems in Germany by another architect, and finally the K2 building was extended in 2005 following a participatory approach, experienced first hand, as a citizen, by us. The participatory approach is what is not present in L' Aquila.

Exceptions to "Zeilenbau" were still made in Western Europe: Sonne (2009) performed a study on the reform of urban high blocks between 1890–1940: dealing with the durability of this type, blocks late avant-garde urban reconfigured by the type of urban fabric, that this tradition, not following Le Corbusier's principles, as in many Siedlungen of so-called "Zeilenbau". Sonne (2009) observed when analyzed the compact urban island housing type, which is demonstrated to be a sustainable one, developing a viable model of everyday city life. Sonne (2009) observed that the atria types ensure green spaces and thus better climate adaptation, even if they showed a differentiation of different climate in the countries concerned. The investigation of Sonne (2009) discussed different types of Western Europe and North America. In Western Europe, innovation was made in social housing in the suburbs, so-called Siedlungen. Exceptions are the Hof-s in Vienna, Austria, but they are a type of housing often overlooked in the history of architecture. The twenty-first century Italian example, the Rationalism of Gruppo 7 (contextualism) that flourished in Milan and Como, but was also represented in other cities, including Rome, is also mentioned by Sonne (2009). If social housing reaches the centre, they become islands, such as the "Hof"s, which are different from "Neues Bauen" Siedlungen.

The Project C.A.S.E. L' Aquila is not social housing, although it was presented as such to the Fiera Milan. This is because it is an expensive project, implementing an option of the most modern seismic protection: base isolation (Calvi and Spaziante 2009). So, from this point of view it is in competition with sustainable housing choices analyzed by Sonne (2009), or blocks of flats for the middle class located in the city center. On the other hand, it was decided to build in the suburbs, following the type of "Zeilenbau" in the interwar period (Le Corbusier—Bauhaus model). This is a new approach, which wasn't followed in Italy, even in the interwar period, a break with tradition and orientation for international style that was made in the twenty-first century in a way that in the twentieth century was not. In terms of architectural language it is similar to that of Kramm + Strigl, noticing thus adopting the new "international style". But in terms of participation it does not follow the German model. The lack of citizen participation is emphasized.

The project C.A.S.E. is innovative above all in the technical part, the seismic protection, and the use of prefabrication and energy saving, and less from the architectural-urban point of view. From an architectural-urban point of view the model of the interwar time or of the years 1960–1970 appeared in Western Europe and Germany, but not in Italy. Since then, Germany has changed its approach, and now building in a participatory manner that in L' Aquila isn't common there. Even for reconstruction after trauma, such as an earthquake, traditional models are better supported to create a "heritage habitat". Provision of such temporary housing is a necessity not only for reconstruction but also for relocation during preventive retrofit, another lesson learned from Progetto C.A.S.E.. We studied whether and how the buildings in the area of L' Aquila, at least one of the Italian models of modernity, was used to design new buildings.

The involvement of architects in the reconstruction of L' Aquila is even less documented, and because of this ways to involve the population could not be designed.

In 2012 the group "L' Aquila- cantiere cultura" (culture building site) called for and launched the first competition for the reconstruction of L' Aquila. It was to be a park at Piazza d'Armi (http://parcourbanolaquila.it/). The competition entries were to include three compartments: A, of 32.132 m^2—a public place and a parking facility. This should include also a memorial for the victims of the earthquake, but otherwise take over the functions of a place (mercato) in a city as a domed place had traditionally done. B of 85.575 m^2—the actual park. This was to include playgrounds, green spaces, alleys, eventually decorative objects (fountains etc). The parking facility was to be separated from the park. Connected to the park should be a theatre, complementary to the function in the historical centre, a traditional theatre including a street theatre function. It was to include a conference hall, offices, a restaurant, bar/coffee shop, library and multimedia services/videotheque/internet point. In the north was the San Bernardino convent that occupied some space from the area, and was to be integrated into the development of the park, even through remodeling. C, 68.100 m^2—sports zone, which did not need to be detailed in this competition. The parking and the theatre should be included in the public place or the sports zone in order not to affect the park. The results, now available, of the competition, won by Giorgio Martocchia, Fabio Cibinel, Roberto Laurenti, Covadonga Rios Diaz (http://www.parcourbanolaquila.it/wc/content/21/img/b07ad7be-2436-4e10-9e09-22af832d5fc0.pdf), show how the memorial can dialogue with green spaces, the aims being described in our contribution to the book. The idea of the park was based on circles, an idea we had followed in a project for reusing a vacant site behind the Parliament Palace in Bucharest (Bostenaru 2010d), where the circles were a parcour to pavilions based on mythology. The centre was left empty and the trees, inexistent at the moment on the site, are on the periphery, suggesting a valley in the mountainous landscape of L' Aquila.

The first competition related to the reconstruction of L' Aquila actually took place in 2009. It was named "Un'idea x la ricostruzione: proposte per l'emergenza" (An idea x for proposed reconstruction for emergency) and aimed at emergency housing. The prizes were given on the 10th of October 2009 (as published by Monica Rossi in the magazine "Detail").

The idea of reconstruction in the form of pavilions is one we supported through research, as it addresses the issue of recalling memory. The theatre envisioned for L' Aquila springs from this idea, and the public space of the park recalls memory in the form of grass as an element to cover a part of a place built for socialization.

Previously Shigeru Ban, the star architect of emergency housing, built a paper concert hall in the spring of 2011, shortly after the Tohoku earthquake affected Japan. It is a temporary construction which can be raised anywhere (http://www.detail.de/daily/Paper-Concert-Hall-von-Shigeru-Ban-fuer-LAquila-2483/).

Another star architect designing a temporary structure for L' Aquila is Renzo Piano. Renzo Piano relates to the "heritage habitat" issue raised in this paper as he is also the architect bringing back to life the early reinforced concrete industrial construction of FIAT Lingotto in Torino. The building by Renzo Piano is an auditorium in the park of the castle, and it has just been approved in winter 2012. This is a timber structure (http://www.detail.de/artikel_auditorium-renzo-piano_26944_It.htm).

Not only pavilions in parks recall memory, but also photography, as shown in the initial hypotheses. http://www.sismycity.com/ L' Aquila 2010 was a project in the framework of the 12th Biennal in Venice (one of 20 collateral events taking place) "A photographic project on the aftermath of the earthquake which struck L' Aquila and its surrounding area" by 8 photographers from different areas of Italy. It was also the year when we took the photos. A related conference "L' Aquila 2010. Place, identity, ethics and reconstruction" took place on the 28th of August. Besides architects it featured philosophers and representatives of the municipality. The motto of the catalogue was "Various themes arose from the silence of the uninhabited city and each photographer developed them independently looking for new forms and relations born with the earthquake". The catalogue includes photographs and texts, mainly related to the ruins but also highlighting the grass which grows over the ruins. However, the major contribution is the role of photography to increase sensibility to the issue.

On the same idea of sensibilisation to the memory of ruins by means of an artistic project, is based Michele de Lucchi's installation Fuorisalone 2011—Mutant Architecture & Design, with Corpo Nazionale Vigili del Fuoco and Riva 1920. Starting from the issue of propping of ruins instead of rebuilding after the earthquake, the installation of the architect Michele de Lucchi together with Philippe Nigro rebuilt the propping in the vaults of the university in Milan, a university specialized in architecture. Numerous architecture journals (such as Domus, Detail) reported on the admired installation, a video of which was also published http://www.youtube.com/watch?v=QQT_GUeJZHY. The idea of having an installation in an arts school is not new, as our installation of the "rediscovered space" was done in the same context. Also the thematisation of timber structures builds on the shapes of the local seismic culture of timber, as shown in a previous paper in the volume, and through this is in dialogue with the structures which did not resist the earthquake. We would like to propose a further installation including also green elements, the X-shaped timber structure building the skeleton on which boxes with green growing on geotextiles can be fixed to cover the walls remaining blank after some buildings collapsed in an earthquake, but also possibly cover the still existing propping subject of ongoing research in 2014.

14.2.3.2 The Reconstruction in Kolontar and Devecser, Hungary

In 2010 a chemical disaster took place at the aluminum factory in Ajka, Hungary, after which the so-called red mud was exuded, which affected a number of villages. In a short time, authorities with the assistance of the star architect Imre Makovecz (who died shortly after the completion of the project) and a group of architects from his group, all volunteers, developed plans for rural reconstruction in Devecser and Kolontar (http://koos.hu/2011/05/26/vorosiszap-milyen-tipusterveket-fogadtak-el-kos-karoly-egyesules/), and their buildings were completed in June 2011. The ecumenical church in Devecser was completed after his death.

If Ajka city is an industrial city, with buildings mostly from the socialist period, including the representative buildings (such as the City Hall), and very rare old buildings (like the baroque church), the surrounding villages have traditional vernacular architecture.

Also here the new architecture will be put in context of

- vernacular housing of Hungary,
- the organic architecture of Imre Makovecz and the anthroposofic movement,
- historic developments in the early twentieth century which created the transition from the vernacular architecture to that of Imre Makovecz (ex. Károly Kós).

The construction of the new buildings is distinguished by respect for the traditional architecture in the area west of the Danube in Hungary, where we meet Swabian houses.

This type of housing presents architectural intermediate space, with the street marked in front by an opening of door type. The street facade has an eardrum. Makovecz's homes (Fig. 14.7) are of this typology, but note the variations in interpretation by incorporating a "neo-Gothic" architecture or elements of wood. In addition to keeping the typology, elements of the witness of the demolished buildings were preserved, for example, marking the entry of the houses. The architect Makovecz designed three types of housing. Two of them have neo-Gothic elements, and elements of wood in the intermediate space: one of them has raw logs and the other wings of the characteristic architecture element of the eagle. The homes designed by Makovecz and the group around him are all ground floor only, whitewashed, and grouped in a street. The street displays, in terms of urban characteristics those of Swabian villages: the sidewalk is located near the street, from which it is separated by a channel for rain water and away from houses, in front of which lies a meadow.

Danube Swabian "is a rural type of popular housing practiced by the German (Swabian) colonists in the Austro-Hungarian Empire of the time. The villages founded by them were located in three different regions of moderate seismicity: the one affected by the Banat earthquakes (encompassing today territories in Romania: Banat, Serbia: Bacska, and Hungary: Bács), the one affected by the earthquakes around lake Balaton (today in Hungary: Dunántul) and the one affected by the Crişana earthquakes or the Maramureş earthquakes (today in Romania: Komithat Sathmar). Historically two variations of this type regarding the functional conformation can be seen: one with the short side to the road and a long wall with no windows on the parcel line (earlier type) and one with long side to the road with windows from the main rooms to it (turn-of-the-century type for wealthy families). The second one could be found in urban environments as well. Contemporary variations of this housing type are still practiced. A functional particularity is that there is a second kitchen, open to the courtyard, at the end of the house to the yard and garden. This one, called "summer kitchen" has only one entrance, from the yard. There is a "winter kitchen" in the main part of the building, forming an ensemble with the other rooms. The "summer kitchen" was sometimes added later on. The load bearing structure consists of masonry walls and timber floors. This kind of building can be found in zones inhabited by the so-called "Danube Swabians", a population which immigrated from south-west Germany to the Eastern limit of the Habsburg Empire around 1700. Today they are spread in Western Romania, West and South Hungary and Northern Serbia, in "Banat", "Bács" and "Dunántul" and "Bacska" respectively and affected by the Romanian Banat earthquakes. This type of housing construction is commonly found in rural areas and has been in practice for less than 100 years. Currently, this type of construction is being built. (Bostenaru 2010c) (Fig. 14.8)

Fig. 14.7 Reconstruction of Kolontar **a** Street with new housing, **b** Housing designed by Imre Makovecz, **c** Housing designed by another architect, **d** Keeping testimonial elements, **e** Chapel in Devecser, **f** The logo of the reconstruction in Devecser, **g** House constructed 2011 displayed on the logo (*sun rays*). (Photos: Bostenaru 2011, 2013)

Fig. 14.8 Typical Swabian architecture **a** intermediate space as seen from inside. Photo: Agota Heinrich 2009, **b** Axonometric projection view of a typical house, **c** Typical Swabian House in the Balaton area. Photo: Bostenaru 2011 **d** Gym in Visegrád, architect Imre Makovecz (1985). (Photo: Bostenaru 2011)

Contemporary Hungarian architecture is internationally known through the works of Imre Makovecz (Bostenaru 2011d) (Fig. 14.9) who reinterpreted tradition in organic architecture, particularly in wood architecture. In September 2011 the architect died and this was his last work. Across the Danube Swabian architecture has received the attention of the architect Imre Makovecz: elements from this, of this, such as the typology of the German Fachwerk, can be found at a sports hall in Visegrád. It relies on early twentieth century architecture with roots in vernacular, like the one by Károly Kós (see figures in the other article by the author in this book), the Transylvan architect after whom the association who published the plans of the reconstruction in Kolontar and Devecser is named.

The construction in Devecser and Kolontar enjoys high acceptance among the population and beyond. The images there that we presented to the public enjoyed the same appreciation as the entire architecture of Makovecz, even of it's minor architecture, an interpretation of the vernacular in worship. And this can not be accidental. Makovecz's architecture was associated with Steiner's anthroposophy in its participative approach (Kuhlmann 1998, p. 135). Makovecz's architecture is open to participatism.

14.2.3.3 The Reconstruction in Corbeni, Romania

The reconstruction in the village of then called Antoneş ti, now Poenari-Nou, Corbeni commune, in northern Argeş county in Romania, done by the architect Bordenache,

Fig. 14.9 Architecture of Imre Makovecz (1935–2011): **a** Camping complex in Visegrad, Hungary (1978), **b** Chapell in Sfântu-Gheorghe, Romania (1998), **c** Church in Miercurea Ciuc, Romania (2001), **d** Chuch in Siofok, Hungary (1986), **e** Auditorium of the University in Piliscsaba, Hungary (1995), **f** Chuch in Százhalombatta, Hungary (1995). (Photos: Bostenaru 2009–2011)

who studied in Italy, mentioned before, followed the tradition of the local vernacular architecture like in the case of Makovecz, an ethnologist being involved in the planning (like in the tradition of participative planning, see Ottokar Uhl at "Wohnen morgen" in the 1970s near Vienna, with a sociologist), more than 70 years ago, after the 1941 flooding of the river Argeş (Popescu 2008, Fig. 14.10). So the houses, although they include concrete elements, present timber loggias and balconies, circular shapes in the openings, and sometimes paintings with vegetal and zoomorphic motives. These paintings served to identify the houses, instead of number (Popescu 2008), similar to the identification of Makovecz's housing. The annexes follow the same tradition. The new village, on a hill, has been foreseen also with a church, to respect tradition, which couldn't be finished until recent years because of World War II and the Communist time after that. The new village is most remarkable for being built during the shortages of the war and dictatorship. The inhabitants participated in the construction process. Public buildings have been foreseen, unlike in L' Aquila. In the same village there is a villa of the architect, listed as a historic monument, in the same tradition of fortified houses from the region. The new village is remarkable since it shows how an architect influenced by studies in Italy built close to this influence a village in interwar time, which should have inspired also the reconstruction in L' Aquila.

In the reconstruction in Corbeni 10 typologies (A-J) have been used, houses being repeated on the about 30 parcels. Structurally there are reinforced concrete slab between the floors and timber sloped roofs. The archive files document the structural computations made by the architect. The costs of the buildings are also thoroughly documented in the archives, and reduction of costs was obtained in war conditions. The houses were sold under the price. The peasant house types feature the so-called "prispa", an intermediary space between interior and exterior, a kind of loggia. If at Makovecz the houses are marked by stone numbers representing remains of the old houses, in case of Antoneşti houses are identified by drawn

Fig. 14.10 Reconstruction after the 1941 flood in Corbeni, Argeş county, Romania: **a** House by architect Richard Bordenache on the main street; **b** New church; **c** The river with concrete after systematization; **d** Different house, with respect for tradition; **e** Identification of a house through painting; **f** Public building. **g**, **h** Archive images of the house by Richard Bordenache (façade and plan). (Photos: Bostenaru 2012)

animals (deer, bird). Like in Devecser a church is the centre of the development. The village is up on a hil, and on its top there is the church. Unlike in L' Aquila and like in Devecser social spaces and buildings are designed. There is a public plazza around which the buildings with public functions are grouped. Drăgoescu (2010) documents the archive records of this urban planning.

Based on the urban plans and using the software Nemetschek, Geo Module, we could model in 3D this development.

14.3 Discussion and Conclusions

Reconstruction after earthquakes is one of the planning stages in disaster management, in fact the first one to be practiced at the beginning of Modernity, as early as the eighteenth century. Preparedness planning such as mitigation and resilience planning came later, and so did emergency planning. It was a planning occasion that allowed major reshaping of the city, with some exceptions (London after the 1666 fire). The paper by Petrea in this book deals with the immediate reaction, the planning of emergency housing, while we considered permanent housing, even if the one in L' Aquila was designed to change function in time.

We considered two completely different approches that have addressed the reconstruction after a disaster: preserving tradition or denying it. These two approaches result in different attitudes of residents, those affected by the measures of reconstruction, subject to the architecture. If the tradition was not followed, the acceptance is low, and affected citizens require a higher level of participation. This is due to what Gociman (2006) called "heritage habitat" which has to be preserved after a disaster. For this reason it is essential that studies of historical construction are included, to identify the elements of tradition that define the "heritage habitat". Failure to exercise "heritage habitat"has led to problems of acceptance and historical reconstruction in other cases such as when the Gibellina, Italy, where a modernist city was created, but abandoned by the residents while it is of tourist interest, and one in San Giuliano di Puglia, where the village was transmuted without respecting neighborhood sociological cells. The situation was different in former socialist cities such as Skopje (architect Kenzo Tange) and Zimnicea (architect Cezar Lăzărescu) where this type of urban "systematisation" was common also for non-earthquake affected cities. Both plans followed the functionalist architecture of the 1960s, which was criticised for its excessive separation through zoning in the participative approach. Differing from this is the reconstruction of Antoneşti, by an architect who adhered to functionalism in some of his main works, but knew to adapt to the context.

The approach was put in context of the growing contributions to participative planning in disaster management. A current started in the 1960s with the students' protests in Western Europe, participative planning won terrain in construction and arrived in the twenty-first century also to the countries of former totalitarian regimes. Therefore the acceptance by citizens is an important measure of the success of a new housing development. Participation can be done to develop the methods to plan the new housing, when large participation in the housing building process is not possible. This has been done in Haiti, in order to gather resources from all around the globe by means of the internet, another innovation of the twenty-first century in participative planning.

Non-engineered housing may not be source for "heritage habitat" inspiration only, but also for seismically resistant local techniques, as shown in the author's other contribution in this book. The World Housing Encyclopedia promotes not only sharing of such knowledge through housing reports, as the Swabian house mentioned here, but also through related projects such as the Confined Masonry Network involved in Haiti rebuilding.

References

Alexander C, Ishikawa S, Silverstein M (1977) A pattern language, Towns. Buildings. Construction. Oxford University Press, New York

Ambraseys N, Finkel C (1987) The seismicity of Turkey and neighbouring regions, 1899–1915. Ann Geophys 5B:701–726

Arslan H, Ünlü A (2006) The evaluation of community participation in housing reconstruction projects after Duzce earthquake. Proceedings of the 3rd International i-rec Conference and Student Competition on post-disaster reconstruction "Meeting stakeholder interests", Florence, Italy, May 17–19 2006

Arslan H, Ünlü A (2008) The role of NGO's in the context of post disaster housing in Turkey, in building resilience achieving effective post-disaster reconstruction. Proceedings of the 4th International i-Rec Conference on post-disaster re-construction "Building resilience: achieving effective post-disaster reconstruction", Christchurch, New Zealand, 30 April–2 May 2008

Asgary A, Badri A, Rafieian M, Hajinejad A (2006) Lost and used post-disaster development opportunities in Bam earthquake and the role of stakeholders. Proceedings of the 3rd International i-rec Conference and Student Competition on post-disaster reconstruction "Meeting stakeholder interests", Florence, Italy, May 17–19 2006

Benko-Medgyaszay I (1909) Über die künstlerische Lösung des Eisenbetonbaues. In: Berichte über den VIII. Internationalen Architektenkongeß Wien 1908, Vienna, pp. 538 ff.

Bostenaru Dan M (2002) Management of Innovation. Innovative Politik in der Stadtentwicklung, Seminar work "Projektmanagement" BOZ, Universität Karlsruhe (TH)

Bostenaru Dan M (2004) Multi-criteria decision model for retrofitting existing buildings. Nat Hazards Earth Sys Sci 4(4)485–499; http://www.copernicus.org/EGU/nhess/4/4/485.htm

Bostenaru Dan M (2005) Multidisciplinary co-operation in building design ac-cording to urbanistic zoning and seismic microzonation. Nat Hazards Earth Sys Sci 5:397–411

Bostenaru Dan M (2007) Von den Partizipationsmodellen der 70er Jahre zu neueren Kommunikationsformen Ende des XXten Jahrhunderts in Architektur und Städtebau. Cuvillier, Göttingen

Bostenaru Dan M (2008a) The rediscovered space, a space of encounter. Proceedings of the 16th ICOMOS General Assembly and International Scientific Symposium "Finding the Spirit of the Place", Quebec, Canada, 29 September–4 October 2008 (CD). Available online at: http://www.international.icomos.org/quebec2008/cd/toindex/77_pdf/77-gU5i-132.pdf

Bostenaru Dan M (2008b) Multi-family urban housing construction with pre-cast concrete: housing models and rationale in Eastern and Western Europe. In: Ravindra KD, Moray DN, Jones MR, Halliday JE (eds) Precast concrete: towards lean construction. IHS BRE Press, Universitatea din Dundee, 98–9 july 2008, Vol. 2 from "Concrete: construction's sustainable option, Volumes 61–6, EP92", section "Housing". p. 297–308

Bostenaru Dan M (2010a) The flat and the knurled space in the reconstruction of cities. In: Panagopoulos T, Noronha T, Beltrao J (eds) Advances in urban rehabilitation and sustainability. WSEAS, pp. 80–85,
Bostenaru Dan M (2010b) Historical use of concrete and innovation in architecture. In: Cruz PJS (ed) Structures and architecture, CRC Press, Leiden, pp. 433–434, (ISBN: 978-0-415-49249-2, Full paper on attached CD ROM)
Bostenaru Dan M (2010c) Vernacular and modernist housing in Germany and Romania. An analysis of vulnerability to earthquakes. Cuvillier, Göttingen
Bostenaru Dan M (2010d) Spaţiul verde redescoperit - Der wiederentdeckte Grünraum. Cuvillier Verlag, Göttingen
Bostenaru Dan M (2011a) The past: space and time. J Appl Eng Sci 1(14):149–157
Bostenaru Dan M (2011b) The use of ontology for digital conservation of architecture works after catastrophes. J Appl Eng Sci 1(14):11–18
Bostenaru Dan M (2011c) Other modernisms in European housing in reinforced concrete (19401920–1940). Bulletin of the Polytechnic Institute of Jassy—Constructions. Architecture Section, Tomme: LVII (LXI)|Fascicle: 1|2011, 161–192
Bostenaru Dan M (2011d) Profil—Imre Makovecz. *Arhitectura* 5/2011
Buckle P (2004) A comparative assessment of community based recovery management in England and Australia. Proceedings of the 2nd International i-rec Conference and Student Competition on post-disaster reconstruction “Planning for reconstruction”, Coventry, UK, April 22–23 2004
Büchner K, Baréz K (1966) Studentenwohnheim in Karlsruhe. Architektur Wohnform 74(7):457–463
Burger J (2007) Gibellina—Il terremoto. Italia e Austria. http://www.film.at/gibellina_il_terremoto/detail.html
Calvi GM, Spaziante V (2009) La ricostruzione tra provvisorio e definitivo: il Progetto C.A.S.E. Progettazione Sismica 03/2009
Cardoso R, Lopes M, Bento R (2004) Earthquake resistant structures of Portuguese old “Pombalino “buildings. Proceedings of the 13th World Conference on Earthquake Engineering, Vancouver, BC, Canada, Paper No. 3329
Coburn A, Spence R (2002) Earthquake Protection. Wiley, West Sussex (2nd edition, 1st edition 1992)
Corbusier Le (2007) Toward an architecture. Translated by John Goodman. Getty Research Institute, Los Angeles
Deleuze G (1980) Mille Plateaux. Minuit, Paris, pp. 592–625
Dikmen N (2006) Relocation or rebuilding in the same area: an important factor for decision making for postdisaster housing projects, Procedings of the 3rd International i-rec Conference and Student Competition on post-disaster reconstruction “Meeting stakeholder interests”, Florence, Italy, May 17–19
Drăgoescu A (2010) Sat model Antoneş ti/Antonesti model village. Urbanismul—Serie nouă 87–8:76–82
Durth W (1999) Kramm + Strigl. Axel Menges, Stuttgart
Duyne Barenstein J (2008) From Gujarat to Tamil Nadu: owner-driven vs. contractor-driven housing reconstruction in India, Building resilience, achieving effective post-disaster reconstruction, Proceeding of the 4th International i-Rec Conference on post-disaster re-construction “Building resilience: Achieving effective post-disaster reconstruction”, Christchurch, New Zealand, 30 April–2 May 2008
Etlin R (1991) Modernism in Italian Architecture, 1890-1940. MIT Press, Cambridge
Findlay RA (2002) Resilience and Sustainable Community Design: The Cases of Pukapuka, Lowa, and La Bahia De Jiquilisco. Proceedings of the 1st International i-rec Conference “Improving post-disaster reconstruction in developing countries”. http://www.grif.umontreal.ca/pages/i-rec%20papers/robert.PDF. Accessed 3 Dec 2013

Forwood E (2004) Community Participation in Honduran Housing Projects After Hurricane Mitch. Proceedings of the 2nd International i-rec Conference and Student Competition on post-disaster reconstruction "Planning for reconstruction", Coventry, UK, April 22–23, 2004

Fox A (2004) Planning for improved resilience. Proceedings of the 2nd International i-rec Conference and Student Competition on post-disaster reconstruction "Planning for reconstruction", Coventry, UK, April 22–23, 2004

Garcia Acosta V (1997) Historia y desastres en America Latina, Red de Estudios en Prevencion de Desastres en América Latina, La Red, http://www.desenredando.org

Gharaati M (2008) An overview of the reconstruction program after the earthquake of Bam, Iran. Proceeding of the 4th International i-Rec Conference on post-disaster re-construction "Building resilience: Achieving effective post-disaster reconstruction", Christchurch, New Zealand, 30 April–2 May 2008

Gharaati M, Davidson C (2008) Who knows best? An overview of reconstruction after the earthquake in Bam, Iran building resilience, achieving effective post-disaster reconstruction. Proceeding of the 4th International i-Rec Conference on post-disaster reconstruction "Building resilience: Achieving effective post-disaster reconstruction", Christchurch, New Zealand, 30 April–2 May 2008

Gibson A (2003) Avant-Garde. In: Nelson R, Schiff R (ed) Critical terms for art history. University of Chicago Press, pp. 202–216

Glavovic BC (2008) Katrina recovery experiences: Imperatives for building sustainable, hazard-resilient communities. Proceeding of the 4th International i-Rec Conference on post-disaster re-construction "Building resilience: Achieving effective post-disaster reconstruction", Christchurch, New Zealand, 30 April–2 May 2008

Gociman CO (2006) Managementul reducerii riscului la dezastre: strategii de arhitectură şi urbanism, Editura Universitară "Ion Mincu", Bucharest

Greene MR (1987) Skopje, Yugoslavia: seismic concerns and land use issues during the first twenty years of reconstruction following a devastating earthquake. Earthq Spectra 3(1):103–117

Jayantha A, Gunasekera R (2006) Building capacity for sustainable and participatory post-tsunami rebuilding. Procedings of the 3rd International i-rec Conference and Student Competition on post-disaster reconstruction "Meeting stake-holder interests", Florence, Italy, May 17–19, 2006

Joedicke J (1976) Angewandte ntwurfsmethodik fur Architekten (Applied Design Methodology for Architects) (in German), Karl Kramer, Stuttgart

Harrison C (2003) Modernism. In: Nelson R, Schiff R (ed) Critical terms for art history. University of Chicago Press, pp. 180–201

Harvey M (2006) Cultivating resilience: the rehabilitation of homestead gar-dens in post-tsunami Sri Lanka. Proceedings of the 3rd International i-rec Conference and Student Competition on post-disaster reconstruction "Meeting stake-holder interests", Florence, Italy, May 17–19, 2006

Häußermann H, Siebel W (1993) Wandel von Planungsaufgaben und Wandel der Planungsstrategie—Das Beispiel der IBA Emscher Park. Arbeitskreis Stadterneuerung an Deutschsprachigen Hochschulen; Institut für Stadt- und Regionalplanung der TU Berlin: Jahrbuch Stadterneuerung 1993: Beiträge aus Lehre und Forschung an deutschsprachigen Hochschulen, 141–151

Hobsbawm E (1994) The Nation as invented Tradition. In: Hutchinson J, Smith A (ed) Nationalism. Oxford University Press, pp. 76–82

Kuhlmann D (1998) Metamorphosen des Organizismus. Zur Formensprache der Lebendigen Architektur von Imre Makovecz, Doctorate thesis, Bauhaus Universität Weimar

Lang H (2008) Community housing in post disaster area on Nias islands, Indonesia: Responding to community needs. Proceeding of the 4th International i-Rec Conference on post-disaster re-construction "Building resilience: achieving effective post-disaster reconstruction", Christchurch, New Zealand, 30 April–2 May 2008

Langenbach R (2009) Don't tear it down, preserving the earthquake resistant vernacular architecture of Kashmir. UNESCO, New Delhi

Langenbach R, Dusi A (2004) On the cross of Sant'Andrea: the response to the tragedy of San Giuliano di Puglia following the 2002 Molise, Italy, earthquake. Earthquake Spectra 20(S1):S341–S358. doi: http://dx.doi.org/10.1193/1.1767162

Lynch K (1960) The image of the city. MIT Press, Cambridge (2000 27th edition 1st edition 1960)

Marsh G, Buckle P, Smale S (2004) The concept of community in community capability and disaster management. Proceedings of the 2nd International i-rec Conference and Student Competition on post-disaster reconstruction "Planning for reconstruction", Coventry, UK, April 22–23, 2004

Özden AT (2006) Developing a model for community involvement in post-disaster housing programmes. Proceedings of the 3rd International i-rec Conference and Student Competition on post-disaster reconstruction "Meeting stake-holder interests", Florence, Italy, May 17–19, 2006

Popescu IL (2008) Un sat istoric: ANTONESTI, Mica enciclopedie AS, Formula AS, Nr. 823

Sivaji K (2002) Community preparedness—an unique experience in post disaster scenario. Proceedings of the 1st International i-rec Conference "Improving post-disaster reconstruction in developing countries". http://www.grif.umontreal.ca/pages/i-rec%20papers/sivaji.pdf. Accessed 3 Dec 2013

Sliwinski A (2006) Social dynamics in participatory reconstruction: an anthropological analysis from El Salvador. Proceedings of the 3rd International i-rec Conference and Student Competition on post-disaster reconstruction "Meeting stakeholder interests", Florence, Italy, May 17–19, 2006

Solnit R (2009) Paradise Built in Hell. Viking Adult, 2009

Sonne W (2009) Dwelling in the metropolis: Reformed urban blocks 1890–1940 as a model for the sustainable compact city. Prog Plann 72(2):53–149

Tobriner S (1982) The Genesis of Noto: An Eighteenth-Century Sicilian City. London A Zwemmer

Turino R (ed) (2010) L' Aquila. Il progetto C.A.S.E. IUSS Press, Pavia

Uhlig G, Cleffa-Rudolph A, van Gool R (1997) Die Stadt in der Zeile. În den 20er Jahren Zeilenbau und Heute—. În Badisches Landesmuseum Karlsruhe der 20er Jahre Neues Bauen, Gropius, Haesler, Schwitters und die Dammerstock-Siedlung în Karslruhe. Info Verlag, Karlsruhe (pp.177–182)

Chapter 15
Earthquake Hazard Impact and Urban Planning—Conclusion and Recommendations for Further Work

Maria Bostenaru Dan, Iuliana Armaş, Alexandru-Ionuţ Petrişor, Alessia Cerqua, Cristina Olga Gociman and Agostino Goretti

Abstract This book, which is the second volume in a new series "Environmental hazards", addresses a topic very much neglected in current research: the role of urban planning in earthquake disaster management. It is not a book about earthquake engineering issues but rather about management issues. Social sciences and arts are often disregarded in earthquake disaster management, the focus being mostly on technical aspects. But the study of the impact is important in order to facilitate appropriate implementation through transferring research knowledge to decision makers. The work is aimed at both junior and senior researchers creating the basis for new university curriculum, as there are very few courses in this field. The volume is organised in four sections, addressing the 'reaction' side of earthquake disaster management: vulnerability, planning, risk assessment and reconstruction. Each section addresses issues for decision makers, some at the level of participatory planning, as well as the role of specialists from urban planning, geography, engineering and seismology/earth

M. Bostenaru Dan (✉) · A.-I. Petrişor
Department of Urban and Landscape Design,
"Ion Mincu" University of Architecture and Urbanism,
Academiei 18–20, 010014 Bucharest, Romania
e-mail: maria.bostenaru@iaim.ro

I. Armaş
Faculty of Geography, University of Bucharest,
1 Balcescu Bd., 010041 Bucharest, Romania

A.-I. Petrişor
National Institute for Research and Development in Constructions,
Urbanism and Sustainable Spatial Development URBAN-INCERC,
Nicolae Filipescu, 53–55, sector 2, 020961 Bucharest, Romania

A. Cerqua
International Society of Biourbanism,
Via G. Giardini, 15b 00133 Rome, Italy

C. O. Gociman
Department of Doctoral Studies, "Ion Mincu" University of Architecture and Urbanism,
Academiei 18–20, 010014 Bucharest, Romania

A. Goretti
Civil Protection Department, Seismic and Volcanic Risk Office,
Via Vitorchiano 4, 00189 Rome, Italy

M. Bostenaru Dan et al. (eds.), *Earthquake Hazard Impact and Urban Planning*,
Environmental Hazards, DOI 10.1007/978-94-007-7981-5_15,
© Springer Science+Business Media Dordrecht 2014

sciences, while technical sciences include civil engineering and remote sensing/geo-information. We present the current state of the art in approaching the problems, the current challenges for further research and issue recommendations based on the results of current research. The work aims to reach a balance between dealing with existing heritage and new planning, because even in the twenty-first century new planning has to rely on lessons from the past. Another balance is reached between dealing with the technical and the humanities/social sciences approach, as reflected in the topic and the research methodology adopted. Case studies are emphasized, mostly from Italy and Romania, two countries that have experienced different types of earthquakes.

Keywords Urban morphology · Urban system · GIS · Participatory planning · Heritage habitat · Restructuring · Economics

15.1 Introduction

The year 2011 was unfortunately rich in seismic events: Christchurch/New Zealand, Tohoku/Japan. Europe remained protected from such events in 2011, but the danger is looming, also in the two countries presented in this book: the possibility of deep earthquakes in Romania and surface earthquakes in Italy. In 2012 a strong earthquake affected the Emilia Romagna region in Italy.

In Christchurch, crowd sourcing data collection was used to assess damage just like in Haiti, the year before, and the importance of this volunteered geographic information is emphasized by a recently started COST action.

In Tohoku we had the unexpected event of a scenario from arts approach turning reality: the historic accident represented by a nuclear disaster triggered by an earthquake and a tsunami, as the disasters which shocked the world in Lisbon in 1755 and in Sumatra in 2004 (Bostenaru 2010).

However, the impact of recent events can only be evaluated as time passes.

Since the 'Natural hazards impact on urban areas and infrastructure' session in the framework of the European Geosciences Union General Assembly has changed the focus to 'Natural hazard impact on technological systems and urban areas' we started a series of books, and in the meantime continued to make contributions on the topic, like the one to the Reviews of Environmental Sciences and Biotechnology (Bostenaru 2011a). Urban planning considered as a social science is the focus of this book, giving also a special emphasis on planning.

15.2 Approach of the Contributors to the Relationship Between the Study of Earthquake Impact and Urban Planning

The case studies presented in the book exhibit a wide variety through the geographical coverage, theoretical and methodological approaches, standpoints influencing the focus of analyses, and scale. Nevertheless, several common elements can easily be detected in more, if not all the papers presented.

The first element is participation. The critical importance of involving relevant stakeholders, particularly local communities, as a basis for building a resilient community, has been stressed looking to the example of post-earthquake reconstruction of L'Aquila, measurement of risk governance and evaluation of performance, involvement in mitigation and recovery in relationship to natural hazards and urban planning, assessment of the urban system of Crotone, Italy in regard to seismic risk, and others situations as well. In addition, the example of Bucharest shows a clear need for educating the population before involving it as a stakeholder.

The second element emphasized by several authors is the importance of multi- and trans-disciplinary approaches to assessing and managing urban risks. This is nothing else than an application of systemic approaches to understanding governance, and measuring public policies in order to strengthen the social capital in terms of social learning and public participation, assessing seismic risks in Tulcea County starting from the geologic and geophysical information, elaborating a GIS holistic model able to handle the seismic behavior of urban centers based on criteria using social, residential, cultural, health, education, energy, economic, emergency and administrative criteria, while taking into account the road infrastructure, or performing an integrated study of Bucharest using data provided by geomorphology, rivers network, geology, tectonics, neotectonics, geophysics and seismology.

Several case studies examined the importance of local and traditional approaches, particularly regarding vernacular architecture. The example of reconstruction in Aquila (Italy) versus Kolontar and Devecser (Hungary) and Corbeni (Romania) revealed differences between the approaches used in Western and Eastern Europe; the later cases indicated a clear preference for the traditions, while the first option for modernity which was not received very well by the community, indicating the need for a participative approach. The local community preference for vernacular architecture is motivated by the sense that in addition to increased resilience, vernacular architecture offers the additional advantage of available local building materials. Multiscale analyses showed that even when assessing an area of the size of Tulcea County, knowing local conditions is very important for mitigating seismic risk. Similarly, in order to make the evaluation processes participatory it is necessary to recognize the multiplicity of stakeholders involved.

The relationship between urban planning and the management of seismic risk has also been explored by several authors. The integration of disaster information in planning is a complex mix of consultation, dialogue and monitoring involving all critical stakeholders, particularly the vulnerable communities, tightly related to both negotiation and participation; a novel cyclic and incremental approach, combining two methods of planning (dynamic construction of future scenarios and Strategic Choice Approach) was proposed in this regard. Such a conclusion must be no surprise at all, since the planning process is essentially participative. In Bucharest, recent residential projects are likely to determine profound changes in the general layout, especially in the peripheral areas. In the author's opinion, the results of the study can represent a decision support framework for policy makers to critically reflect on possible risk mitigation plans. Of course, the statement relates to the real relationship between planning and science—how much of the scientific evidence is actually used to support the final decisions?

The papers in this book can be divided in two broad categories: those dealing with technical issues and those dealing with creative spatial design.

We will first deal with the technical side, focusing on GIS for analysis and visualisation.

15.3 GIS and Urban Risks

Geographical Information Systems (GIS) continue to be at the forefront of science even though GIS was first used in 1781 (Iosub 2008) and the first computer-based GIS in 1969 (Letham 2007). During the 40 years of computer-based GIS, many definitions appeared. While Redlands, a Californian based Environmental Systems Research Institute (ESRI), founded also in 1969 (Letham 2007) made its way to becoming the world's GIS leader, its definition of a GIS is most likely the best known: a GIS is a "system for the management, analysis and display of geographic information" (ESRI 2001), which "integrates hardware, software, and data for capturing, managing, analyzing, and displaying all forms of geographically referenced information" (ESRI 2011). The statement does not mention two other implicit components—the user and specific methods (Ioniţă and Moise 2001). Even though comprehensive, these descriptions do not point towards a very important side of GIS, the management-oriented one, emphasized in a much earlier definition: "decision support system involving the integration of spatially referenced data in a problem solving environment" (Cowen 1988). Its importance consists in pointing to the use of a GIS in a management context. Basically, a GIS joins two features: a database and a map, embedded in the concept of '*geodatabase*', defined as "spatial database containing datasets that represent geographic information" (ESRI 2001).

From this perspective, it must be stressed that even though a GIS has increased mapping abilities, it is not just 'another mapping software', similar to AutoCad, Corel etc. This perspective is important for the Romanian planning practice, where the analytical and managerial capacity of GIS are insufficiently used, with very few exceptions. Utilizing GIS for mapping only is similar to assigning to a multi-purpose Swiss knife only the role of a letter opener. In fact, users of graphic software are disadvantaged when using GIS, as mapping is a little bit more complicated, being influenced by choosing the proper projection, availability of geographically referenced data, different coordinate systems and many others.

Last but not least, it is important to indicate the source of GIS data. While data from the table of attributes can be provided by specialized agencies or censuses, the map must be built in a different way. The first digital maps were created, as shown in the next chapter, by digitizing existing paper-based maps. However, for updated and accurate information, other sources must be used instead. Remote sensing is the process of acquiring "information about an object without being in direct contact with it" (Jensen 2000). Even if arguable that the role of remote sensing is to provide data for a GIS (Cowen 1988), it became clear that remote sensing outputs—digital images—are the easiest way to produce accurate and up-to-date digital maps.

Remote sensing techniques include satellite imagery, aerial photography, RADAR and LIDAR images, and other technologies. Simple images acquired at a specific wavelength were replaced by hyper-spectral ones covering more wavelengths and the resolution increased continuously, as well as the frequency of acquisition.

In terms of methodology, the contributions indicated without doubts that the use of GIS is a must. Nevertheless, a GIS requires a geospatial database, and as much data as possible. In many cases, the scarcity of data was the main limitation of analyses. Sources of data include historical earthquakes, assessed using historical documents, with the caveat of the need to look whether they are affected by voluntary or involuntary 'falsification' using intelligent methods, such as computational procedures based on 'expert judgment'. Another source of data is high resolution satellite imagery (IKONOS) used to model population in urban areas and supplement existing building inventory data. The knowledge of the spatial distribution of urban population can be used in relationship to other hazards such as tsunamis, floods, or volcanic eruptions, but also terrorist attacks and epidemiological events.

15.4 Urban Planning and Risks

The concept of morphology, which means the study of form, was initially proposed by Goethe (1891); the term was first used in bioscience, which connects to the approaches in this project, regarding the urban organism or the evolutionary structural organisation inspired by organic forms. Alexander's participatory and phenomenology approach (1977) leads to organic shapes. It is based on tradition, which is something we envisage also at urban scale when relating the reconstruction language to the 'heritage habitat' (Gociman 2006). Under this concept, reconstruction should call upon the memory of the affected population to rebuild using urban and architectural structures compatible with their memory map. The idea of a 'heritage habitat' is based on the approach of Lynch (1960) regarding the perception of the urban image, which we previously used for urban analysis in decision for retrofit (Bostenaru 2004). For the purpose of defining the 'heritage habitat' in reconstruction it is important to study the existing built substance in the area. The structures in the city renew themselves continuously, just like living organisms, as structure forming forces have been acting since the founding of the city. Therefore the task of urban planners is the most difficult ones if their intervention is asked for, such as due to an imminent dangerous disaster. Starting from urban topology and the topological transformations given by movement, projection and deformation, Florescu proceeds to urban morphogenesis, a result of the dynamics of urban forms and generator of their complexity. Discontinuity is caused by risks and vulnerability at an urban scale. Discontinuities can be static or dynamic, the latter relating the urban form to the context. Risks are a relating element between form and context, relating continuity and discontinuity in the evolution of the urban organism. Florescu, in another chapter of this volume developed formulas regarding the risk in the context, in the static urban form and in the metabolic urban form, thus relating the risk of the individual element to that of the urban form, depending on discontinuities, which affect the vulnerability. This is different for buildings, which enjoy a certain degree of autonomy for change by their owners.[1]

Several papers deal with reconstruction issues exemplifying these principles. In the following we include some detailed viewpoints on the L'Aquila reconstruction, in line with the participatory approach, contributed by Alessia Cerqua (but used in the book as an example for the heritage habitat). Further studies will focus on the destructured area in central Bucharest, endangered by earthquake and by flood, to name the major hazards, but affected by 'memory erasing' due to the communist demolitions.

15.5 Case study: L'Aquila City Recovery Program

The Strategic Choice Approach was adopted in an international workshop held in 2009 organized by Roma Tre University, Tipus Lab[2], under the coordination of Professor A. Giangrande and E. Mortola, for the recovery program of L'Aquila city. In April 2009, a big earthquake occurred in the region of Abruzzo, in central Italy, with epicenter near L'Aquila city. The main shock occurred at 3:32 local time on 6 April 2009, and was rated 5.8 on the Richter scale (6.3 on the magnitude scale). The earthquake caused damage to between 3,000 and 11,000 buildings; 308 people were killed and approximately 1,500 people were injured, making this the deadliest earthquake to hit Italy since the 1980 Irpinia earthquake. Around 65,000 people were made homeless, and 40,000 of them found accommodation in tented camps and a further 10,000 were housed in hotels on the coast.

Roma Tre University organized an international workshop to study a participative recovery program, using the Strategic Choice Approach, and involving community.

During the workshop, the tutor worked with students and community in a brainstorming session: members of the working group were invited by the tutor to build the scenario (*Visioning*) of L'Aquila city as a sort of 'tale from the future', referring to a distant time horizon. The tutor asked the citizens and students to describe the activities, thoughts and emotions of an inhabitant of L'Aquila that returns to his or her place of origin after being away for many years.

The Visioning was built as a 'collage' of the sentences pronounced in turn by individual participants. The tutors took notes faithfully of these sentences and rearranged them properly to build the scenario, where possible alternative and not compatible solutions were foreseen.

The working group proceeded to interpret the acquired information as problems (*decision areas*) and alternative design solutions to resolve them (*options*). With the help of Strategic Choice Approach the whole design problem was then divided into problem foci: a 'focus' is a subset of decision areas and relative options, and the

---

[1] Based on a project proposal submitted by Boştenaru (2012).

[2] Tipus Lab: www.pism.uniroma3.it.

criteria used to identify each focus is its level of internal cohesion, as measured by the total quantity of links among its decision areas, and its level of isolation, measured by the number of links among the areas of the focus and all other areas. For this participative recovery planning, three foci have been identified: the first, about the future of the historical center and the new social spaces in Collemaggio, the second, about the quality of life and participation in new towns, and the last, concerning a river park, traffic congestion and sustainable mobility.

Let us have a look at them in detail.

Focus 1. '*The future of the historical center and the new social spaces in Collemaggio'*.

This scenario is based on the expectation that not all buildings in the old town will be restored: in the space already occupied by buildings destroyed or heavily damaged by the earthquake, new buildings will be built. In the historical center, old and new buildings will alternate. Some parts of the city center will not be rebuilt, but all recovered buildings will be used. Some spaces and social activities in the Collemaggio psychiatric hospital will expand. A large auditorium in wood will be realized, and a stone amphitheater where the protagonists are the dreams of those who participate in the drama activities that will take place in it. Finally, the possibility of overcoming the sense of uncertainty for many residents will largely depend on the ability and willingness of the public administration to ensure some recovery time and the securing of their old homes.

Focus 2. '*The quality of life and participation in the new towns'*.

In the historical center a suitable space will be arranged to accommodate the general assembly of the inhabitants. A laboratory of participative design will be realized for each group of new towns. To overcome the current poor social life of the new towns, some specific activities will be organized for the residents. Where possible, the terraces on the platforms will be linked together to create ways that favor interaction between people living in different modules. In the open spaces of the new towns there will be new facilities and services: kindergartens, playgrounds, fountains, ecological common gardens, etc.

Interventions of sustainable mobility will be implemented in order to connect the new towns to each other and with other interesting parts of the territory, including both the services implemented in central position and the historical town. Some cycle and pedestrian paths will be created. In any case it will be necessary to strengthen public transport, increasing the number of bus trips, ensuring compliance with the scheduled times, etc.

Focus 3. '*The river park. Traffic congestion and sustainable mobility*'.

The project involves construction of the river Aterno park, with all the necessary redevelopment and regeneration projects. Only a few houses in the park will be rebuilt: the rest will be demolished to create a special park where the ruins of buildings will coexist with a large garden space. Mobility will be realized by an extensive network of cycle and pedestrian paths, crossing the whole area of the park and other parts of the municipality. It will be nice to walk along these paths in the green, away from noise and traffic hazards.

The project also includes interventions aimed at improving the existing road network of L'Aquila. In particular it will be necessary to achieve those new features that will reduce traffic congestion. The reduced congestion will also be due to the increasing use of public transport and the new tram, which will use two tracks: one realized by recovering the existing track and the other completely new, which will connect the city center with the western part of the township, passing through some new towns.

From this developmental perspective, the first Strategic Choice workshop at l'Aquila was not an academic exercise, but a serious attempt in engaging an already serious community in a participatory planning process for its future.

15.6 Urban Restructuration for Disaster Resilience

A theoretical approach followed by Gociman (2006) is the security zoning approach. While there are known types of zoning, such as microseismic zoning in seismology, and functional zoning in urban planning, security zoning proposes to react to risk by means of architecture and urbanism. Thus in planning green spaces or other empty spaces such as parking facilities, areas have to be foreseen to assure the minimal distance of access without being covered by rubble in case of building collapse. In case that the disaster strikes, security centres have to be identified to evacuate population: first in buildings such as schools then in green spaces to provide emergency housing. Such analyses for subzones have to be connected, as in other more central dense areas, subject to recent research projects, which were not affected by the demolitions in Ceausescu's Era, there might be lack of such spaces and the emergency housing zones have to be coupled to neighbouring areas. The security centres are priority areas in the pre-disaster phase to be retrofited so they can accomplish their function. An organisation plan must be done for each of them so they can be easily converted from the current function to a security centre in case disaster strikes.

The following mapping is to be performed:

- Hazard mapping, to identify multihazard
- Exposed elements (what has been called "elements at risk" in the Urban System Exposure approach in Masure and Lutoff, 2003, analyzed in two chapters of the book) mapping to a specific hazard (buildings, urban system, public space, heritage value, institutional value, environmental value, occupational structure),
- Vulnerability mapping of exposed elements to one specific local hazard. The specific hazard considered is again earthquake, but also secondary events, such as, for example, fire and epidemics from earthquake. The vulnerability of buildings and population is considered, mapping differently the degree of use during the day and the night. Also measures are considered that reduce vulnerability, such as presence of firefighting. A vulnerability score is given.
- Global risk mapping to superpose local hazards (ex. fire, flood, earthquake).

The result is the building-up of a poly-nuclear system of space security centers, able to relocate an affected population after disasters occurrence, named support system of emergency habitat and the identification of post-disaster reconstruction opportunities generated by the system of the emergency habitat, having as support, the specific components of heritage habitat.

One of the tasks of the urban planner is protection from disasters, and this may be achieved through participative means, with the support of the population. This kind of approach is used by the innovative teaching module called 'Prospective planning' by Florescu, Gociman and Bostenaru, encompassing both the study of vulnerability and of risk in urban planning, at the 'Ion Mincu' University of Architecture and Urbanism. Mapping, definition of evacuation ways and centres, and emergency housing definition are parts of the project work to be performed by master students in 'Urban design'. Teaching by research involves students in several research projects by way of workshops. Emergency housing is also a topic of the doctoral school, as shown by other works included in the volume. Recently also an ERASMUS LLP network has been launched on teaching disaster resilience 'Academic Network for Disaster Resilience to Optimise educational Development' (http://www.disaster-resilience.net/). It is an example of teaching excellence in planning, an innovative approach in an emerging field where these competences are important, as shown by the book contributions, but rare.

15.7 Outlook to Further Studies

Research still has to be performed regarding the supportive role of information and communication technology. Such efforts are currently performed in the framework of the Global Earthquake Model (http://www.globalquakemodel.net/). They include transforming classifications done for vulnerability classes into taxonomy and subsequently ontology in order to transform them into a software to compute risk, which will make use of databases regarding hazard, exposure, earthquake consequences. The orientation towards the employment of ontology in civil engineering and urban planning is not new. Former Europe-wide networking in the framework of COST actions included such initiatives, such as the COST action C21 'Towntology—Urban Ontologies for an Improved Communication in Urban Civil Engineering Projects' (http://www.towntology.net/), and the TU0801 action, 'Semantic enrichment of 3D city models for sustainable urban development', in which we participated, aimed at a further modeling, regarding the aggregation of information in 3D, for decision processes. We included some related issues in this volume. 3D representation permits analysing the function of city zones, in relationship to urban morphology, aiming at sustainable development. As the conclusions of the action show, urban morphology, a subject of the book in terms of disaster resilience, is also a key factor in defining usefulness, usability and usage of 3D city models.

A research endeavour to be soon completed aims to look at the 3D representation of Lisbon, pre-1755 earthquake, using a recent instrument for research developed for Second Life and thus for participative planning, and the contemporary Google Earth model, using techniques like the L'Aquila model about which we wrote, to identify change and conservation of the memory represented by the ruins, so continuity and discontinuity. We started looking at the issue of datascapes (landscapes of data) which can also be 3D models (http://www.dataappeal.net/, see representations on http://www.nedimah.eu/, an early browser being done in frame of the CRC461 at the University of Karlsruhe according to Zikas and Gehbauer 2007), or even the representation of vulnerability without using GIS by means of the SELENA software (authored by NORSAR) and aim to analyze these alternative Google Earth models in a chapter in book dedicated to landscape, and even attempt to build teaching modules dedicated to this kind of digital landscape (project application to EEA grants). Photography and 3D modeling of a 'mythic' space or just a simulation can be put in dialogue. The memory of disaster ruins in photography will be the subject of another book at this publisher. It is about representation of mythical space and time, an approach also followed by WG1 of the Network for Digital Methods in Arts and Humanities. The mythological dimension of disasters has also been identified by Gociman (2006) and we spent some time to develop artistic installations to display this memory, both in scenography and on computer (Bostenaru 2011b). Sand, water, vegetation can cover ruins, and the mythological significance of the elements used to create a symbolic door to the memory, scenografically represented through door models or, in a computer simulation, through pavilions creating the feeling of the respective spaces and also giving architectural language solutions for retrofit methods, for the morphology of these. Such a pavilion was shown in the contribution about L'Aquila. Green spaces can be the first step for reconstruction, not only through providing space for emergency centres, but also through the symbolic-scenographic potential, as in the same reconstruction, for example in our contribution about L'Aquila, and contributing at the same time also to covering of blind walls remaining after collapses in earthquakes through green walls, a contribution against climate change, a recent research we started. The scenography and innovative materials for this have to be investigated. This shows again that the scale of the architecture object and the urban scale are in permanent dialogue when talking about planning disaster resilience measures.

Earthquake protection is a long term goal. Technical prerequisites are well known, implementation is stagnating. In both seismic retrofit as well as in seismic resistant new construction, in this particular case in the framework of reconstruction efforts, other aspects of sustainability have to be incorporated. For example, in the C.A.S.E. project in L'Aquila solar panels were used. This enhances the acceptability of the measures. Emergency housing is an interdisciplinary field, where architecture can make a significant contribution. Since the arrival of the Modern Movement, architecture has been concerned with the employment of new materials for dealing with social challenges such as industrial production. Today the challenge is dealing with environmental problems. Resilience to natural hazards is such a problem, but the larger framework of reaction to climate change has to be kept in mind. Geotextiles are being employed in stabilising earth in order to be able to plant on oblique or even vertical surfaces. Mostly they are employed in large landscapes, such as ash or garbage deposits, or on river sides, but an employment on building scale is possible. Such geotextiles should be also employed in l'Aquila to stabilise inclined ground around the housing in this mountaineous area. Green roofs and walls are helping also in coping with climate change. They are also assuring the link with tradition, as shown in Imre Makovecz's organic architecture, and thus assuring the 'heritage habitat' in order to socially integrate the new development in the 'heads of people'. Green walls allow integrating the parcels remaining empty after the collapse of a building through reusage such as Pocket Parks. In this way the destructive impact of an earthquake becomes an occasion for urban restructuration, providing green in urban areas without having the negative urban image of brand walls. It is possible to integrate the elements also with propping systems.[3]

Another aspect of research in need of attention is related to the definition of the term 'environmental planner'. A recent proposal is to make this a branch of landscape planning, specialised in disaster risk reduction. In several hazards, not so much for earthquake mitigation, this plays a role, dealing for example with the ambivalence of water, between heritage and danger (Bostenaru 2012).

As in the case of special issues dedicated to Natural Hazard Impact, we aim to compile a volume on all hazards, separate from the one focused on earthquake hazard only. In a succeeding volume, aspects regarding other hazards (such as floods, including tsunami, fire) will be covered. However, it should be noted that, in case of those hazards, computer technology is much more supportive of management due to the surface spread of the effects. Also this volume made reference in some of the papers to multihazards.

Another aspect of our research that needs attention is the economic impact of earthquakes. Research so far, still scarce, has been done mainly by the group around Kappos from the Aristoteles University of Thessaloniki (Lekidis et al. 2005; Kappos and Dimitrakopoulos 2008; Kappos et al. 1998, 2007) apart from the first editor of this volume (Bostenaru 2001, 2006). Also the research of the group of Iervolino from the University of Naples addresses decision systems for retrofitting buildings to earthquakes, including the investor (Caterino et al. 2006, 2007, 2009). A recently initiated COST action (IS1104) aims at investigating economic aspects and we hope progress will be made in this direction. It aims to also employ game theory to investigate tools, since socio-economic models lie at the basis of most games. Also an EU project aims at investigating the "Costs of natural hazards", but not earthquakes. ICOMOS has a working group on 'Economics of conservation' (ISCEC) to which we are candidates, and which aims to coordinate research on the topic. It was a core topic of the General Assembly in 2011. Also the ESF-COST high level conference on Geohazards in Sant Feliu (http://www.geohazcop.org/workshops/Sant_Feliu_2011/) to which we contributed dealt with and released a declaration on such neglected aspects in the cycle where not every progress in prediction leads to one in mitigation due to socio-economic issues of implementation of measures (including corruption) or lack of economic research (few papers were dedicated to that). The GEO community that organized the conference brought further these topics for discussion to the European Geosciences General Assembly in Vienna in 2012, where, although there is no longer a session continuing our series, the topic of communication between urban planning and earthquake engineering/seismology will be further addressed by the 'resilient cities' session, and the topic of 'Costs of natural hazards' is addressed in a new session to which we contributed.

[3] From a project proposal Boştenaru (2011), working towards a patent for such a system.

References

Bostenaru Dan M (2001) Calculation of costs for seismic rehabilitation of historical buildings. In: Brebbia CA, Corz A (eds) Earthquake resistant engineering structures III. WIT Press, Southampton, pp 515–524

Bostenaru Dan M (2004) Multi-criteria decision model for retrofitting existing buildings. Nat Hazards Earth Sys Sci 4(4):485–499

Bostenaru Dan M (2006) Wirtschaftlichkeit und Umsetzbarkeit von Gebäudeverstärkungsmaßnahmen zur Erdbebenertüchtigung, 1st edition, Shaker Verlag, Aachen

Bostenaru Dan M (2010) The earthquakes and the Tsunami of 1755 and 2004—historic accidents? Bulletin of the Polytechnic Institute of Jassy—CONSTRUCTIONS. ARCHITECTURE Section, Tomme: LVI (LX) I Fascicle: 3:107–120

Bostenaru Dan M (2011a) The session series on 'Natural Hazards'. Impact on Urban Areas and Infrastructure'. Rev environ sci biotech 10(1):9–24. doi: 10.1007/s11157-011-9231-1

Bostenaru Dan M (2011b) The past: space and time. J Appl Eng Sci 1(14) 3 2011, pp 149–157

Bostenaru Dan M (2012) Water ambivalence or natural hazards' impact on riverine urban areas. In: Sârbu C, Popa A (eds) Development opportunities for areas related to Danube river and Danube Delta. Editura Universitară "Ion Mincu", Bucharest, p 112–121

Caterino N, Iervolino I, Manfredi, G, Cosenza, E (2006) Multi-criteria decision making for seismic retrofitting of an underdesigned rc structure. First European conference on earthquake engineering and seismology, Geneva, Switzerland, 3–8 September 2006

Caterino N, Iervolino I, Occhiuzzi A, Manfredi G, Cosenza E (2007) Dissipazione passiva nella selezione dell'intervento di adeguamento sismico di un edificio in c.a. mediante analisi decisionale, L'Ingegneria Sismica in Italia–XII Convegno Nazionale dell'ANIDIS, Pisa, Italy

Caterino N, Iervolino I, Manfredi G, Cosenza E (2009) Comparative analysis of multi-criteria decision-making methods for seismic structural retrofitting, Computer-Aided Civil Infrastructure Eng 24:432–445

Cowen DJ (1988) GIS versus CAD versus DBMS: what are the differences? Photogramm Eng Remote Sensing 54:1551–1555

ESRI (2001) ArcGIS 9. What is ArcGIS? Environmental Systems Research Institute, Redlands, California. Environmental Systems Research Institute Press, Redlands

ESRI (2011) What is GIS? http://www.esri.com/what-is-gis/index.html. Accessed 2 Oct 2011

Gociman CO (2006) Managementul reducerii riscului la dezastre: strategii de arhitectură şi urbanism. Editura Universitară "Ion Mincu", Bucharest

Goethe JW (1891) Schriften zur Naturwissenschaft—Zur Morphologie

Ioniţă A, Moise E (2001) Bilingual English-Romanian GIS & TIC dictionary. Research Institute for Artificial Intelligence. Romanian Academy, Bucharest

Iosub F (2008) Evolution of Geographical Information Systems (GIS) [in Romanian]. http://earth.unibuc.ro/articole/evoluia-sistemelor-informaionale-geografice-gis. Accessed 2 Oct 2011

Jensen JR (2000) Remote sensing of the environment. An earth resource perspective. Prentice Hall, Upper Saddle River

Kappos AJ, Dimitrakopoulos EG (2008) Feasibility of pre-earthquake strengthening of buildings based on cost-benefit and life-cycle cost analysis, with the aid of fragility curves. Nat Hazards 45(1):33–54

Kappos AJ, Stylianidis KC, Pitilakis K (1998) Development of seismic risk scenarios based on a hybrid method of vulnerability assessment. Nat Hazards 17:177–192

Kappos A, Lekidis V, Panagopoulos G, Sous I, Theodulidis N, Karakostas C, Anastasiadis T, Salonikios T, Margaris B (2007) Analytical estimation of economic loss for buildings in the area struck by the 1999 Athens earthquake and comparison with statistical repair costs. Earthquake Spectra 23(2):333–355

Lekidis VA, Karakostas CZ, Sous II, Anastasiadis A, Kappos A, Panagopoulos G (2005) Evaluation of economic loss for structures in the area struck by the 7/9/1999 Athens earthquake and comparison with actual repair costs. In: Brebbia CA, Beskos DE, Manolis GD, Spyrakos CC (eds) Earthquake resistant engineering structures V, pp 301–312
Letham G (2007) ESRI, Arc/Info, ArcGIS, ArcView . . . 25 years in the making—a time line. http://www.gisuser.com/content/view/11991/53/. Accessed 2 Oct 2011
Lynch K (1960) The image of cities, MIT Press, Cambridge
Masure P, Lutoff C (2003) Methodology of Urban System Exposure (USE). Assessment to natural disasters, report, European commission
Zikas T, Gehbauer F (2007) Decision process and optimisation rules for seismic retrofit programs. In: SFB 461 (ed) Proceedings of the international symposium on strong Vrancea earthquakes and risk mitigation. Bucharest, Romania, p 472–484

Index

M. Bostenaru Dan et al. (eds.), *Earthquake Hazard Impact and Urban Planning*, Environmental Hazards, DOI 10.1007/978-94-007-7981-5,
© Springer Science+Business Media Dordrecht 2014

MIX
Papier aus verantwortungsvollen Quellen
Paper from responsible sources
FSC® C105338

If you have any concerns about our products,
you can contact us on
ProductSafety@springernature.com

In case Publisher is established outside the EU,
the EU authorized representative is:
Springer Nature Customer Service Center GmbH
Europaplatz 3, 69115 Heidelberg, Germany

Printed by Libri Plureos GmbH
in Hamburg, Germany